AF779090

Electrolyte and Electrode Design for Next-Generation Rechargeable Batteries

Electrolyte and Electrode Design for Next-Generation Rechargeable Batteries

Guest Editor

Shaokun Chong

MDPI

Basel • Beijing • Wuhan • Barcelona • Belgrade • Novi Sad • Cluj • Manchester

Guest Editor
Shaokun Chong
Institute of Flexible
Electronics
Northwestern Polytechnical
University
Xi'an
China

Editorial Office
MDPI AG
Grosspeteranlage 5
4052 Basel, Switzerland

This is a reprint of the Special Issue, published open access by the journal *Batteries* (ISSN 2313-0105), freely accessible at: https://www.mdpi.com/journal/batteries/special_issues/4K7J54J2BV.

For citation purposes, cite each article independently as indicated on the article page online and as indicated below:

Lastname, A.A.; Lastname, B.B. Article Title. *Journal Name* **Year**, *Volume Number*, Page Range.

ISBN 978-3-7258-7024-0 (Hbk)
ISBN 978-3-7258-7025-7 (PDF)
https://doi.org/10.3390/books978-3-7258-7025-7

© 2026 by the authors. Articles in this reprint are Open Access and distributed under the Creative Commons Attribution (CC BY) license. The reprint as a whole is distributed by MDPI under the terms and conditions of the Creative Commons Attribution-NonCommercial-NoDerivs (CC BY-NC-ND) license (https://creativecommons.org/licenses/by-nc-nd/4.0/).

Contents

About the Editor

Shaokun Chong

Shaokun Chong is an Associate Professor at Northwestern Polytechnical University, and a member of the China Association for Science and Technology's Young Talent Support Program. His research focuses on advanced nano-energy materials and battery technologies. He has published 45 SCI papers as the first/corresponding author in international high-level journals such as *Angew. Chem. Int. Ed., Matter, Adv. Energy Mater., ACS Nano, Adv. Funct. Mater., Nano Lett.*, etc. He has also led 14 research projects.

batteries

MDPI

Article

Bismuth Nano-Rods Wrapped with Graphene and N-Doped C as Anode Materials for Potassium- and Sodium-Ion Batteries

Shuangyan Qiao [1,2], Yongning Liu [3,*], Kai Wang [1,2,*] and Shaokun Chong [1,2,*]

1 Research & Development Institute, Northwestern Polytechnical University in Shenzhen, Shenzhen 518063, China; iamsyqiao@nwpu.edu.cn
2 Frontiers Science Center for Flexible Electronics, Xi'an Institute of Flexible Electronics, Xi'an Institute of Biomedical Materials & Engineering, Northwestern Polytechnical University, Xi'an 710072, China
3 State Key Laboratory for Mechanical Behavior of Materials, School of Material Science and Engineering, Xi'an Jiaotong University, Xi'an 710049, China
* Correspondence: ynliu@mail.xjtu.edu.cn (Y.L.); iamkwang@nwpu.edu.cn (K.W.); iamskchong@nwpu.edu.cn (S.C.)

Abstract: Alloying-type anode materials have considerably promoted the development of potassium-ion batteries (PIBs) and sodium-ion batteries (SIBs), enabling them to achieve high-energy-density. However, large volume expansion and sluggish dynamic behavior have become key issues affecting electrochemical performance. Herein, bismuth (Bi) nano-rods are anchored on reduced graphene (rGO) and encapsulated via N-doped C (NC) to construct Bi@rGO@NC architecture as anode materials for SIBs and PIBs. The hierarchical confinement effect of three-dimensional conductive networks can not only improve electrode stability upon cycling via suppressing the large volume variation, but also eliminate the band gap of Bi and accelerate ion diffusion, thereby exhibiting favorable electrochemical reaction kinetics. Thus, Bi@rGO@NC contributes an ultra-long lifetime, over 1000 cycles, and an outstanding rate property to SIBs and PIBs. This work can pave the way for the construction of high-performance alloying-type anode materials for SIBs and PIBs.

Keywords: potassium-ion batteries; sodium-ion batteries; anode materials; bismuth; composites

1. Introduction

Lithium-ion batteries (LIBs) have been widely applied in the fields of portable electronic devices, electric vehicles, and large-scale electric energy storage (EES) systems, with the remarkable advantages of higher energy density and longer cycle life. In this context, the limited lithium resources in the Earth's crust and their uneven distribution make it difficult to meet the demand for both electric vehicles and grid-scale energy storage, which has gradually aroused concern. There is an urgent need for low-cost, resource friendly, high-energy-density alternative batteries to satisfy the rapidly increasing need [1–3]. Potassium ion batteries (PIBs) and sodium ion batteries (SIBs) provide potential solutions because of the relatively abundant reserves (2.09 wt.% for K and 2.36 wt.% for Na vs. 0.0017 wt.% for Li) and low cost, and similar storage mechanisms to LIBs [4]. Low redox potential, close to lithium's (−2.93 V vs. K^+/K, −2.71 V vs. Na^+/Na, −3.04 V vs. Li^+/Li), allows PIBs and SIBs to achieve high energy density. Because of the weaker Lewis acidity, smaller Stokes' radius, and higher mobility of K^+ and Na^+ in electrolyte, SIBs and PIBs can achieve better electrochemical kinetic behavior than LIBs [5–7]. However, the issue with PIBs and SIBs lies in the large ion radius (1.38 Å of K^+, 0.98 Å of Na^+), which would easily disrupt the electrode materials, leading to sluggish dynamic and poor cyclability [8,9].

Significant progress has been made in the research of electrode materials for PIBS and SIBs over the past years. In particular, the structure–property relationship, ion storage mechanism, and optimization strategy of different cathode materials (represented by layered oxides, polyanion compounds, Prussian blue analogs, and organic materials) have been thoroughly explored, and achieved encouraging performance results [10–14]. As for layered oxides (A_xMO_2, A = Na and K, M = transition metal), studies mainly focus on O3- and P2-type structures (alkaline metal ions coordinated in the octahedral and prismatic sites, respectively), where O3-type oxides can contribute to high theoretical specific capacity due to high content alkaline metal ions, while P2-type oxides can boost excellent rate capability, benefitting the large channels for ion migration [15,16]. Polyanionic compounds (fluorophosphates, pyrophosphates, etc.) can exhibit high working voltage and superior cyclic stability owing to the stable crystal structure [17]. Prussian blue analogues ($A_xM[Fe(CN)_6]_y{\cdot}zH_2O$, where A represents alkali ions, and M stands for transition metals) show great promise because their open framework structure and large interstitial sites can accommodate reversible de/intercalation of Na-/K-ion, thus guaranteeing high specific capacity, good rate properties, and a long lifetime [18]. Organic cathode materials possess flexible frameworks and large primary interlayer spacing, and are suitable as host materials with a high capacity [19].

In contrast, there is still a shortage of suitable anode materials to match these mature cathode materials. Among the emerging anode materials, carbonaceous materials (such as hard carbon, soft carbon, graphene, etc.) provide excellent chemical stability, but their low capacity cannot be ignored [20]. Although organic compounds possess controllable structures and high electrochemical performance due to multi-electron reversible redox reaction, their low electronic/ionic conductivity and high solubility are serious obstacles [21]. Generally, conversion-type anode materials (transition-metal oxide, chalcogenides, nitride, phosphide, etc.) deliver high theoretical specific capacity, yet high redox potential and huge volume change reduce the long cycles of the electrode [22]. Thus, low working voltage coupled with high theoretical capacity thrust alloy-based electrode materials (Sn-based, Sb-based, Bi-based, P-based, and so on) to the forefront [23]. As an alloy-based anode material, bismuth (Bi) has attracted increasing interest from researchers, thanks to its nontoxic nature, large interlayer spacing, excellent electrical conductivity (7.8×105 $S{\cdot}m^{-1}$), high volumetric capacity (3750 $mAh{\cdot}cm^{-3}$), and suitable reaction potential [24,25]. Nevertheless, the large volume expansion (≈244% for Na_3Bi, ≈411% for K_3Bi) caused by alloying processes and poor kinetic behavior have become the key scientific challenges given their adverse impact on Na/K-ion storage properties [26–28]. These issues could be properly addressed by the strategies of nano-structuring, engineering morphology, and composite constructure with C-based materials [29–34]. The Bi nanoparticles wrapped in graphene to form a Bi@graphene nanocomposite delivered a high capacity of 561 mAh g^{-1} for SIBs; however, the cycle life requires improvement [35]. Later, Zhang et al. improved the cycle life of Bi@graphene to 200 cycles via optimizing the electrolyte with a more stable solid electrolyte interphase layer [36]. In spite of these advances, there are still huge challenges in achieving high capacity during long cycling ranges with stable structures.

Herein, Bi nano-rods encapsulated by reduced graphene oxide and nitrogen-doped carbon (Bi@rGO@NC), constructed through solvothermal and calcination processes, serve as alloying anode materials for PIBs and SIBs. The electronic structure of Bi is adjusted through the synergistic effect of NC and rGO, and thus exhibits conductor behavior. In addition, the hierarchical encapsulation effects of NC and rGO can assist in a high electron/ion transport ability, and relieve the huge lattice stress generated by volume expansion, thereby improving electrochemical dynamic processes and achieving excellent electrode stability upon cycling. Hence, Bi@rGO@NC electrodes deliver superior rate capability and cyclic stability over 1000 cycles in PIBs and SIBs.

2. Materials and Methods

2.1. Materials Synthesis

A total of 4 mmol of thioacetamide (TAA) was dissolved in a mixture of ethanol (15 mL, >99.7%) and glycerol (5 mL, 99%). Then, 2 mmol $BiCl_3$ and a certain amount of graphene oxide was added under magnetic stirring to form a homogeneous solution, which was heated at 180 °C for 12 h in a Teflon-lined stainless-steel autoclave. A Bi_2S_3@rGO sample was collected using freeze-drying after washing with ethanol and water several times.

Then, 121 mg of trihydroxy methyl aminomethane was dissolved in 100 mL of deionized water and 15 μL of HCl (36~38%) under stirring to form a homogeneous solution. Subsequently, 200 mg of dopamine hydrochloride and 200 mg of Bi_2S_3@rGO were added with ultrasonic and stirring, and stirred for 24 h at room temperature. Finally, Bi@rGO@NC composite was acquired through annealing at 500 °C for 3 h in an Ar atmosphere.

2.2. Materials Characterizations

An X-ray Diffraction (XRD, D8 Advance, Bruker, Karlsruhe, Germany) with Cu Kα radiation (λ = 1.5406 Å) was employed to examine the crystal structure of the synthesized material. The microscopic morphology of the sample was observed using field-emission scanning electron microscopy (FESEM, ZEISS-300, ZEISS, Oberkochen, Germany). An energy dispersion spectrum (EDS) mapping was also performed during SEM observation. The Raman spectrum was recorded using a WITec-Alpha300R spectrometer with a laser excited at 532 nm. Thermogravimetric analysis (TGA, TG 209 F3, NETZSCH, Freistaat Bayern, Germany) was carried out in air atmosphere at a heating rate of 10 °C/min from room temperature to 800 °C to evaluate the carbon content and Bi loading of materials. The surface chemical composition and electronic state of the sample were obtained via X-ray photoelectron spectroscopy (XPS, Shimadzu Kratos Axis Supra, Al Ka radiation, Shimadzu Kratos, Manchester, UK). The pore size distribution and the Brunauer–Emmett–Teller (BET) specific surface area were measured using N_2 adsorption/desorption method on a BELSORP-MAX tester.

2.3. Electrochemical Tests

The working electrodes were prepared by mixing active material with acetylene black and polyvinylidene difluoride (PVDF), in a weight ratio of 8:1:1, in 1-Methyl-2-pyrrolidinone (NMP), which was coated onto Cu foil and dried at 80 °C for 12 h. The electrode was cut into discs with a diameter of 12 mm, and the mass loading of active material in each electrode was controlled at the range of 1.5–2.0 mg. Coin-type cells (CR-2025) were assembled in a glove box ($H_2O \leq 0.01$ ppm, $O_2 \leq 0.01$ ppm) for the electrochemical measurement, where K/Na metal (Aladdin) and glass microfiber were used as counter electrode and separator, respectively. Then, 1 M $NaClO_4$ in a mixture of EC and DEC (volume ratio of 1:1) with 5% fluoroethylene carbonate (FEC) was employed as the electrolyte for SIBs. A total of 1 M KPF_6 in ethylene carbonate (EC), diethyl carbonate (DEC), propylene carbonate (PC), with a volume ratio of 1:1:1, was utilized as the electrolyte for PIBs. Galvanostatic charge/discharge tests were performed on a Neware battery test system at a voltage range of 0.01~3.0 V. Cyclic voltammetry (CV) was determined on a CHI 660E electrochemical workstation. A Galvanostatic intermittent titration technique (GITT) test was applied via charging/discharging at 50 $mA \cdot g^{-1}$ for 600 s and then relaxation for 3600 s at open circuit voltage.

2.4. Computational Methods

Theoretical calculations based on density functional theory (DFT) methods were performed in the CASTEP package. For electron exchange–correlation, the Perdew–Burke–Ernzerhof generalized gradient approximation (GGA-PBE) was used. The energy cutoff of plane wave basis was set to 500 eV, which was adequate to achieve energy convergence. The super cell of $3 \times 3 \times 1$ was used to calculate the adsorption energy of K/Na ions to avoid the interaction energy between ions and obtain more accurate results. The K-ion adsorption energy was obtained by the formula of $E_{ads.} = E_{Bi\text{-}K} - E_{Bi} - E_K$.

3. Results and Discussion

3.1. Morphology and Structure Analysis

The Rietveld refinement X-ray Diffraction (XRD) pattern of Bi@rGO@NC composite is displayed in Figure 1a. It can be indexed as a rhombohedral Bi structure with the space group of R-3m (JCPDS NO. 85-1329), whose sharp diffraction peaks show good crystallinity with no impurities [37]. The detailed refinement parameters are listed in Table 1, whose unit cell parameters are determined as a = b = 4.56 Å and c = 11.90 Å, and the fitting results are satisfactory (R_{wp} = 5.76%, R_p = 4.43%). The crystal structure diagram of Bi is shown in Figure 1b, in which the inherent layered structure is accompanied by a large interlayer distance of 3.95 Å, which provides favorable conditions for the de/intercalation of sodium- and potassium-ions [38]. The field-emission scanning electron microscopy (FESEM) image of Bi@rGO@NC is displayed in Figure 1c. It can be observed that the Bi nanorods are tightly anchored on graphene substrate and wrapped with an N-doped C layer, and the diameter of Bi nanorods is about 10~20 nm. Energy-dispersive spectroscopy (EDS) mapping images are displayed in Figure 1d to further illustrate the elemental composition. It can be clearly observed that Bi, C, and N elements are distributed uniformly throughout the Bi@rGO@NC architecture. Such a hierarchical structure can provide a buffer area to accommodate the large volume expansion of Bi electrodes during alkaline-ion insertion and extraction, and the nanorod morphology of Bi can shorten the ion transport paths to enhance the rate capability. In Raman spectra (Figure 1e), two distinct characteristic bands at ~1350 and ~1580 cm^{-1} are related to D-band (sp^3 hybridized disordered carbon) and G-band (sp^2 hybridized graphitic carbon), respectively [8]. The intensity ratio of D-band and G-band (I_D/I_G) is 1.09, implying a high amorphous degree of carbon in Bi@rGO@NC, which could offer better electron conductivity and is conducive to improving electrochemical kinetic behavior. Moreover, two peaks can be observed at ~2685 and ~2930 cm^{-1}, which correspond to the characteristic 2D band and D+G band of rGO, respectively, confirming the successful introduction of rGO [39]. The N_2 adsorption/desorption profile in Figure 1f exhibits type-IV isotherm with H3 hysteresis loops, revealing a hierarchically porous structure, which contributes to the rapid Na- and K-ion diffusion capability. It is also confirmed by the pore size distribution curve (inset of Figure 1f, pore size of <10 nm). Furthermore, Bi@rGO@NC possesses a high specific surface area of 70.87 $m^2 \cdot g^{-1}$, which offers efficient buffering space for volume change during alloying/de-alloying processes. As shown in the TGA curve (Figure 1g), the final weight percentage of Bi_2O_3 was 44.69%, and the mass loss after 200 °C is ascribed to the oxidation from Bi to Bi_2O_3 and the consumption of carbon in the air. The loading of Bi in Bi@rGO@NC can be calculated to be 40.08% using the equation Bi wt.% = 100 × (4 × M_{Bi} × weight percentage of Bi_2O_3)/(2 × M_{Bi2O3}) [40,41].

Table 1. Lattice parameters of Bi@rGO@NC obtained by Rietveld refinement XRD.

a = b = 4.56128, c = 11.90056, α = β = 90°, γ = 120°				
	X	Y	Z	Occupation
Bi	0.66667 (0)	0.33333 (0)	0.56724 (0)	1
R_p = 4.43%, R_{wp} = 5.76%				

X-ray photoelectron spectroscopy (XPS) was performed to investigate the surface composition and electronic state of elements in Bi@rGO@NC. The survey spectrum in Figure S1a indicates the existence of Bi, C, N, and O elements. In Bi 4f fitting spectra (Figure 1h), two peaks located at 164.6 and 159.4 eV can be assigned to Bi $4f_{5/2}$ and Bi $4f_{7/2}$ of Bi-Bi bonds, demonstrating that Bi exists in metal form [42]. The high-resolution C 1s spectrum (Figure 1i) can be fitted into three peaks, at 284.7, 285.7, and 289.1 eV, which belong to C-C/C=C, C-O/C-N, and O-C=O, respectively [43]. The O 1s spectrum (Figure S1b) is divided into three peaks, corresponding to C-O (533.5 eV), C-OH (532.2 eV), and C=O (530.8 eV), respectively [44]. Figure 1j presents N 1s spectra, where the peaks at 400.3 and 398.5 eV are related to pyrrolic N and pyridinic N [45], respectively, which is evidence of successful N doping into the carbon layer. It is well known that the introduction of N can increase defects in carbon layers, which is conducive to improved electronic conductivity and surface wettability, and thus leads to rapid ion storage dynamic process and good electrolyte wetting [46].

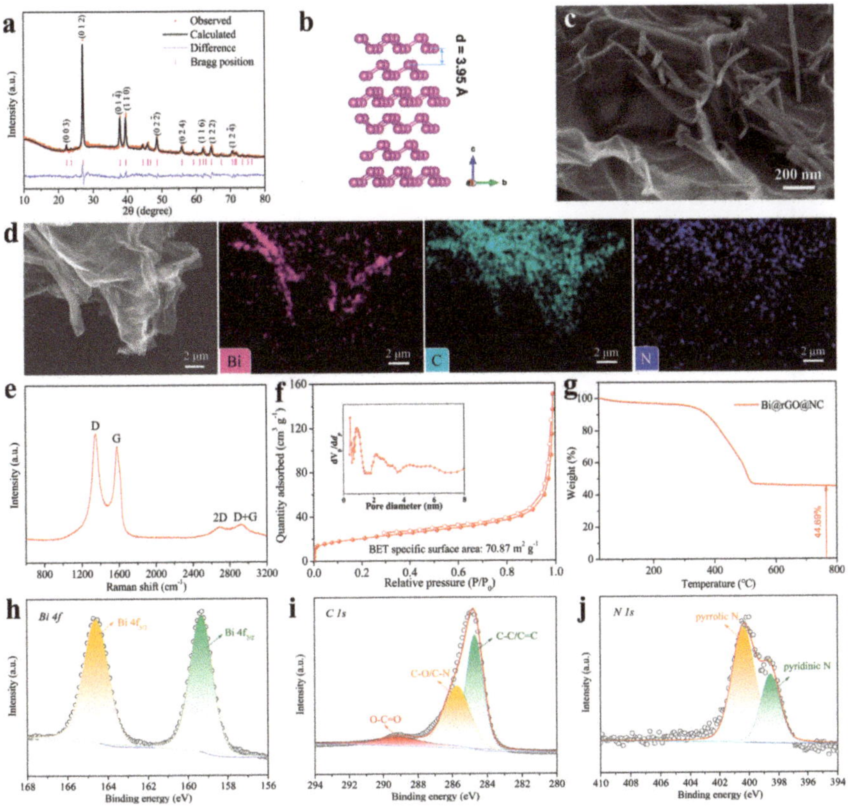

Figure 1. Structural and morphological characterization: (**a**) Rietveld refinement XRD pattern of Bi@rGO@NC; (**b**) crystal structure diagram of Bi; (**c**) FESEM image; (**d**) EDS mapping images of Bi@rGO@NC; (**e**) Raman spectrum; (**f**) N_2 adsorption-desorption isotherm and pore size distribution (inset); (**g**) TGA curve; (**h**) high resolution Bi 4f; (**i**) C 1s; and (**j**) N 1s spectra of Bi@rGO@NC.

3.2. Electrochemical Performances

In SIBs, cyclic voltammetry (CV) was conducted in the potential window of 0.01–3.0 V to investigate the redox behavior of Bi@rGO@NC. Figure 2a gives CV curves of the initial third cycles at 0.1 mV s^{-1} in SIBs, where a peak at 1.05 V can be judged as the formation of a solid–electrolyte interface (SEI) during the first cathodic process. In the first sodiation process, two peaks can be observed, at 0.50 and 0.24 V. These are attributed to the two-step alloying reactions of Bi → NaBi → Na_3Bi [47]. The anodic sharp peaks around 0.75 and 0.86 V are ascribed to the dealloying process of Na_3Bi to form Bi. As a result of the activation process of the electrode, the reduction peaks of subsequent cycles slightly move to higher voltage, while the oxidation peaks show no obvious shift, indicating outstanding electrochemical reversibility and small polarization of the alloying-dealloying process between Bi and Na_3Bi [31]. Figure 2b depicts the galvanostatic charge/discharge curves of Bi@rGO@NC at 50 mA·g^{-1} for SIBs, contributing high initial charge/discharge capacities of 285.5/501.5 mAh·g^{-1}. Such large irreversible specific capacity originates from the formation of SEI film and the decomposition of electrolytes.

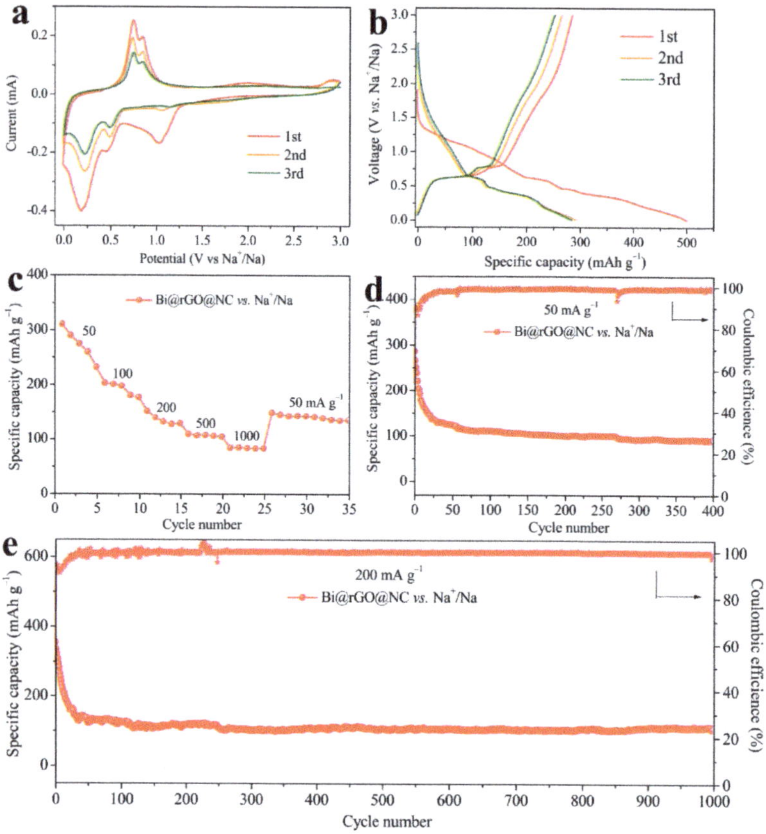

Figure 2. Electrochemical performances of Bi@rGO@NC for SIBs: (**a**) CV curves at 0.1 mV ·s^{-1} and (**b**) discharge/charge profiles of the first three cycles at 50 mA·g^{-1}; (**c**) rate capability; (**d**) cycle performance at 50 mA·g^{-1}; and (**e**) at 200 mA·g^{-1}.

Rate capability analysis was performed to understand the fast and stable redox kinetics. The finding was that Bi@rGO@NC electrodes in SIBs provide reversible specific capacities of 203.4, 151.7, 109.6, and 84.5 mAh·g^{-1} at 100, 200, 500, and 1000 mA·g^{-1} (Figure 2c), respectively. When the current density returns to 50 mA·g^{-1}, the reversible capacities of 149.0 mAh·g^{-1} for SIBs can be obtained. Figure 2d displays the cyclic property at 50 mA·g^{-1}, showing great cycling stability over 400 cycles for SIBs, with a high Coulombic efficiency of around 99% after several initial cycles, where Bi@rGO@NC maintains high-capacity retention value of 90.5 mAh·g^{-1} after 400 cycles. Long-term cycling performance of Bi@rGO@NC electrodes was further evaluated at 200 mA·g^{-1} (Figure 2e). Superior cyclability for Na-ion storage can be achieved for Bi@rGO@NC, showing ultra-long lifetime over 1000 cycles with the capacity retention of 109.7 mAh·g^{-1}, and a capacity decay rate of only 0.069% per cycle. The good electrochemical performance of Bi@rGO@NC architecture may be derived from the synergistic effects of the N-doped carbon layer and graphene. Figure S3 displays the cycling performances of electrode materials with and without N-doped carbon to illustrate its effect on cycling stability. Obviously, Bi@rGO@NC consistently exhibits much better cycling stability with than without N-doped C electrode at 50 and 200 mA·g^{-1}, demonstrating that N-doped C can effectively enhance electrode stability via accommodating the large volume expansion of Bi during cycling.

In PIBs, two pairs of oxidation/reduction peaks at 0.69/1.32 and 0.16/0.70 V (vs. K^+/K) in Figure 3a correspond to the two-step alloying/dealloying process between Bi and K. Overlapping CV profiles manifest highly reversible K-ion insertion/extraction behavior. There are two pairs of plateaus located at 0.74/1.29 and 0.19/0.66 V in Figure 3b, which are consistent with the CV result. For K-ion storage, Bi@rGO@NC delivers high first reversible specific capacity of 348.3 mAh·g^{-1}. Meanwhile, the charge specific capacities of 179.6, 156.6, 66.2, and 31.9 mAh·g^{-1} can be acquired at current densities of 100, 200, 500, and 1000 mA·g^{-1}, respectively (Figure 3c). When the current density turns back to 50 mA·g^{-1}, the reversible specific capacity of 176.4 mAh·g^{-1} can be obtained. By comparison, it can be observed that superior rate capability can be achieved for SIBs compared with PIBs, demonstrating better electrochemical kinetics for reversible Na-ion storage. Figure 3d shows the cycling performance of Bi@rGO@NC at 50 mA·g^{-1}, showing that it maintains a specific capacity of 52.1 mAh·g^{-1} after 400 cycles. Furthermore, the Columbic efficiency of the cell increases continuously upon cycling, eventually reaching around 99%, suggesting great electrochemical reversibility. The long-term cycling property was further measured at 200 mA·g^{-1}, as displayed in Figure 3e, exhibiting a long lifespan over 1000 cycles, while the capacity retention is slightly lower than that in SIBs.

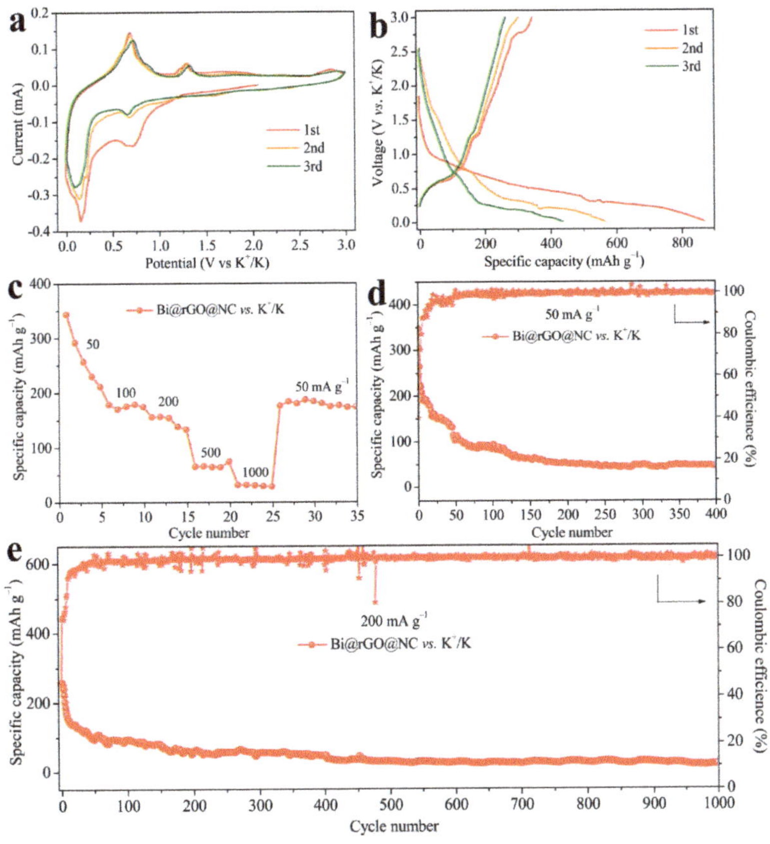

Figure 3. Electrochemical performances of Bi@rGO@NC for PIBs: (**a**) CV curves at 0.1 mV·s^{-1} and (**b**) discharge/charge profiles of the first three cycles at 50 mA·g^{-1}; (**c**) rate capability; (**d**) cycle performance at 50 mA·g^{-1} and (**e**) at 200 mA·g^{-1}.

All these results confirm that Bi@rGO@NC is a promising anode material for Na- and K-in storage. It is worth mentioning that the Bi@rGO@NC composite shows better cyclic stability and rate property in SIBs than PIBs because the larger ionic radius of K in PIBS adversely impacts the kinetic behavior and stability of the electrode material. Ex situ FESEM images of Bi@rGO@NC after 20 cycles for both SIBs and PIBs were collected to understand the electrode integrity. Although Bi@rGO@NC suffers from volume expansion to some degree, as shown in Figure S2a, its rod-like morphology is well maintained, indicating great morphology robustness during repeated sodiation/desodiation processes. By contrast, Bi@rGO@NC could not withstand the huge stress induced by de/intercalation of large-sized K-ion, leading to the formation of bulky agglomerates (Figure S2b), which results in relatively faster capacity decay.

3.3. Electrochemical Kinetics

Density functional theory (DFT) based on first principles calculations was employed to evaluate the kinetics of Bi@rGO@NC for Na-/K-ion storage at the atomic level. As evidenced by the electronic density of states (DOS, Figure 4a), Bi@rGO@NC exhibits metallic behavior. Such a result demonstrates that the introduction of a double carbon layer makes Bi change from a semiconductor to a conductor [48], thus effectively adjusting the electronic structure of Bi and improving its electronic conductivity, which provides favorable conditions for alkaline-ion storage. The band structures in Figure 4b also clearly evidence its metallicity. Furthermore, the calculated adsorption energy of K^+ and Na^+ on Bi@rGO@NC are −1.63 eV and −1.64 eV, respectively, based on the model of K/Na-embedded Bi@rGO@NC crystal (Figure 4c), indicating the favorable tendency of thermodynamic adsorption. Therefore, it is concluded that N-doped carbon and rGO matrices enhance the electronic conductivity and Na-/K-ion adsorption capacity of Bi.

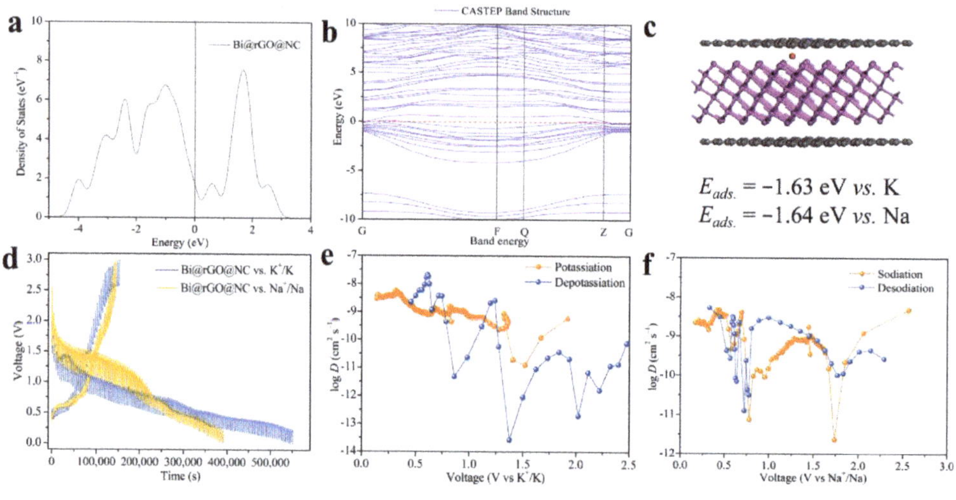

Figure 4. First-principle calculations and electrochemical kinetics: (**a**) DOS and (**b**) band structure; (**c**) structural model for calculation (K/Na atom in red, N atom in blue, C atom in black); (**d**) GITT curves; (**e**) the calculated D_{K+}; and (**f**) D_{Na+} plots of Bi@rGO@NC from GITT.

The ion transport dynamic behavior was analyzed using galvanostatic intermittent titration techniques (GITT) at 50 mA·g^{-1} (Figure 4d). Na-/K-ion diffusion coefficient (D_{Na+} and D_{K+}) was calculated based on the formula of $D_{Na+/K+} = 4/\pi\tau\ (n_m VM/S)^2 \cdot (\Delta E_s/\Delta E_\tau)^2$, where τ, n_m, V_m, S, ΔE_s and ΔE_τ can be ascribed to the time of the pulse, molar number of the active material, molar volume of electrode material, active surface area of electrode in contact with electrolyte, change in battery voltage during charging/discharging, and change in voltage in equilibrium state, respectively [49]. The calculated diffusion coefficients (D_{Na+} and D_{K+}) are in the range of 10^{-14}~10^{-8} $cm^2 \cdot s^{-1}$ for K-ion storage (Figure 4e) and 10^{-12}~10^{-8} $cm^2 \cdot s^{-1}$ for Na-ion storage (Figure 4f). Such results illustrate that hierarchical carbon encapsulation plays a key role in the electrochemical dynamics. The K-/Na-ion diffusion and capacitive behavior were evaluated by CV curves. Analogous shapes at different scanning rates (Figure 5a) suggest good response capability to high-rates for Na-ion storage. The proportion of capacitance at different scan rates can be quantified with the formula $I = k_1 v + k_2 v^{1/2}$, where $k_1 v$ and $k_2 v^{1/2}$ correspond to the contributions of capacitive and diffusion behavior, respectively [50]. The equation can be adjusted to $I/v^{1/2} = k_1 v_{1/2} + k_2$, where k_1 (slope) and k_2 (intercept) are obtained by linear fitting of $I/v^{1/2}$ *vs.* $v^{1/2}$. Finally, the capacitance and diffusion contributions can be determined by averaging the values of $k_1 v/(k_1 v + k_2 v^{1/2})$ and $k_2 v^{1/2}/k_1 v + k_2 v^{1/2})$ under each voltage. The diffusion-controlled ratio is 85.7% for SIBs at 0.1 $mV \cdot s^{-1}$ (Figure 5b). As displayed in Figure 5c, the capacitance contributions at scanning rates of 0.2, 0.3, 0.5, 1.0, 2.0 and 3.0 mV s^{-1} are calculated to be 18.09%, 20.65%, 24.26%, 29.8%, 36.01% and 36.9%, respectively. In terms of PIBs, the polarization is more severe at large sweep speeds, as shown in Figure 5d, which explains why rate performance of PIBs is worse than that of SIBs. In addition, K-ion diffusion contribution ratio is calculated to be 84.11% at 0.1 $mV \cdot s^{-1}$ (Figure 5e), illustrating that the capacity is determined by diffusion behavior to a significant extent. Similarly, the proportion of capacitance shows an upward trend with the increase in sweep rates. The high proportion of capacitive behavior contributes to outstanding rate ability and cycling stability. Figure S4 shows the electrochemical impedance spectroscopy (EIS) plots of Bi@rGO@NC electrode after 10 cycles, where a much lower charge transfer resistance (R_{ct}) value can be achieved for SIBs compared with that for PIBs, demonstrating fast electron and Na-ion transfer dynamic behavior.

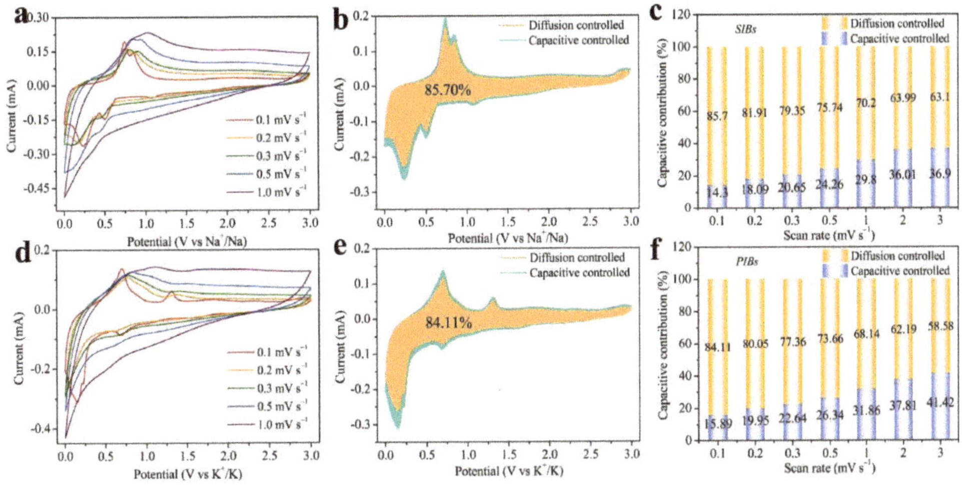

Figure 5. Electrochemical kinetics: (**a**) CV profiles at different scan rates, CV curves displaying capacity distribution (**b**) at 0.1 $mV \cdot s^{-1}$ and (**c**) at various scan rates for SIBs; (**d**) CV profiles at different scan rates, CV curves displaying capacity distribution (**e**) at 0.1 $mV \cdot s^{-1}$ and (**f**) at various scan rates for PIBs.

4. Conclusions

In summary, we propose THE Bi@rGO@NC architecture as anode material for Na- and K-ion storage. The robust hierarchical encapsulation engineering from graphene and N-doped C can effectively relieve the large lattice stress induced by huge volume variation, and is therefore conducive to the preservation of electrode integrity. Experimental characterization and theoretical calculation verify that the electron conductivity, Na/-K-ion adsorption, and diffusion ability of Bi anode can be enhanced by double C to present great electrochemical kinetic behavior. Therefore, outstanding alkaline-ion storage properties with a long-term lifespan of over 1000 cycles and superior rate capability can be achieved for Bi@rGO@NC electrode. This work provides an important reference for research into the Bi-based alloying composite anode material.

Supplementary Materials: The following supporting information can be downloaded at https://www.mdpi.com/article/10.3390/batteries9100505/s1. Figure S1: (a) XPS survey spectrum; (b) O 1s spectrum of Bi@rGO@NC. Figure S2: Ex situ FESEM images of Bi@rGO@NC after 20 cycles for (a) SIBs and (b) PIBs. Figure S3: Cycle performances of composite electrodes with and without N-doped C for SIBs at (a) 50 $mA{\cdot}g^{-1}$ and (b) 200 $mA{\cdot}g^{-1}$. Figure S4: EIS plots of Bi@rGO@NC after 10 cycles.

Author Contributions: Investigation, formal analysis, and writing—original draft, S.Q.; conceptualization and supervision, Y.L.; methodology and writing—original draft, K.W.; conceptualization, methodology, supervision, writing—review and editing, S.C. All authors have read and agreed to the published version of the manuscript.

Funding: This work was supported by State Key Laboratory for Mechanical Behavior of Materials (20212315), Guangdong Basic and Applied Basic Research Foundation (2021A1515110164, 2022A1515010208, 2020A1515110353 and 2023A1515010941), National Natural Science Foundation of China (52207248), Natural Science Basic Research Program of Shaanxi (2022JQ-113), and China Postdoctoral Science Foundation (2022M722606 and 2021TQ0266).

Data Availability Statement: Data will be made available on request.

Conflicts of Interest: The authors declare no conflict of interest.

References

1. Zhong, S.; Zhang, X.; Liu, J.; Sui, Y. Study on $xLiVPO_4F{\cdot}yLi_3V_2(PO_4)_3/C$ composite for high-performance cathode material for lithium-ion batteries. *Front. Chem.* **2020**, *8*, 361. [CrossRef] [PubMed]
2. Guo, X.; Guo, S.; Wu, C.; Li, J.; Liu, C.; Chen, W. Intelligent monitoring for safety-enhanced lithium-ion/sodium-ion batteries. *Adv. Energy Mater.* **2023**, *13*, 2203903. [CrossRef]
3. Lei, Y.J.; Liu, H.W.; Yang, Z.; Zhao, L.F.; Lai, W.H.; Chen, M.; Liu, H.; Dou, S.; Wang, Y.X. A review on the status and challenges of cathodes in room-temperature Na-S batteries. *Adv. Funct. Mater.* **2022**, *33*, 2212600. [CrossRef]
4. Chang, C.-B.; Lu, Y.-R.; Tuan, H.-Y. High-entropy NaCl-type metal chalcogenides as K-ion storage materials: Role of the cocktail effect. *Energy Storage Mater.* **2023**, *59*, 102770. [CrossRef]
5. Luo, P.; Zheng, C.; He, J.; Tu, X.; Sun, W.; Pan, H.; Zhou, Y.; Rui, X.; Zhang, B.; Huang, K. Structural engineering in graphite-based metal-ion batteries. *Adv. Funct. Mater.* **2021**, *32*, 2107277. [CrossRef]
6. Zhao, W.; Ma, X.; Yue, L.; Zhang, L.; Luo, Y.; Ren, Y.; Zhao, X.-E.; Li, N.; Tang, B.; Liu, Q.; et al. A gradient hexagonal-prism $Fe_3Se_4@SiO_2@C$ configuration as a highly reversible sodium conversion anode. *J. Mater. Chem. A* **2022**, *10*, 4087–4099. [CrossRef]
7. Wu, X.; Leonard, D.P.; Ji, X. Emerging non-aqueous potassium-ion batteries: Challenges and opportunities. *Chem. Mater.* **2017**, *29*, 5031–5042. [CrossRef]
8. Zheng, J.; Yang, Y.; Fan, X.; Ji, G.; Ji, X.; Wang, H.; Hou, S.; Zachariah, M.R.; Wang, C. Extremely stable antimony-carbon composite anodes for potassium-ion batteries. *Energy Environ. Sci.* **2019**, *12*, 615–623. [CrossRef]
9. Chong, S.; Ma, M.; Yuan, L.; Qiao, S.; Dong, S.; Liu, H.; Dou, S. Hierarchical encapsulation and rich sp^2N assist Sb_2Se_3-based conversion-alloying anode for long-life sodium- and potassium-ion storage. *Energy Environ. Mater.* **2023**, *0*, e12458. [CrossRef]
10. Zhao, S.; Guo, Z.; Yan, K.; Guo, X.; Wan, S.; He, F.; Sun, B.; Wang, G. The rise of Prussian blue analogs: Challenges and opportunities for high-performance cathode materials in potassium-ion batteries. *Small Struct.* **2020**, *2*, 2000054. [CrossRef]
11. Zhang, X.; Yang, X.; Sun, G.; Yao, S.; Xie, Y.; Zhang, W.; Liu, C.; Wang, X.; Yang, R.; Jin, X.; et al. Hydration enables air-stable and high-performance layered cathode materials for both organic and aqueous potassium-ion batteries. *Adv. Funct. Mater.* **2022**, *32*, 2204318. [CrossRef]
12. Hong, Y.; Hu, J.; Tang, W.; Wei, B.; Guo, M.; Jia, S.; Fan, C. A universal small-molecule organic cathode for high-performance Li/Na/K-ion batteries. *Energy Storage Mater.* **2022**, *52*, 61–68. [CrossRef]
13. Li, H.; Xu, M.; Long, H.; Zheng, J.; Zhang, L.; Li, S.; Guan, C.; Lai, Y.; Zhang, Z. Stabilization of multicationic redox chemistry in polyanionic cathode by increasing entropy. *Adv. Sci.* **2022**, *9*, e2202082. [CrossRef] [PubMed]
14. Gao, X.; Liu, H.; Chen, H.; Mei, Y.; Wang, B.; Fang, L.; Chen, M.; Chen, J.; Gao, J.; Ni, L.; et al. Cationic-potential tuned biphasic layered cathodes for stable desodiation/sodiation. *Sci. Bull.* **2022**, *67*, 1589–1602. [CrossRef] [PubMed]
15. Zhang, J.; Kim, J.-B.; Zhang, J.; Lee, G.-H.; Chen, M.; Lau, V.W.-H.; Zhang, K.; Lee, S.; Chen, C.-L.; Jeon, T.-Y.; et al. Regulating pseudo-jahn-teller effect and superstructure in layered cathode materials for reversible alkali-ion intercalation. *J. Am. Chem. Soc.* **2022**, *144*, 7929–7938. [CrossRef] [PubMed]

16. Sha, M.; Liu, L.; Zhao, H.; Lei, Y. Review on recent advances of cathode materials for potassium-ion batteries. *Energy Environ. Mater.* **2020**, *3*, 56–66. [CrossRef]
17. Peng, B.; Wan, G.; Ahmad, N.; Yu, L.; Ma, X.; Zhang, G. Recent progress in the emerging modification strategies for layered oxide cathodes toward practicable sodium ion batteries. *Adv. Energy Mater.* **2023**, *13*, 202300334. [CrossRef]
18. Liu, X.; Cao, Y.; Sun, J. Defect engineering in Prussian blue analogs for high-performance sodium-ion batteries. *Adv. Energy Mater.* **2022**, *12*, 2202532. [CrossRef]
19. Li, C.; Wang, S. Improving strategies for the molecular structure of organic anode/cathode materials in potassium-ion batteries. *EcoMat* **2022**, *4*, e12246. [CrossRef]
20. Lei, H.; Li, J.; Zhang, X.; Ma, L.; Ji, Z.; Wang, Z.; Pan, L.; Tan, S.; Mai, W. A review of hard carbon anode: Rational design and advanced characterization in potassium ion batteries. *InfoMat* **2022**, *4*, e12272. [CrossRef]
21. Hu, J.; Hong, Y.; Guo, M.; Hu, Y.; Tang, W.; Xu, S.; Jia, S.; Wei, B.; Liu, S.; Fan, C.; et al. Emerging organic electrodes for Na-ion and K-ion batteries. *Energy Storage Mater.* **2023**, *56*, 267–299. [CrossRef]
22. Soltani, N.; Bahrami, A.; Giebeler, L.; Gemming, T.; Mikhailova, D. Progress and challenges in using sustainable carbon anodes in rechargeable metal-ion batteries. *Prog. Energy Combust.* **2021**, *87*, 100929. [CrossRef]
23. Wang, X.; Tang, S.; Guo, W.; Fu, Y.; Manthiram, A. Advances in multimetallic alloy-based anodes for alkali-ion and alkali-metal batteries. *Mater. Today* **2021**, *50*, 259–275. [CrossRef]
24. Lv, Z.; Xu, H.; Xu, W.; Peng, B.; Zhao, C.; Xie, M.; Lv, X.; Gao, Y.; Hu, K.; Fang, Y.; et al. Quasi-topological intercalation mechanism of $Bi_{0.67}NbS_2$ enabling 100 C fast-charging for sodium-ion batteries. *Adv. Energy Mater.* **2023**, *13*, 2300790. [CrossRef]
25. Liu, X.; Zhu, J.; Wang, X.; Yue, L.; Wang, W.; Wang, B.; Shen, D.; Li, Y. Boosting potassium storage kinetics, stability, and volumetric performance of honeycomb-like porous red phosphorus via in situ embedding self-growing conductive nano-metal networks. *Adv. Funct. Mater.* **2022**, *33*, 2209388. [CrossRef]
26. Xiong, P.; Bai, P.; Li, A.; Li, B.; Cheng, M.; Chen, Y.; Huang, S.; Jiang, Q.; Bu, X.H.; Xu, Y. Bismuth nanoparticle@carbon composite anodes for ultralong cycle life and high-rate sodium-ion batteries. *Adv. Mater.* **2019**, *31*, e1904771. [CrossRef] [PubMed]
27. Sun, Z.; Liu, Y.; Ye, W.; Zhang, J.; Wang, Y.; Lin, Y.; Hou, L.; Wang, M.S.; Yuan, C. Unveiling intrinsic potassium storage behaviors of hierarchical nano Bi@N-doped carbon nanocages framework via in situ characterizations. *Angew. Chem. Int. Ed.* **2021**, *60*, 7180–7187. [CrossRef] [PubMed]
28. Hirahara, T.; Nagao, T.; Matsuda, I.; Bihlmayer, G.; Chulkov, E.V.; Koroteev, Y.M.; Hasegawa, S. Quantum well states in ultrathin Bi films: Angle-resolved photoemission spectroscopy and first-principles calculations study. *Phys. Rev. B* **2007**, *75*, 035422. [CrossRef]
29. Liang, Y.; Song, N.; Zhang, Z.; Chen, W.; Feng, J.; Xi, B.; Xiong, S. Integrating Bi@C nanospheres in porous hard carbon frameworks for ultrafast sodium storage. *Adv. Mater.* **2022**, *34*, e2202673. [CrossRef]
30. Yang, H.; Chen, L.W.; He, F.; Zhang, J.; Feng, Y.; Zhao, L.; Wang, B.; He, L.; Zhang, Q.; Yu, Y. Optimizing the void size of yolk-shell Bi@void@C nanospheres for high-power-density sodium-ion batteries. *Nano Lett.* **2020**, *20*, 758–767. [CrossRef]
31. Zhang, F.; Liu, X.; Wang, B.; Wang, G.; Wang, H. Bi@C nanospheres with the unique petaloid core-shell structure anchored on porous graphene nanosheets as an anode for stable sodium- and potassium-ion batteries. *ACS Appl. Mater. Interfaces* **2021**, *13*, 59867–59881. [CrossRef] [PubMed]
32. Qiao, S.; Zhou, Q.; Ma, M.; Liu, H.K.; Dou, S.X.; Chong, S. Advanced anode materials for rechargeable sodium-ion batteries. *ACS Nano* **2023**, *17*, 11220–11252. [CrossRef] [PubMed]
33. Ma, M.; Chong, S.; Yao, K.; Liu, H.K.; Dou, S.X.; Huang, W. Advanced anode materials for potassium batteries: Sorting out opportunities and challenges by potassium storage mechanisms. *Matter* **2023**, *6*, 1–54. [CrossRef]
34. Chen, M.; Zhou, L.; Wang, T.; Xia, H.; Liu, H.K.; Dou, S.X.; Chou, S. Nitrogen as an anionic center/dopant for next-generation high-performance lithium/sodium-ion battery electrodes: Key scientific issues, challenges and perspectives. *Adv. Funct. Mater.* **2023**, *33*, 2214786. [CrossRef]
35. Su, D.; Dou, S.; Wang, G. Bismuth: A new anode for the Na-ion battery. *Nano Energy* **2015**, *12*, 88–95. [CrossRef]
36. Zhang, Q.; Mao, J.; Pang, W.K.; Zheng, T.; Sencadas, V.; Chen, Y.; Liu, Y.; Guo, Z. Boosting the potassium storage performance of alloy-based anode materials via electrolyte salt chemistry. *Adv. Energy Mater.* **2018**, *8*, 1703288. [CrossRef]
37. Cai, S.; Yan, F.; Zhao, Y.; Li, M.; Chen, Y.; He, X.; Wang, C. Hierarchical micro-composite assembled from Bi spheres and expanded graphite flakes as anodes for sodium-ion half/full cells with excellent comprehensive electrochemical performance. *Chem. Eng. J.* **2022**, *430*, 132938. [CrossRef]
38. Wang, A.; Hong, W.; Yang, L.; Tian, Y.; Qiu, X.; Zou, G.; Hou, H.; Ji, X. Bi-based electrode materials for alkali metal-ion batteries. *Small* **2020**, *16*, 2004022. [CrossRef]
39. Ngaloy, R.T.; Fontanilla, A.M.; Soriano, M.S.R.; Pascua, C.S.; Matsushita, Y.; Agulo, I.J.A. Highly efficient photocatalysis by zinc oxide-reduced graphene oxide (ZnO-rGO) composite synthesized via one-pot room-temperature chemical deposition method. *J. Nanotechnol.* **2019**, *2019*, 1895043. [CrossRef]
40. Hong, W.; Wang, A.; Li, L.; Qiu, T.; Li, J.; Jiang, Y.; Zou, G.; Peng, H.; Hou, H.; Ji, X. Bi dots confined by functional carbon as high-performance anode for lithium ion batteries. *Adv. Funct. Mater.* **2020**, *31*, 2000756. [CrossRef]
41. Cen, Y.; Dong, J.; Zhu, T.; Cai, X.; Wang, X.; Hu, B.; Xu, C.; Yu, D.; Liu, Y.; Chen, C. Bi nanorods anchored in N-doped carbon shell as anode for high-performance magnesium ion batteries. *Electrochim. Acta* **2021**, *397*, 139260. [CrossRef]

42. Cheng, M.; Liu, J.; Wang, X.; Li, Y.; Xia, W.; Liu, Q.; Hu, J.; Wei, T.; Ling, Y.; Liu, B.; et al. In-situ synthesis of Bi nanospheres anchored in 3D interconnected cellulose nanocrystal derived carbon aerogel as anode for high-performance Mg-ion batteries. *Chem. Eng. J.* **2023**, *451*, 138824. [CrossRef]
43. Chong, S.; Qiao, S.; Wei, X.; Li, T.; Yuan, L.; Dong, S.; Huang, W. Sb_2S_3-based conversion-alloying dual mechanism anode for potassium-ion batteries. *iScience* **2021**, *24*, 103494. [CrossRef] [PubMed]
44. Topçu, E.; Dağcı Kıranşan, K. Flexible gold nanoparticles/rGO and thin film/rGO papers: Novel electrocatalysts for hydrogen evolution reaction. *J. Chem. Technol. Biotechnol.* **2019**, *94*, 3895–3904. [CrossRef]
45. Wang, Y.; Zhou, Y.; Shen, Y.; Zhu, S. Pyridinic N: A special group for enhancing direct decomposition reaction of NO over N-doped porous carbon. *Micropor. Mesopor. Mat.* **2018**, *265*, 98–103. [CrossRef]
46. Pan, Y.; Cheng, X.; Gao, M.; Fu, Y.; Feng, J.; Gong, L.; Ahmed, H.; Zhang, H.; Battaglia, V.S. Cagelike $CoSe_2$@N-doped carbon aerogels with pseudocapacitive properties as advanced materials for sodium-ion batteries with excellent rate performance and cyclic stability. *ACS Appl. Mater. Interfaces* **2020**, *12*, 33621–33630. [CrossRef] [PubMed]
47. Xue, P.; Wang, N.; Fang, Z.; Lu, Z.; Xu, X.; Wang, L.; Du, Y.; Ren, X.; Bai, Z.; Dou, S.; et al. Rayleigh-instability-induced bismuth nanorod@nitrogen-doped carbon nanotubes as a long cycling and high rate anode for sodium-ion batteries. *Nano Lett.* **2019**, *19*, 1998–2004. [CrossRef] [PubMed]
48. Kowalczyk, P.J.; Mahapatra, O.; Mccarthy, D.N.; Kozlowski, W.; Klusek, Z.; Brown, S.A. STM and XPS investigations of bismuth islands on HOPG. *Surf. Sci.* **2011**, *605*, 659–667. [CrossRef]
49. Liu, H.; He, Y.; Cao, K.; Wang, S.; Jiang, Y.; Liu, X.; Huang, K.J.; Jing, Q.S.; Jiao, L. Potassium-ion batteries: Stimulating the reversibility of Sb_2S_3 anode for high-performance potassium-ion batteries. *Small* **2021**, *17*, 2170044. [CrossRef]
50. Zhou, Q.; Yuan, L.; Li, T.; Qiao, S.; Ma, M.; Wang, Y.; Chong, S. Boosting cobalt ditelluride quantum-rods anode materials for excellent potassium-ion storage via hierarchical physicochemical encapsulation. *J. Colloid. Interface Sci.* **2023**, *646*, 493–502. [CrossRef]

Disclaimer/Publisher's Note: The statements, opinions and data contained in all publications are solely those of the individual author(s) and contributor(s) and not of MDPI and/or the editor(s). MDPI and/or the editor(s) disclaim responsibility for any injury to people or property resulting from any ideas, methods, instructions or products referred to in the content.

batteries

MDPI

Article

Deciphering Electrolyte Degradation in Sodium-Based Batteries: The Role of Conductive Salt Source, Additives, and Storage Condition

Mahir Hashimov and Andreas Hofmann *

Institute for Applied Materials, Karlsruhe Institute of Technology, Hermann-von-Helmholtz-Platz 1, D-76344 Eggenstein-Leopoldshafen, Germany; mahir.hashimov@kit.edu
* Correspondence: andreas.hofmann2@kit.edu; Tel.: +49-(0)721-60825920

Abstract: This work investigates the stability of electrolyte systems used in sodium-ion-based batteries. The electrolytes consist of a 1:1 (*v:v*) mixture of ethylene carbonate (EC) and propylene carbonate (PC), a sodium-conducting salt (either $NaPF_6$ or NaTFSI), and fluoroethylene carbonate (FEC), respectively, sodium difluoro(oxalato) borate (NaDFOB), as additives. Through systematic evaluation using gas chromatography coupled with mass spectrometry (GC-MS), we analyze the formation of degradation products under different conditions including variations in temperature, vial material, and the presence or absence of sodium metal. Our results reveal the significant influence of the conductive salt's source on degradation. Furthermore, we observe that FEC's stability is affected by the storage temperature, vial material, and presence of sodium metal, suggesting its active involvement in the degradation process. Additionally, our results highlight the role of NaDFOB as an additive in mitigating degradation. The study provides crucial insights into the complex network of degradation reactions occurring within the electrolyte, thus informing strategies for improved electrolyte systems in sodium-based batteries. Since the production, material selection and storage of electrolytes are often insufficiently described, we provide here an insight into the different behavior of electrolytes for Na-ion batteries.

Keywords: sodium-based batteries; electrolyte degradation; GC-MS analysis

1. Introduction

Sodium-based batteries (SIBs) have been gathering substantial attention in recent years as potential replacements for lithium-ion batteries, taking advantage of sodium's abundance and low cost [1–3]. However, realizing the full potential of these batteries requires a profound understanding of all system components, particularly the electrolyte, which is central to high performance and cell aging [4–6]. These batteries typically employ carbonate-based solutions as electrolytes, chosen for their high ionic conductivity, stability, and compatibility with both, the anode and cathode materials [7–9].

However, the storage of these carbonate solutions has proven to be of particular challenge [10–12]. These electrolyte solutions can decompose over time, producing various degradation products that can compromise the performance and safety of the battery [13,14]. The decomposition process may accelerate depending on the storage conditions, such as temperature and exposure to light or air, leading to the formation of volatile degradation products [15–17].

Gas chromatography (GC) coupled with mass spectrometry (MS) has emerged as a suitable analytical method for evaluating these decomposition products [18]. The ability of GC-MS to separate and identify different compounds makes it ideal for characterizing the complex mixtures arising from electrolyte degradation. This provides critical insights into the decomposition process, facilitating the development of strategies for improving electrolyte stability [19–21].

The degradation of electrolytes is significantly influenced by temperature, with various studies examining the effects within specific thermal ranges. Observations indicate that within the moderate thermal confines of 30 °C to 60 °C, there is an observable acceleration in the decomposition of certain electrolyte constituents, especially $LiPF_6$ [22]. N. Gauthier et al. demonstrated that $LiPF_6$ salt decomposition at 60 °C induces the formation and deposition of (fluoro)phosphate species in higher proportions at the LTO surface [23]. As the thermal spectrum escalates beyond 60 °C, the effects become more severe. Empirical data suggest a severe degradation of electrolyte components as temperatures approach and exceed 80 °C [22,24,25]. Above 80 °C it is triggering the deterioration of the solid electrolyte interphase (SEI) layer [26]. This degradation enables an exothermic reaction involving the lithiated carbon and the electrolyte, commencing around 100 °C [27]. Further thermal increases cause the electrolyte to decompose above 110 °C, and above 135 °C, the integral separator begins to melt [28]. The maximum thermal degradation is observed at approximately 140 °C, where the electrolyte evaporates and its vapors, in the presence of oxygen, are highly combustible [27,29].

Fluoroethylene carbonate (FEC) and sodium difluoro(oxalato)borate (NaDFOB) play important roles as additives in sodium-ion battery electrolytes, enhancing the electrochemical performance and safety of these systems [30–32]. FEC is renowned for its capability to form a stable solid electrolyte interphase (SEI) on electrode surfaces, a crucial factor in extending the cycle life and efficiency of SIBs [33–35]. This SEI layer inhibits detrimental reactions between the electrode and electrolyte, thus promoting stable cycling and reducing capacity fading [36–39]. Additionally, FEC's ability to stabilize electrode interfaces and decrease electrode–electrolyte reactivity enhances the overall durability and reliability of SIBs [40,41].

NaDFOB, on the other hand, stands out as a highly effective conducting salt for SIB electrolytes due to its exceptional ionic conductivity and electrochemical stability [42–44]. Its robust nature makes it a suitable choice for achieving stable electrolyte–electrode interfaces, even at high voltages. This property is critical for preserving the structural integrity of both electrodes and electrolytes, thereby enhancing the long-term performance and safety of SIBs [45,46]. NaDFOB's unique molecular structure contributes to its remarkable compatibility with various electrode materials and supports efficient ion transport within the battery system, facilitating high-rate capability and prolonged cycle life [42,43,46,47].

Several decomposition products are known that form during the storage of carbonate solutions. Commonly, these include alcohols such as ethylene glycol and propylene glycol, formed from the reduction in the corresponding carbonate solvents [48,49]. Other degradation products such as ethers, esters, and various organic carbonates have also been reported [49–53].

The degradation of electrolytes in sodium-ion batteries (SIBs) has significant consequences, impacting the sustainability and longevity of these energy storage systems. One of the primary concerns is based on the instability of carbonate-based electrolytes within SIBs, leading to a problematic solid electrolyte interface (SEI) due to the reactivity of the electrolyte solution, which directly contributes to capacity fading [54,55] and the inevitable generation of hydrofluoric acid (HF) during battery cycling [56–58]. This issue is exacerbated by the presence of water, which may be introduced during manufacturing or generated within the battery, particularly during instances of overcharging [59,60]. Degradation products also have a pronounced impact on the cathode side, where dissolution of transition metals can occur, especially under high voltage and elevated temperature conditions (see also above) [61,62] as well as on cell cycling [63,64]. Moreover, the presence of these degradation products raises significant concerns regarding the recyclability and environmental sustainability of SIBs, as the extraction and treatment of these materials are energy-intensive and potentially environmentally hazardous [65,66].

In addition to the previously mentioned aspects, the cathode material, respectively, the state of charge (SOC) of the cathode material, has an important impact on the electrolyte degradation behavior of batteries. In sodium-ion batteries, cycling aging effects show both

SOC-independent and SOC-dependent behaviors [67]. In addition, a changed SOC affects the electrochemical and thermal dynamics within the cell. At higher SOCs, irreversible heat generation increases, which can accelerate electrolyte degradation [68]. These findings underscore the importance of understanding and managing SOC to optimize battery performance, longevity, and safety.

The primary aim of our experiment (Scheme 1) is to investigate the effect of storage conditions and how the purity of sodium hexafluorophosphate ($NaPF_6$) salts affects the formation of degradation products in carbonate-based electrolytes. Two different $NaPF_6$ salts of varying purities are used to evaluate their impact on electrolyte stability and degradation. We hypothesize that the level of impurities in these salts is a significant factor in the rate and extent of electrolyte decomposition. Identifying any links between salt purity and electrolyte stability could offer new pathways to improve cell aging and the operational lifetime of alkali metal batteries.

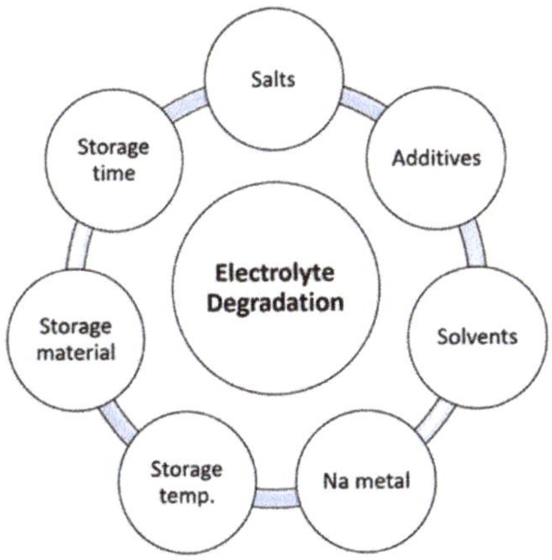

Scheme 1. Experimental design overview.

Furthermore, we compare sodium bis(trifluoromethanesulfonyl)imide (NaTFSI) as an alternative conducting salt with $NaPF_6$. NaTFSI has been proposed as a promising substitute for $NaPF_6$ in some battery systems, owing to its exceptional thermal stability and compatibility with various electrode materials, despite its corrosive tendency [69]. By comparing the performance and degradation patterns of electrolytes based on NaTFSI and $NaPF_6$, we aim to evaluate whether NaTFSI can offer any advantages in terms of electrolyte stability and battery performance.

2. Materials and Methods

2.1. Chemicals and Sample Preparation

The electrolytes were prepared using 1 M $NaPF_6$, sourced from Alfa Aesar (99.5%, Ward Hill, MA, USA) and ChemFish Tokyo Co., Ltd. (99.9%, Tokyo, Japan) or 1 M NaTFSI (TCI Europe, 98%, Zwijndrecht, Belgium) dissolved in a 1:1 volume-to-volume (*v:v*) ratio of ethylene carbonate (EC) and propylene carbonate (PC), with or without the addition of 2 wt% FEC (Gotion, Fremont, CA, USA) or 2 wt% NaDFOB (ChemFish Tokyo Co., Ltd.) additives. The chemical and physico-electrochemical properties of the carbonates and salts are presented in Tables 1 and 2, respectively. High purity sodium metal (Sigma-Aldrich, Taufkirchen, Germany, contains mineral oil, 99.9% trace metals basis) was manually cut into discs approximately half a centimeter in diameter and then submerged in the corresponding electrolytes. Each plastic vial was filled with 0.5 mL of the prepared electrolyte solution, whereas aluminum vials contained 1 mL.

Table 1. Chemical and physico-electrochemical properties of the carbonates used in this study. M_w is the molar weight, ρ the density, ε the dielectric constant at 1 kHz, T_b the boiling point and η the viscosity.

Solvent	M_w [$g \cdot mol^{-1}$]	ρ at 25 °C [$g \cdot cm^{-3}$]	T_b [°C]	ε at 40 °C	η at 40 °C [mPa·s]	Refs.
EC	88.06	1.321 [a]	248	89.78	1.93	[70]
PC	102.09	1.205 [b]	242	66.14 [b]	2.53 [c]	[70]
FEC	106.05	1.454	212	78.4	4.10	[71,72]

[a] at 39 °C. [b] at 20 °C. [c] at 25 °C.

Table 2. Chemical and physico-electrochemical properties of the salts used in this study. M_w is the molar weight, T_m is the melting point and σ is the conductivity.

Salt	Structure of Anion	M_w [$g \cdot mol^{-1}$]	T_m [°C]	σ in 1 M PC Solution [$mS \cdot cm^{-1}$]	Anodic Stability/V vs. Na^+/Na^0	Refs.
$NaPF_6$	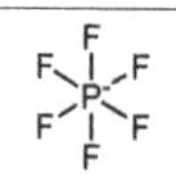	167.95	dec. 300	7.98	5 [a]	[41,73]
NaTFSI	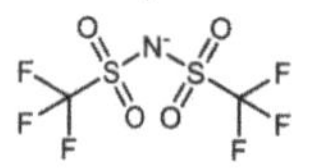	303.13	257	6.20	3.4/5 [a,b]	[41,73]
NaDFOB	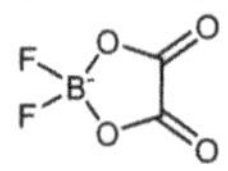	159.82	-	4.27	5.51 [c]	[30,41]

[a] Sodium metal cube slice as counter and reference electrodes, and an aluminum plunger as the working electrode. [b] limited to 3.4 V by Al dissolution, but up to 5 V with 5% $NaPF_6$. [c] stainless steel as the working electrode and Na as the counter and reference electrodes.

It should be mentioned that the purity of the sodium metal, its precise dimensions, and the method used for cutting the metal impact the results. Therefore, care was taken to maintain consistency and accuracy during this process.

All electrolyte solutions were divided and stored in polyethylene (PE) and aluminum (Al) vials, then kept at varying temperatures (25 °C, 40 °C, 50 °C, 58 °C, and 65 °C) for a period extending up to 18 weeks. This range of temperatures was chosen to study the effect of heat on the stability and reactivity of the electrolytes. All vials were dried at under vacuum 80 °C before sample preparation. The electrolyte preparation was carried out inside a glovebox under an inert argon gas atmosphere to prevent any oxidation or moisture-induced reactions, as these could significantly influence the results of the experiment. Glass vials were not used in the study because the glass surface can easily react with HF, which is typically formed in PF_6-containing electrolytes by traces of water over time or at higher temperatures.

For GC-MS measurements, aliquots of 15 µL of each electrolyte were extracted and diluted in 1.5 mL of dichloromethane (DCM) solvent inside of the glove box. This solvent was chosen due to its excellent miscibility with a variety of substances and its relatively low boiling point, facilitating subsequent analysis. Afterwards, centrifugation at a rate of 8000 rpm for 5 min was conducted to separate the salt layer in order to protect the capillary column of the GC. The speed and duration of the centrifugation were carefully optimized to ensure effective separation without causing degradation or alterations to the samples. After centrifugation, 1 mL of the supernatant solution from each sample was transferred into glass vials. These vials were then immediately sealed to prevent the loss or contamination of the sample and used for GC-MS analysis to identify degradable products. The entire process, from sample preparation to the final GC-MS analysis, was repeated twice to ensure the reliability and reproducibility of the data.

In Tables 3–5, a comprehensive overview of the electrolyte compositions is presented, encompassing the respective salt names, solvent names, types of additives used, as well as the specific conditions under which the electrolytes were stored, namely the temperature and type of vial used. In addition, the unique identifiers assigned to each sample are also specified in these tables for reference throughout the study. This detailed organization allows for a streamlined comparison and assessment of the electrolyte properties and their behavior under varying conditions.

Table 3. Summary of electrolyte compositions and conditions ($NaPF_6$ from Alfa Aesar).

Solvent	EC/PC																	
Salt	$NaPF_6$; **99.5%**																	
Additive	-						NaDFOB						FEC					
***T* [°C]**	40			65			40			65			40			65		
Vial	PE	PE	Al	PE	PE	Al	PE	PE	Al	PE	PE	Al	PE	PE	Al	PE	PE	Al
Na	+	-	-	+	-	-	+	-	-	+	-	-	+	-	-	+	-	-
name/code	a1	a2	a3	a4	a5	a6	b1	b2	b3	b4	b5	b6	c1	c2	c3	c4	c5	c6

Table 4. Summary of electrolyte compositions and conditions ($NaPF_6$ from ChemFish Tokyo Co., Ltd.).

Solvent	EC/PC											
Salt	$NaPF_6$; 99.9%											
Additive	-								NaDFOB		FEC	
***T* [°C]**	25		40	50		58		65	40	65	40	65
Vial	PE	Al	Al	PE	Al	PE	Al	Al	Al	Al	Al	Al
Na	+	+	+	+	+	+	+	+	+	+	+	+
name/code	a7	a8	a9	a10	a11	a12	a13	a14	b7	b8	c7	c8

Table 5. Summary of electrolyte compositions and conditions (NaTFSI from TCI Europe).

Solvent	EC/PC											
Salt	NaTFSI											
Additive	-		NaDFOB		FEC							
***T* [°C]**	65		65		40				65			
Vial	PE	Al	PE	Al	PE	Al	PE	Al	PE	Al	PE	Al
Na	+	+	+	+	+	+	-	-	+	+	-	-
name/code	d1	d2	e1	e2	f1	f2	f5	f6	f3	f4	f7	f8

2.2. *Instrumentation*

GC experiments were conducted using established procedures as detailed in literature reference [74]. Specifically, a Clarus 690 GC instrument from PerkinElmer Inc. (Waltham, MA, USA) was employed, which was equipped with an autosampler, a flame ionization detector (FID) and a mass detector (SQ 8T). Turbomass (version 6.1.2) and TotalChrom (version 6.3.4) were utilized for both, data acquisition and subsequent data analysis.

The setup during the measurements included the following parameters: the gases employed were He 6.0 (Air Liquide, Düsseldorf, Germany), H_2 from a PG+160 hydrogen generator (Vici DBS, Schenkon, Switzerland), and Air (Air Liquide). The GC column utilized was an Optima 5MS with a length of 30 m, an interior diameter of 0.25 mm, and a film thickness of 0.5 µm. Injection parameters involved a split flow of 20 mL/min, an inlet temperature of 250 °C, an injection volume of 0.5 µL, an initial pressure of 175 kPa, and a pressure-controlled mode. The oven temperature was maintained at 40 °C. Oven and pressure parameters were set as follows: an initial temperature of 40 °C for 1.5 min, followed by heating at a rate of 20 °C/min up to a final temperature of 320 °C. The initial pressure was maintained at 175 kPa for 2 min, then increased at a rate of 7.8 kPa/min up to a final pressure of 300 kPa. For the MS setup, the filament voltage was set at 70 kV, with an ion source temperature of 200 °C and a transfer line temperature of 200 °C. For the FID setup, the flow rates were 450 mL/min for synthetic air, and 45 mL/min for hydrogen gas, with a detector temperature of 280 °C. Post-separation, the gas flow was divided using a SilFlow™ (Crownhill, UK) GC Capillary Column 3-port Splitter to allow signal detection in both the MS and the FID. The FID was utilized for quantification, while the MS was used for compound identification. Thus, the MS was operated in scan mode, scanning a range from 33 u to 350 u with an event time of 0.3 s and an interscan delay of 0.02 s. The FID signals were used to calculate peak areas.

All samples, including electrolytes and mixtures, were compared and corrected against pure dilution solvent as well as pure electrolyte solvent. Whenever possible, impurities in the electrolyte solvents were identified based on NIST searches (using electron ionization fragmentation match), and by measuring the pure substance separately. Chromatograms of pure solvents can be found at https://zenodo.org/, accessed on 28 September 2023 (see below).

3. Results and Discussion

In this study, the effect of storage conditions and conducting salt source is investigated in terms of the formation of volatile degradation products in liquid electrolytes. The choice of appropriate storage conditions and conducting salt sources is crucial for maintaining the stability and performance of liquid electrolytes, especially in energy storage systems such as batteries. In order to make a comparison, two $NaPF_6$ salts were selected for the study, one with purity of 99.5% and the other with purity of 99.9%. In addition, NaTFSI was directly compared with the corresponding $NaPF_6$ conducting salt. As indicated in the introduction, an increased degradation of PF_6^- occurs above about 60 °C. Although this has been shown so far, especially for $LiPF_6$, such a degradation is also plausible for $NaPF_6$. For this reason, the maximum test temperature was limited to 65 °C.

3.1. $NaPF_6$ with 99.5% Purity

A purity of 98.5–99.5% is the purity typically available commercially, so this provides a basis for the study.

3.1.1. Degradation Product Formation

Initially, the storage of electrolytes without sodium was studied in polyethylene and aluminum vials. The chromatograms after a storage time of 18 weeks at T = 65 °C are depicted in Figure 1. Within the study, a number of components were found to be present in the electrolyte mixtures. The chemical structures of all compounds are depicted in Figure 2. In addition to the solvents and additives EC (**5**), PC (**6**) and FEC (**10**), decomposition products are in particular 1,2-ethanediol (ED, **1**), 1,2-propanediol (PD, **2**), 2-ethyl-1,3-dioxolane (**3**), 2-ethyl-4-methyl-1,3-dioxolane (**4**), amylene hydrate (**7**), 1.4-dioxane (**8**), ethanediol monoformate (**9**), diethylene glycol (**11**) and triethylene glycol (**12**). An overview of all compounds is presented in Table 6. Identification of these compounds was achieved by comparing both, retention time and specific mass fragmentation patterns with known reference standards, underscoring the multifaceted nature of the degradation process. The numbering was chosen according to the sequential position in the chromatograms. It was observed that some more very weak peaks were present in the chromatogram, indicating the formation of additional compounds. However, it was not possible to identify them confidently, so these peaks will not be discussed in this work.

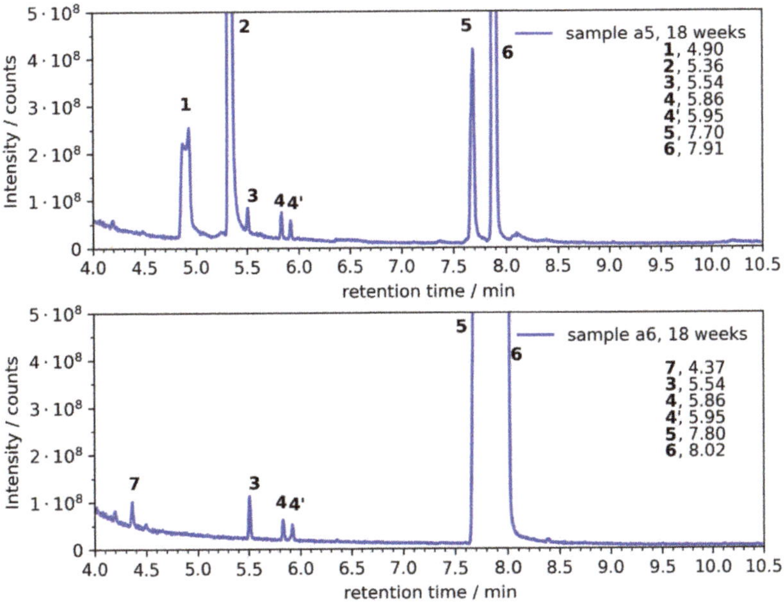

Figure 1. Chromatographic separation of degradation products (EC/PC + 1 M $NaPF_6$; **a5**: plastic vial, **a6**: aluminum vial; T = 65 °C; without sodium): 1,2-ethanediol (ED, **1**), 1,2-propanediol (PD, **2**), 2-ethyl-1,3-dioxolane (**3**), 2-ethyl-4-methyl-1,3-dioxolane (**4** and **4′**), ethylene carbonate (EC, **5**), propylene carbonate (PC, **6**), 2-methyl-2-butanol (**7**). Due to the higher concentration of sample a6, the EC and PC peak exhibit higher signal intensity.

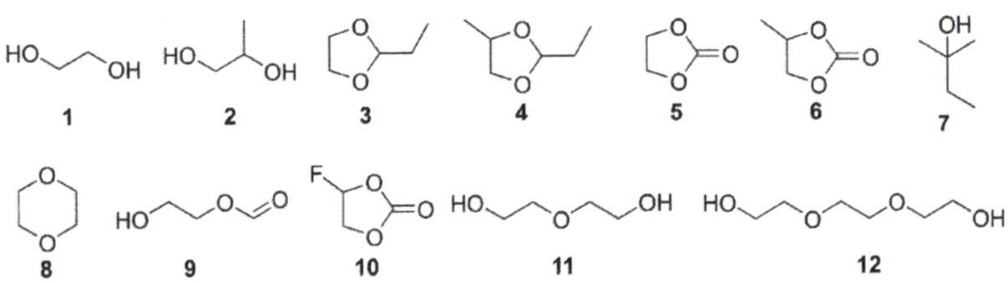

Figure 2. Chemical structures of compounds discussed in the manuscript.

The elution order, as a consequence of the respective volatilities and interactions with the GC column, commenced with amylene hydrate (2-methyl-2-butanol (**7**)), followed by 1,2-ethanediol (**1**), 1,2-propanediol (**2**), 2-ethyl-1,3-dioxolane (**3**), and finally 2-ethyl-4-methyl-1,3-dioxolane (**4**). This order is significant as it can provide insights into the physicochemical characteristics of these compounds, particularly their volatility and polarity, which are fundamental in their interaction with the column and thus their separation. (**1**) and (**2**) could be formed from the reduction in EC and PC, respectively.

Interestingly, the chromatogram showed that 2-ethyl-4-methyl-1,3-dioxolane (**4**) was characterized by two distinct peaks (**4** and **4′**). This result implies the presence of two separate species, possibly due to structural isomers or different conformations. It is critical to note this aspect as it highlights the complexity of the degradation process and suggests the existence of multiple pathways or mechanisms. 2-Methyl-2-butanol (**7**), for instance, suggests the occurrence of reduction reactions involving alkyl carbonate solvents.

It was observed that in aluminum vials no formation of the diols (**1**) and (**2**) took place (see also Tables 7 and 8). In addition, both components (**3**) and (**4**/**4′**), which were also formed in lower concentration relative to EC and PC than in the plastic vials, only (7) was found. This was also confirmed at a storage temperature of T = 40 °C (Tables 7 and 8).

Table 6. Overview of electrolyte components and degradation products.

№	Compound	CAS	Ret. Time [a] [min]	RI [b]	RI [c] NIST	Mass Frag. m/z	Confirmed [d]
1	1,2-ethanediol	107-21-1	4.90	692	702 ± 10	43, 33, 42, 62, 61	yes
2	1,2-propanediol	57-55-6	5.35	739	740 ± 18	45, 43, 61, 75, 76	yes
3	2-ethyl-1,3-dioxolane	2568-96-9	5.54	753	780 ± 7	73, 45, 57, 72, 101	yes
4	2-ethyl-4-methyl-1,3-dioxolane	4359-46-0	5.86, 5.95	790, 799	NA	87, 59, 41, 57, 72, 115	yes
5	EC	96-49-1	7.70	978	NA	43, 88, 44, 58, 42	yes
6	PC	108-32-7	7.91	1000	NA	57, 43, 87, 58, 42	yes
7	amylene hydrate	75-85-4	4.37	637	615 ± 16	59, 73, 55, 43, 41	yes
8	1,4-dioxane	123-91-9	5.08	710	675 ± 19	88, 58, 57, 87, 89	yes
9	ethanediol monoformate	628-35-3	5.63	764	NA	60, 44, 43, 45, 61	yes
10	FEC	1144335-02-8	6.55	859	NA	62, 44, 58, 106, 73	yes
11	diethylene glycol	111-46-6	7.65	978	980	45, 75, 76, 43, 44	yes
12	triethylene glycol	112-27-6	9.77	1256	1255	45, 89, 58, 75, 43	yes
13	1,3-dioxolane-2-methanol	5694-68-8	6.72	883	881	73, 45, 43, 44, 74	yes

[a] retention time at peak maximum; [b] RI values are provided based on n-alkanes; RI values are determined at peak onset; [c] experimental, semi-standard non-polar and estimated non-polar retention index (n-alkane scale) were used; [d] confirmed by the retention time of the pure compound.

Table 7. Identification and formation of degradation products in electrolytes without additives. Values in brackets account for very low signal intensities (e.g., seen by mass extraction only).

Sample	Weeks			
	3	6	9	18
a1	---	(1), (2)	1, 2	1, 2, (8)
a2	---	---	7	3, 4/4′, 7
a3	---	---	(7)	7
a4	1, 2	1, 2, (3), (4/4′)	1, 2, (3), (4/4′)	1, 2, 3, 4/4′, (8)
a5	3, 4/4′	1, 2, 3, 4/4′	1, 2, 3, 4/4′	1, 2, 3, 4/4′, (8)
a6	---	(4/4′), (7)	(3), (4/4′), 7	3, 4/4′, 7

Figure 3 compares the electrolyte after storage for 18 weeks with sodium metal at T = 40 °C and T = 65 °C. It was observed that in the case of T = 40 °C the diol formation is significantly lower (compound **(1)** and **(2)**) and the dioxolane adducts **(3)** and **(4/4′)** do not occur at all. Compared to the electrolytes without sodium, the formation of the diol components **(1)** and **(2)** is enhanced. This is indicative of Na-induced decomposition, whereas the formation of the dioxolane components **(3)** and **(4/4′)** is rather inhibited in the case when sodium is present.

Table 7 shows an overview about the degradation product formation identified in the analysis, specifying the particular electrolytes in which they were formed, along with the corresponding week of detection. This information sheds further light on the temporal and compositional complexity of the degradation process, illustrating the dynamic nature of the electrolyte system.

These findings offer a deeper understanding of the stability and degradation behavior of the electrolyte system, with the presence and identification of these degradation products signifying key information on the possible reactions or pathways taking place in the system. This can be further used to optimize and improve the stability and performance of sodium-based batteries in future studies. The time evolution of the diol and dioxolane formation can be seen clearly, which is considerably accelerated at higher temperature and already shows significant decomposition in the electrolyte after 3 weeks. The results are also important for half-cell experiments with sodium, which are carried out at higher temperatures, since they show that electrolyte degradation already occurs without cell potential in the presence of sodium metal.

Table 8. Identification and formation timeline of degradation products in electrolytes including additives. Values in brackets account for very low signal intensities (e.g., seen by mass extraction only).

Sample	Weeks			
	3	**6**	**9**	**18**
b1	---	---	(1), (2)	1, 2
b2	(7)	(7)	7, (9)	1, 2, 3, 4/4′, 7, 9
b3	---	(7)	(7)	7
b4	---	(1), 2, (3), (4/4′), 9	1, 2, 3, 4/4′, 9	1, 2, 3, 4/4′, 9
b5	(3)	1, 2, 3, 4/4′, 9	1, 2, 3, 4/4′, 9	1, 2, 3, 4/4′, 9
b6	---	---	7	7
c1	---	(1), (2)	1, 2	1, 2
c2	10	10	7, 10	3, 4/4′, 7, 10
c3	10	10	(7), 10	7, 10
c4	1, 2	1, 2	1, 2, 3, 4/4′	1, 2, 3, 4/4′
c5	10	1, 2, 3, 4/4′, (10)	1, 2, 3, 4/4′	1, 2, 3, 4/4′
c6	10	10	7, 10	3, (4/4′), 7, 10

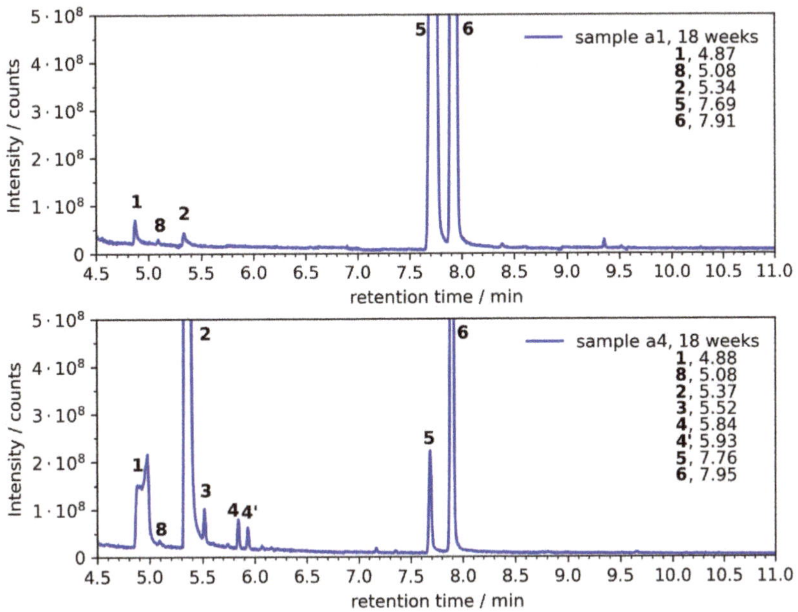

Figure 3. Chromatographic separation of degradation products (EC/PC + 1 M $NaPF_6$ + Na metal; **a1**: plastic vial, T = 40 °C, **a4**: plastic vial; T = 65 °C): 1,2-ethanediol (ED, **1**), 1,2-propanediol (PD, **2**), 2-ethyl-1,3-dioxolane (**3**), 2-ethyl-4-methyl-1,3-dioxolane (**4** and **4′**), ethylene carbonate (EC, **5**), propylene carbonate (PC, **6**), 1,4-dioxane (**8**). Due to the higher concentration of sample **a1**, the EC and PC peak exhibit higher signal intensity.

3.1.2. Degradation Product Formation in the Presence of NaDFOB and FEC

In the next step, the electrolytes were investigated with both additives, NaDFOB and FEC. These additives are suitable for cycling sodium half cells and were selected for this reason. An overview about all detected decomposition products as well as the additive FEC is presented in Table 8. Corresponding chromatograms are shown in Figure 4 for NaDFOB and in Figure 5 for the FEC additive. It is obvious that the storage in aluminum vials reduces the formation of diols as well as dioxolanes significantly (Figures 4 and 5). Table 8 shows the time-dependent formation of degradation products. In the case of NaDFOB when stored in PE vials, the product (**9**) is detected, which is formed first at higher temperature (65 °C), but then at 40 °C when the reaction time is longer. The detection of (**9**) implies partial degradation of ethylene carbonate (EC), one of the primary components of the electrolyte.

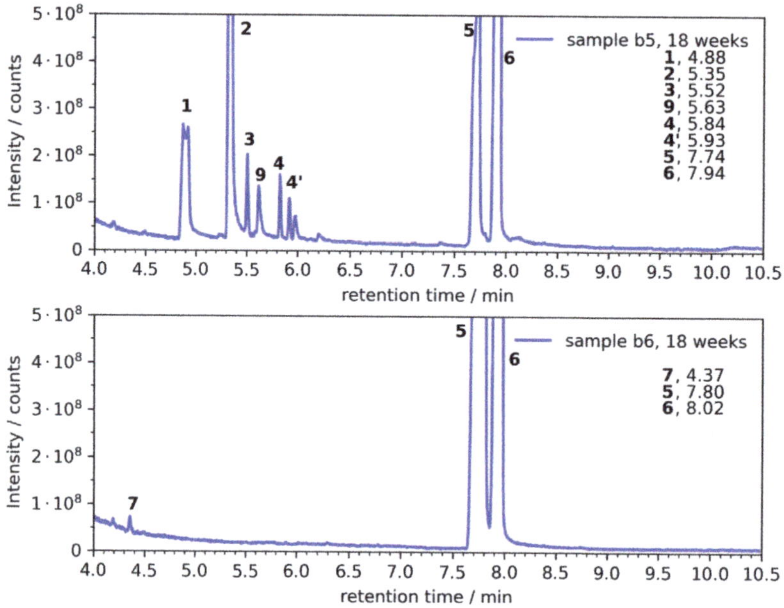

Figure 4. Chromatographic separation of degradation products (EC/PC + 1 M $NaPF_6$ + NaDFOB; **b5**: plastic vial, **b6**: aluminum vial; T = 65 °C): 1,2-ethanediol (ED, **1**), 1,2-propanediol (PD, **2**), 2-ethyl-1,3-dioxolane (**3**), 2-ethyl-4-methyl-1,3-dioxolane (**4** and **4′**), ethylene carbonate (EC, **5**), propylene carbonate (PC, **6**), 2-methyl-2-butanol (**7**) and ethandiol monoformate (**9**).

The storage tests containing FEC as an additive show that FEC is consumed and no longer detectable when the electrolyte is brought into contact with sodium metal (**c1**, **c4**), even after 3 weeks at T = 40 °C. This means that FEC reacts readily with Na metal. This highlights the suitability as an additive because FEC forms a boundary phase which then enables cell cycling. However, this also means that FEC is continuously consumed and explains the well-known rapid deterioration that occurs in the case of cell tests as soon as FEC is no longer present, since the newly forming Na surface is then no longer protected. Surprisingly, it was also found that FEC decomposes over time, especially at higher temperatures when stored in PE vials, and is no longer detectable in the electrolyte (Figure 5, sample **c5**). In this case, storage in Al vials is strongly recommended.

Storage in Al vials is confirmed by the results shown in Tables 7 and 8 where the resilience of electrolytes stored in aluminum vials is clearly visible. The reason for the superior characteristics of aluminum vials vs. PE or PP vials could not be clarified by the study. It is plausible that there is a difference in the sealing properties, especially at higher temperatures (even if the containers were stored again in sealed glass containers under argon), or that plastic components were released, which then accelerated the degradation reactions. In essence, the experiments show that PE vials should not be used for storage of the electrolytes.

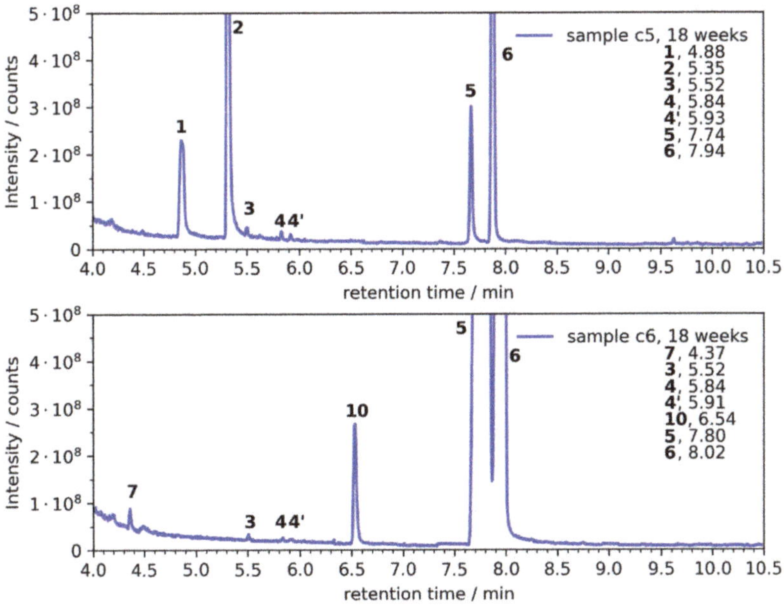

Figure 5. Chromatographic separation of degradation products (EC/PC + 1 M $NaPF_6$ + FEC; **c5**: plastic vial, **c6**: aluminum vial; T = 65 °C): 1,2-ethanediol (ED, **1**), 1,2-propanediol (PD, **2**), 2-ethyl-1,3-dioxolane (**3**), 2-ethyl-4-methyl-1,3-dioxolane (**4** and **4′**), ethylene carbonate (EC, **5**), propylene carbonate (PC, **6**), 2-methyl-2-butanol (**7**) and FEC (**10**).

3.1.3. Analysis of Diol Formation

The GC-MS analysis, conducted on the samples stored in plastic vials to identify degradation products, showed that 1,2-ethanediol (**1**, ED) and 1,2-propanediol (**2**, PD) demonstrated higher relative intensities. Therefore, these both compounds were selected for further evaluation. To accomplish this, mass area ratios of both ED and PD relative to propylene carbonate (PC) were calculated. Figure 6 illustrates the formation of ED and PD in electrolytes, both with and without additives (NaDFOB and FEC) over time, specifically at temperatures of 40 °C and 65 °C. A clear observation from Figure 6 is that the formation of both diol degradation products is significantly higher at 65 °C than at 40 °C. This phenomenon is consistent with chemical reactions, where elevated temperatures increase reaction rates and, consequently, expedite the degradation of the electrolyte.

At 40 °C, both graphs in Figure 6 depicted a trend of increasing ED and PD formation over time with the FEC additive (**c1**). However, in the electrolyte **a1** without any additive, PD formation began to plateau after nine weeks. At 65 °C, both products exhibited a similar trend in formation, but the electrolyte **a4** without any additive demonstrated a more substantial increase in both, ED (**1**) and PD (**2**) over time. This significant increase could be due to an enhanced reaction rate at elevated temperatures, leading to faster degradation of the electrolyte.

Interestingly, the formation of ED in the presence of the FEC additive (**c4**) was less pronounced at the higher temperature, suggesting that the FEC additive may have a stabilizing effect at elevated temperatures (Figure 6). Meanwhile, for PD formation at 65 °C, the electrolyte c4 with the FEC additive exhibited an equal amount of PD after 9 weeks compared to the electrolyte **a4** without any additives. However, after 18 weeks, the effect of the FEC additive became more pronounced, leading to a reduced amount of PD. This suggests that the FEC additive might exert its stabilizing effect over a longer time frame, demonstrating its potential in enhancing the long-term stability of electrolytes.

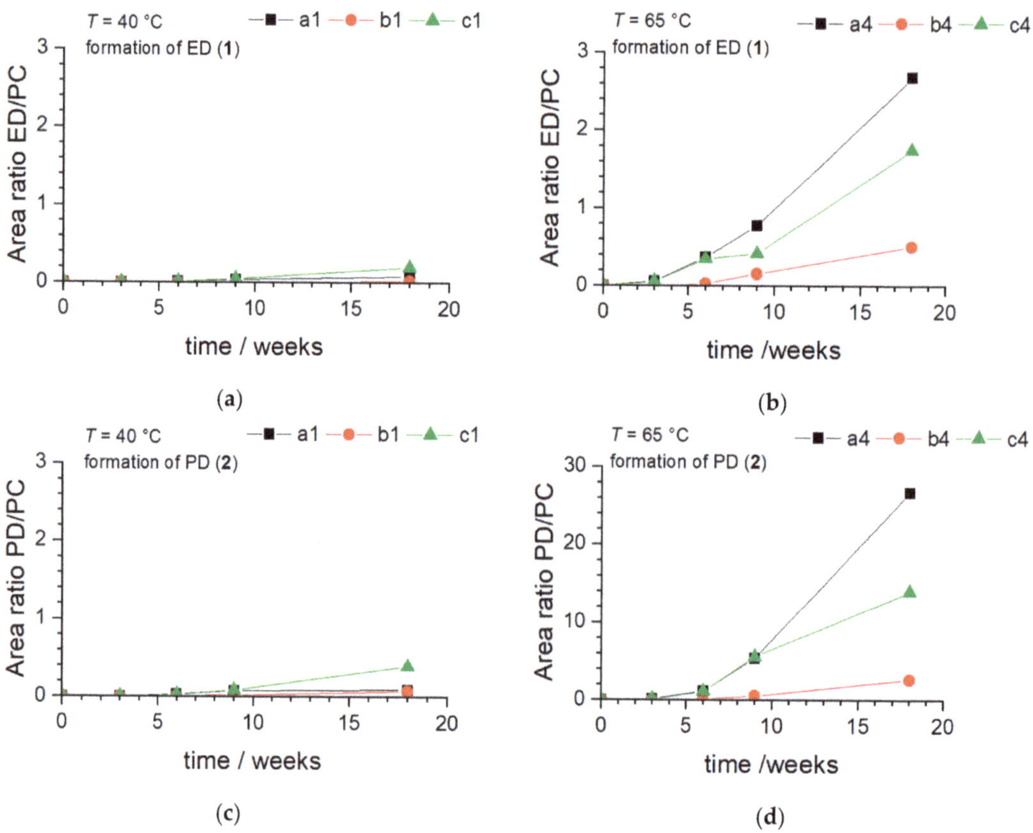

Figure 6. Time-dependent formation of ED (**1**) and PD (**2**) at T = 40 °C (**a**,**b**) and T = 65 °C (**c**,**d**). All samples were consistently stored in PE vials and contain sodium metal. Please note that the scale in (**d**) is ten times higher than in (**c**) for a better visualization.

Remarkably, in both temperature conditions, the formation of ED and PD was minimal in the electrolytes **b1** and **b4** with the NaDFOB additive. This suggests that NaDFOB could be interacting with potential reactants or intermediates, mitigating the reaction pathways leading to ED and PD. In addition, the electrolytes **b1** and **b4** (Figure 6) containing the NaDFOB additive outperform the other electrolyte formulations across all conditions. With significantly lower levels of ED and PD formation, this additive shows strong potential as a stabilizing agent in the electrolyte system. The NaDFOB additive's efficacy aligns with earlier observations, reinforcing the hypothesis that it might mitigate degradation reactions, thereby enhancing electrolyte stability [75].

The impact of sodium metal immersed in the electrolytes was also evaluated more carefully. In Figure 7, a direct comparison with and without sodium metal is depicted for T = 65 °C. The electrolytes without additives showed tendentially slightly lower formation of ED and PD when Na metal was absent, suggesting that the presence of sodium metal may catalyze the electrolyte degradation process to some extent. However, such an effect cannot be observed in case of both additives. In case of FEC, the formation of (**1**) and (**2**) is almost identical with and without Na which further supports the readily surface reaction of FEC with Na metal. In this case the promotion of the Na metal regarding the diol formation is inhibited. It is quite interesting that in case of NaDFOB the formation of the diol derivatives is reduced. It is assumed that the NaDFOB additive directly influence the mechanism of the diol formation in the electrolyte. Nevertheless, NaDFOB is not able to fully inhibit the formation of diols. When subjected to a temperature of 40 °C (not shown here), the samples (**a2**, **b2**, and **c2**) that did not undergo immersion in Na metal exhibited negligible EO or PO formation after both, 9-week and 18-week intervals. This observation underscores the role of Na metal immersion in facilitating the EO and PO formation processes within the designated time frames.

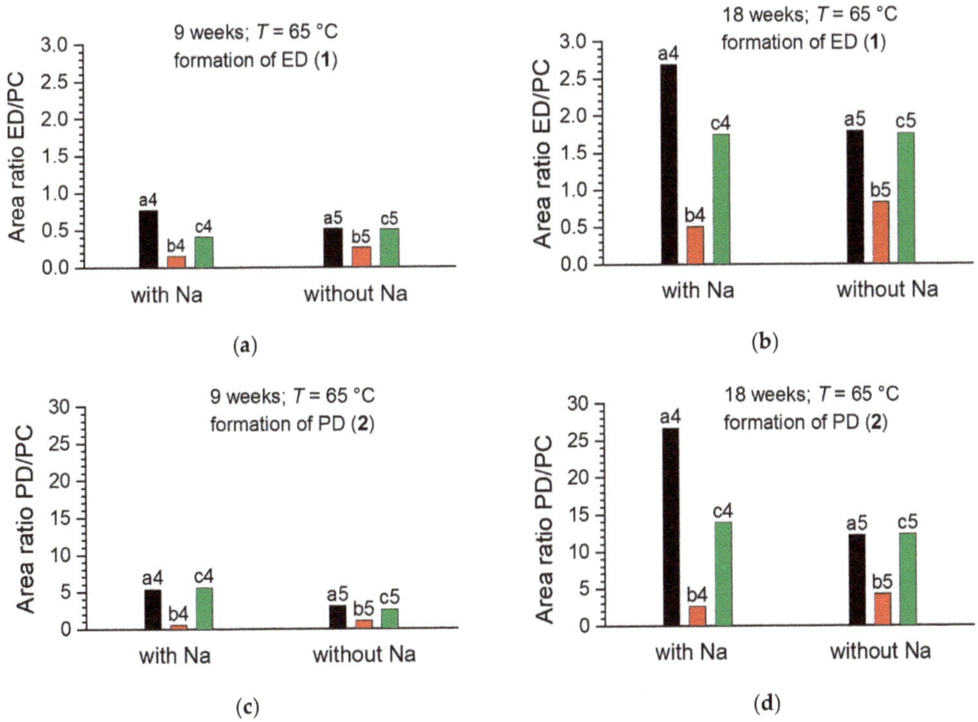

Figure 7. Comparison of electrolytes with and without sodium metal after 9 weeks and 18 weeks of storage. (**a**,**b**): 1,2-ethanediol, (**c**,**d**): 1,2-propanediol. Sample "a" denote electrolytes without additives, "b" denotes NaDFOB as additive, and "c" denotes FEC as additive.

3.2. $NaPF_6$ with 99.9% Purity

The electrolytes incorporating $NaPF_6$ conductive salt with a purity of 99.9% demonstrated significantly different degradation behaviors compared to those using the less-pure salt (see above). Remarkably, the electrolytes with the high-pure salt showed considerably lower amounts of degradation product formation. Specifically, the degradation product PD (**2**) only started forming after an extended duration of 15 weeks, with the lone exception being sample **a12**, where PD (**2**) formation was observed after 9 weeks. In terms of ED (**1**) formation, it was detected only in the samples of **a10** and **a12**, after a prolonged storage duration of 15 weeks. It is important to note that these samples were stored even at relatively higher temperatures of 50 and 58 °C, respectively. Figure 7 illustrates the influence of storage materials on the PD formation. It can be observed that in PE vials, the amount of PD increased with temperature significantly, highlighting the temperature dependence of the degradation process (Figure 8). In contrast, the PD formation in aluminum vials remained constant over the temperature range between 25 °C and 65 °C, suggesting an enhanced thermal stability offered by these vials.

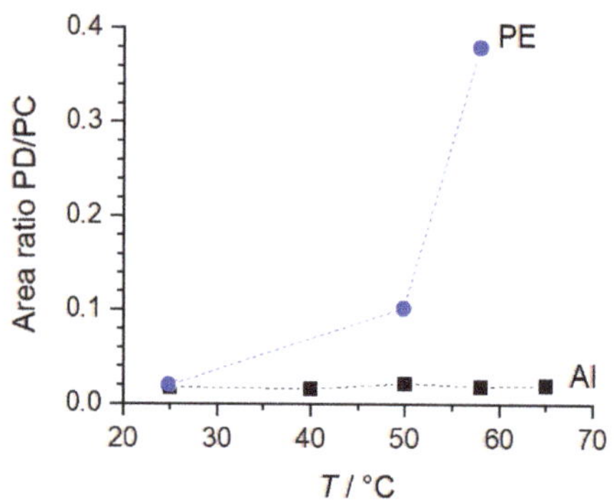

Figure 8. Comparison of temperature dependence and thermal stability of PD formation in PE and aluminum vials for sodium metal immersed electrolytes after 18 weeks of storage. The smaller measurement interval in the case of Al vials is based on the fact that it should be checked whether no PD/PC increase can actually be observed.

The observations with $NaPF_6$ source point to a notable influence of the source of the conductive salt on electrolyte stability. This difference could be attributed to variations in impurity profiles or minute differences in physical properties of the $NaPF_6$ salt quality. The results underscore the importance of raw material quality in determining the performance and stability of the electrolytes. Moreover, the distinct behaviors of PE and aluminum vials further emphasize the role of storage material in electrolyte stability. Despite the same electrolyte composition, the material of the vial plays a significant role, potentially due to its thermal properties or its interaction with the electrolyte components.

Figure 9 offers a comparative analysis of the degradation products formed in electrolytes of samples **a4**, **a12**, and **a14**, all stored at T = 65 °C. The reduced intensity of degradable components in sample **a12**, which uses $NaPF_6$ with 99.9% purity, compared to sample **a4** (NaPF6 purity of 99.5%), is a confirmation to the positive impact of higher purity on electrolyte stability. This observation reinforces the earlier discussion on the importance of raw material purity in determining the performance and stability of the electrolytes. The absence of component **4** (2-ethyl-4-methyl-1,3-dioxolane) in sample **a12** is noteworthy. This could be attributed to the absence or reduced concentration of specific impurities in the higher purity $NaPF_6$ that might catalyze or initiate the formation of this compound. The detection of diethylene glycol (**11**) and triethylene glycol (**12**) in sample **a12** only after 18 weeks suggests a slow degradation process. These compounds might be formed through secondary reactions or the further breakdown of primary degradation products. Their delayed appearance underscores the enhanced stability offered by the higher purity $NaPF_6$. The observation that only 1,2-propanediol (**2**) was formed in sample **a14**, stored in aluminum, further emphasizes the protective role of aluminum vials. The metal's thermal properties and chemical inertness likely contribute to this observed stability.

Figure 10 provides a detailed comparison of the degradation products formed in electrolytes of samples **a14**, **b8**, and **c8**, all stored at T = 65 °C in aluminum vials over sodium metal. These samples utilized the higher purity $NaPF_6$, and the study aimed to understand the influence of different additives, namely NADFOB and FEC, on the degradation behavior of the electrolytes. A striking observation from the figure is the consistent formation of only 1,2-propanediol (**2**) across all three samples, irrespective of the presence or absence of additives. This consistency suggests that the formation of 1,2-propanediol is a dominant degradation pathway under these conditions and is not significantly influenced by the additives in question.

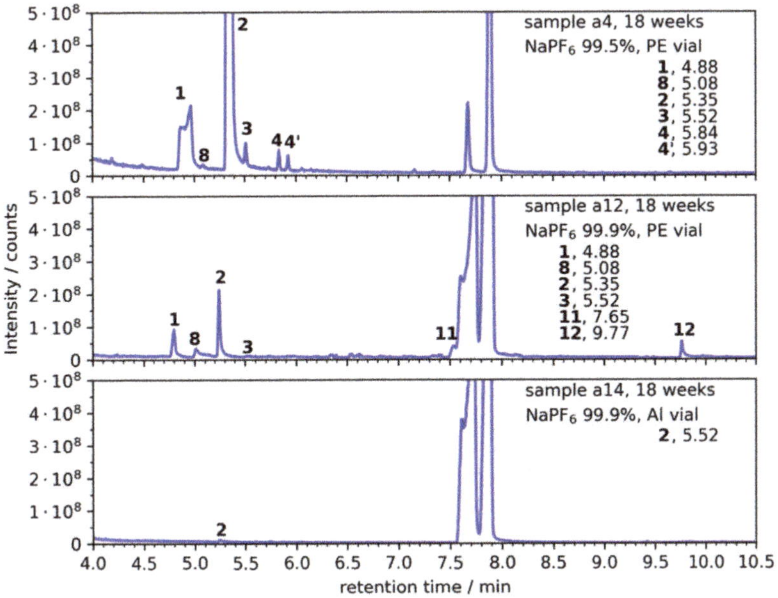

Figure 9. Chromatographic separation of degradation products from electrolyte EC/PC + 1 M $NaPF_6$ at T = 65 °C.

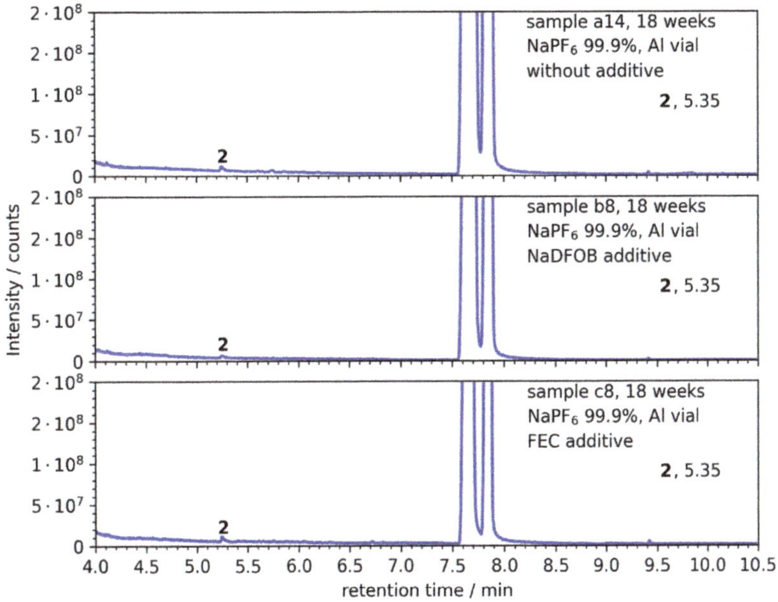

Figure 10. Chromatographic separation of degradation products from electrolyte EC/PC + 1 M $NaPF_6$ at T = 65 °C with selected additives in Al vials stored over sodium metal.

The use of aluminum vials, known for their thermal properties and chemical inertness, might also play a role in this consistent observation. The vials could be limiting the degradation pathways by offering a more stable environment, further narrowing down the degradation products to just 1,2-propanediol.

When the temperature is reduced to 40 °C (not shown here), a notable observation is the significantly reduced intensity of 1,2-propanediol (**2**). This suggests that the degradation process, leading to the formation of this compound, is notably subdued at this lower temperature. Another intriguing observation is the persistent presence of FEC in the electrolyte of sample **c7**. This contrasts with the behavior observed in sample **c8** from Figure 10, where FEC, despite being used as an additive, had disappeared when the electrolyte was stored at 65 °C. This difference underscores the temperature-dependent reactivity of FEC. At the elevated temperature of 65 °C, FEC seems to undergo rapid degradation or consumption, leading to its absence in the electrolyte. However, at the milder temperature of 40 °C, FEC remains stable and does not degrade as quickly, allowing for its detection in the electrolyte even after the storage period.

3.3. NaTFSI with 99.9% Purity

The results obtained from the analysis of electrolytes using NaTFSI as the conductive salt illustrate a similar behavior to that observed with high-pure 99.9% $NaPF_6$. Across most samples, the formation of PD is detected after a lengthy storage duration of 15 weeks, with the exceptions being samples **d1**, **e1**, **f1**, and **f3**. These samples were stored in PE vials, and notably, they had sodium metal immersed in them.

Figure 11 illustrates the formation of ED and PD over 18 weeks in PE vials with different additives, and the superior performance of the NaDFOB additive becomes evident. Both PD and ED formation were the least in electrolytes with this additive, suggesting that NaDFOB can effectively mitigate degradation processes in these electrolytes. The graphical representation (Figure 12) of the area ratio of ED/PC and PD/PC against time provides a comprehensive view of the degradation behavior of the electrolytes **f1** and **f3** over an 18-week period. Both these electrolytes were stored in PE vials and contained the FEC additive, but they were subjected to different temperature conditions: **f1** at 40 °C and **f3** at 65 °C.

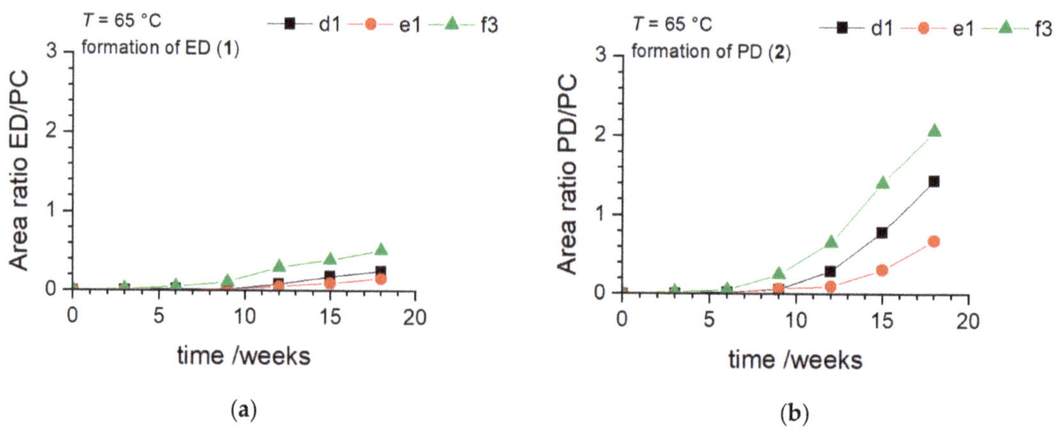

Figure 11. Time-Dependent Formation of ED (**a**) and PD (**b**) at 65 °C with additive-free electrolyte (**d1**), with NaDFOB additive (**e1**) and with FEC in the electrolyte (**f3**). All samples contain sodium metal and are stored in PE vials.

From the Figure 12a, which plots the formation of ED against time, a clear trend emerges. The amount of ED in both electrolytes increases over time, but the rate of increase is markedly different between the two. The electrolyte **f3**, stored at the higher temperature of 65 °C, exhibits a significantly accelerated formation of ED compared to its counterpart, **f1**, stored at 40 °C. This observation aligns with the fundamental understanding of chemical kinetics, where reaction rates, and consequently degradation rates, are known to increase with temperature. The elevated temperature likely provides the necessary energy to overcome activation barriers more readily, leading to faster degradation.

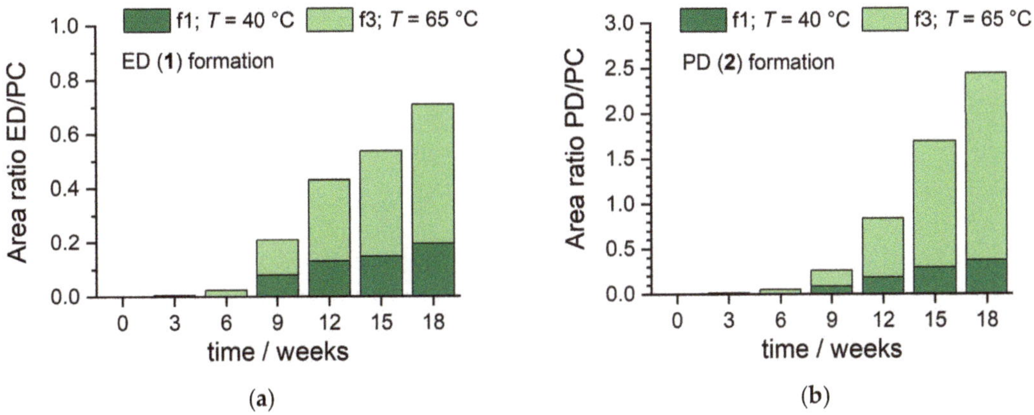

Figure 12. Comparative Analysis of ED (**a**) and PD (**b**) Formation in Electrolytes **f1** and **f3** over 18 Weeks at Different Temperatures.

The Figure 12b, focusing on PD formation, mirrors the trend observed for ED. The temporal increase in PD is evident for both electrolytes, but once again, the electrolyte **f3** demonstrates a more pronounced increase. The significant difference in PD formation between both temperature conditions further underscores the temperature's pivotal role in influencing the degradation behavior of these electrolytes.

The presence of the FEC additive in both electrolytes was expected to offer some level of stabilization. While the additive does play a role, as evidenced by the formation of degradation products over an extended period rather than immediately, its efficacy seems to be compromised at higher temperatures. The significantly higher formation of ED and PD in the **f3** electrolyte suggests that while FEC might offer protection, its stabilizing effect is temperature-dependent and diminishes at elevated temperatures.

The degradation behavior of the electrolytes listed in Table 5 is consistent in terms of the formation of PD. The majority of these electrolytes began to show the presence of PD after an extended storage duration of 15 weeks. This consistent observation across multiple samples underscores the stability of these electrolytes, especially when considering the relatively long-time frame required for the onset of degradation. The low intensity of PD, even when it does form, further emphasizes the resilience of these electrolytes against degradation.

However, a unique observation was made with sample **f4** (Figure 13). This sample, stored in an aluminum vial at 65 °C with the FEC additive, exhibited the formation of a new degradation product: 1,3-dioxolane-2-methanol. While the presence of this compound was detected, its intensity was notably low, suggesting that its formation is not a dominant degradation pathway under the given conditions.

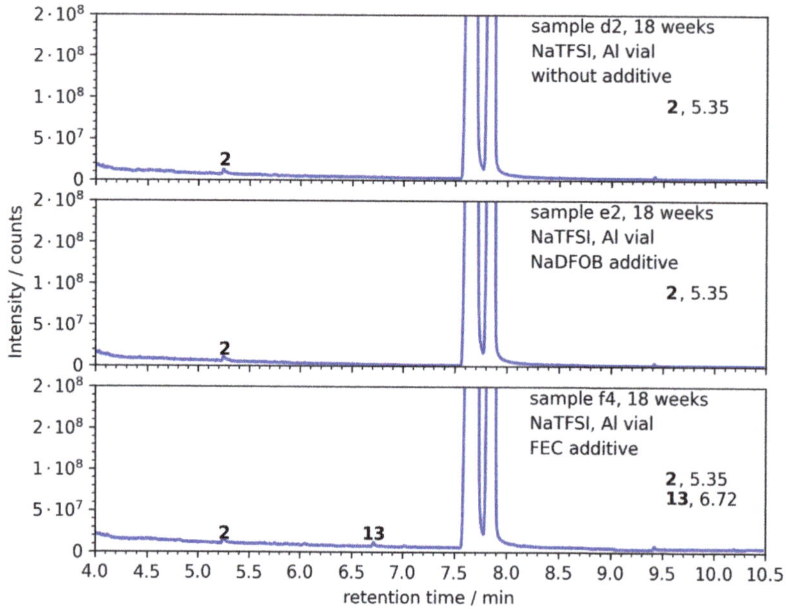

Figure 13. Chromatographic separation of degradation products from electrolyte EC/PC + 1 M NaTFSI at T = 65 °C with selected additives in Al vials stored over sodium metal.

3.4. FEC Reactivity and Color Changes

The influence of various conditions—storage temperature, type of vial material, presence or absence of sodium metal, and the conductive salt used—on the reactivity of FEC was carefully examined over a period of 18 weeks using GC-MS analysis.

For the samples containing $NaPF_6$ as the conductive salt, the following trends are as follows (refer to Tables 2 and 3 for detailed data):

1. In PE vials at 40 °C, FEC disappears after 3 weeks when sodium metal is present (**c1**) but remains stable over 18 weeks when sodium metal is absent (**c2**).
2. In aluminum vials at 40 °C, regardless of the presence (**c7**) or absence (**c3**) of sodium metal, FEC remains stable over the 18 weeks duration.
3. At 65 °C in PE vials, FEC disappears after 3 weeks when sodium metal is present (**c4**), and after 6 weeks when sodium metal is absent (**c5**).
4. In aluminum vials at 65 °C, FEC remains stable over 18 weeks when sodium metal is absent (**c6**), but it disappears after 12 weeks when sodium metal is present (**c8**).

For the samples containing NaTFSI as the conductive salt, similar patterns emerge (see also Table 4 for detailed data):

1. In PE vials at 40 °C, FEC disappears after 9 weeks when sodium metal is present (**f1**) but remains stable over 18 weeks when sodium metal is absent (**f5**).
2. In aluminum vials at 40 °C, regardless of the presence (**f2**) or absence (**f6**) of sodium metal, FEC remains stable over the 18 weeks duration.
3. At 65 °C in PE vials, FEC disappears after 3 weeks when sodium metal is present (**f3**), and after 12 weeks when sodium metal is absent (**f4**).
4. In aluminum vials at 65 °C, FEC remains stable over 18 weeks, irrespective of the presence (**f8**) or absence (**f7**) of sodium metal.

The color change in electrolytes over time offers additional evidence for FEC reactivity and possible degradation pathways in various conditions. The electrolyte sample c1, stored at 40 °C in a PE vial with $NaPF_6$ as a conductive salt and sodium metal present, turns dark after 6 weeks, coinciding with the disappearance of sodium metal. This suggests a reaction involving FEC, sodium metal, and possibly the conductive salt under these conditions. This process seems to involve the reduction of FEC and possibly the formation of insoluble compounds causing the darkening. According to Dugas et al., CO_2 was also found in our study when FEC in EC/PC is stored over Na [34]. Additionally, NaF was identified in our samples. Concludingly, a reductive decomposition of FEC leading to NaF as well as CO_2, which also promotes the formation of the corresponding diols from EC and PC, respectively, can be deduced.

In contrast, samples **c2**, **c5**, **c8**, **f1**, and **f4**, which change color progressively from pale yellow to brown over weeks, might suggest slow degradation processes that produce colored compounds. The rusting of the sodium metal in samples **c8**, **f1**, and **f4** could be attributed to the formation of insoluble sodium salts possibly due to reactions with FEC degradation products.

It is interesting to note that samples **c3**, **c7**, **f2**, **f5**, **f6**, **f7**, and **f8** showed no color change throughout the 18 weeks. This suggests the stability of FEC in these conditions and aligns with our previous observations about FEC's reactivity.

For the sample **f3**, the dark color and precipitation formed after 3 weeks might indicate the rapid degradation of FEC and the formation of insoluble compounds under these specific conditions (65 °C, PE vial, NaTFSI, sodium metal present).

4. Conclusions

This study presents a comprehensive investigation into the stability and degradation behavior of various electrolyte systems intended for use in sodium-based batteries. The electrolytes comprised a 1:1 (*v*:*v*) mixture of ethylene carbonate (EC) and propylene carbonate (PC), with sodium salts (either $NaPF_6$ or NaTFSI) and FEC additive.

The findings support the hypothesis that the type of conductive salt, presence of sodium metal, choice of additive, and vial material all exert significant influences on the degradation pathways of the electrolyte. NaDFOB continues to demonstrate superior performance in minimizing the formation of degradation products, underpinning its role as a potential stabilizer in these systems. Moreover, the results strengthen the case for the use of aluminum as a preferred material for vials, as it appears to provide a protective effect even in the presence of sodium metal.

Significant variations were observed in the formation of degradation products based on the choice of the conductive salt. Notably, electrolytes utilizing very pure $NaPF_6$ (99.9%) showed a markedly lower amount of degradation product formation compared to those using $NaPF_6$ in pure quality (99.5%), underlining the influence of raw material quality on electrolyte stability.

Further, the choice of vial material for storage was found to play a crucial role in the degradation behavior, with aluminum vials demonstrating superior performance over PE vials. This observation points towards the potential thermal properties or surface interactions of aluminum impacting the stability of the electrolyte system.

The presence of sodium metal was found to accelerate degradation, particularly at higher temperatures, indicating its participation in degradation reactions. However, the NaDFOB additive exhibited a mitigating effect on this degradation, suggesting its potential utility in enhancing the stability of these electrolytes.

The trend in the reactivity of the FEC additive under various conditions observed in this study provides valuable insights into its role in the electrolyte system and the factors influencing its stability. The following effects can be summarized:

1. Temperature effects: In both sets of samples, FEC appears to degrade faster at higher temperatures. This behavior aligns with the general principle of chemical kinetics that reaction rates typically increase with temperature. This is usually explained by the Arrhenius equation, which states that a higher temperature increases the fraction of molecules possessing energy greater than the activation energy, leading to an increased rate of reaction. In the context of our study, higher temperatures could facilitate degradation reactions involving FEC, thereby causing its faster consumption.

2. Effects of sodium metal presence: The presence of sodium metal seems to accelerate the disappearance of FEC. This suggests that FEC reacts with sodium, possibly through a reductive decomposition mechanism. The resulting products could contribute to the formation of a stable SEI layer, which could help improve the overall stability of the electrolyte system.
3. Effects of vial material: FEC tends to last longer in aluminum vials than in PE vials, indicating that the material of the storage vial can impact the stability of the additive. The reason behind this observation could be the better thermal conductivity of aluminum compared to PE, which could help dissipate heat more efficiently, thereby reducing the rate of degradation reactions.
4. Effects of conductive salt: Although both types of salts studied ($NaPF_6$ and NaTFSI) are sodium salts, differences in the reactivity of FEC were observed between them. This suggests that the anion part of the salt might play a role in the reactions involving FEC. Some reports in the literature suggest that $TFSI^-$ is less reactive and more thermally stable than PF_6^-, which could explain the observed behavior [76,77].

Author Contributions: Conceptualization, M.H. and A.H.; formal analysis, M.H. and A.H.; funding acquisition, A.H.; investigation, M.H.; methodology, M.H.; project administration, A.H.; resources, A.H.; supervision, A.H.; validation, M.H. and A.H.; visualization, M.H. and A.H.; writing—original draft preparation, M.H. and A.H.; writing—review and editing, M.H. and A.H. All authors have read and agreed to the published version of the manuscript.

Funding: This work contributes to the research performed at CELEST (Center for Electrochemical Energy Storage Ulm-Karlsruhe) and was funded by the German Research Foundation (DFG) under Project ID 390874152 (POLiS Cluster of Excellence). We acknowledge support by the KIT-Publication Fund of the Karlsruhe Institute of Technology.

Data Availability Statement: Raw data of the GC measurements are available at https://zenodo.org, accessed on 28 September 2023 (https://doi.org/10.5281/zenodo.8388636).

Conflicts of Interest: The authors declare no conflict of interest.

References

1. Yabuuchi, N.; Kubota, K.; Dahbi, M.; Komaba, S. Research development on sodium-ion batteries. *Chem. Rev.* **2014**, *114*, 11636–11682. [CrossRef] [PubMed]
2. Kim, H.; Hong, J.; Park, K.Y.; Kim, H.; Kim, S.W.; Kang, K. Aqueous rechargeable Li and Na ion batteries. *Chem. Rev.* **2014**, *114*, 11788–11827. [CrossRef] [PubMed]
3. Wang, Y.; Song, S.; Xu, C.; Hu, N.; Molenda, J.; Lu, L. Development of solid-state electrolytes for sodium-ion battery—A short review. *Nano Mater. Sci.* **2019**, *1*, 91–100. [CrossRef]
4. Slater, M.D.; Kim, D.; Lee, E.; Johnson, C.S. Sodium-Ion Batteries. *Adv. Funct. Mater.* **2013**, *23*, 947–958. [CrossRef]
5. Bin, D.; Wang, F.; Tamirat, A.G.; Suo, L.M.; Wang, Y.G.; Wang, C.S.; Xia, Y.Y. Progress in Aqueous Rechargeable Sodium-Ion Batteries. *Adv. Energy Mater.* **2018**, *8*, 1703008–1703038. [CrossRef]
6. Gupta, P.; Pushpakanth, S.; Haider, M.A.; Basu, S. Understanding the Design of Cathode Materials for Na-Ion Batteries. *ACS Omega* **2022**, *7*, 5605–5614. [CrossRef]
7. Hirsh, H.S.; Li, Y.X.; Tan, D.H.S.; Zhang, M.H.; Zhao, E.Y.; Meng, Y.S. Sodium-Ion Batteries Paving the Way for Grid Energy Storage. *Adv. Energy Mater.* **2020**, *10*, 2001274–2001281. [CrossRef]
8. Tian, Z.; Zou, Y.; Liu, G.; Wang, Y.; Yin, J.; Ming, J.; Alshareef, H.N. Electrolyte Solvation Structure Design for Sodium Ion Batteries. *Adv. Sci.* **2022**, *9*, e2201207. [CrossRef]
9. Liu, Z.; Lu, Z.; Guo, S.; Yang, Q.H.; Zhou, H. Toward High Performance Anodes for Sodium-Ion Batteries: From Hard Carbons to Anode-Free Systems. *ACS Cent. Sci.* **2023**, *9*, 1076–1087. [CrossRef]
10. Rafie, A.; Kim, J.W.; Sarode, K.K.; Kalra, V. A review on the use of carbonate-based electrolytes in Li-S batteries: A comprehensive approach enabling solid-solid direct conversion reaction. *Energy Stor. Mater.* **2022**, *50*, 197–224. [CrossRef]
11. Wang, J.; Xu, Z.; Zhang, Q.; Song, X.; Lu, X.; Zhang, Z.; Onyianta, A.J.; Wang, M.; Titirici, M.M.; Eichhorn, S.J. Stable Sodium-Metal Batteries in Carbonate Electrolytes Achieved by Bifunctional, Sustainable Separators with Tailored Alignment. *Adv. Mater.* **2022**, *34*, e2206367. [CrossRef]
12. Qin, M.; Zeng, Z.; Cheng, S.; Xie, J. Challenges and strategies of formulating low-temperature electrolytes in lithium-ion batteries. *Interdiscip. Mater.* **2023**, *2*, 308–336. [CrossRef]

13. Barnes, P.; Smith, K.; Parrish, R.; Jones, C.; Skinner, P.; Storch, E.; White, Q.; Deng, C.J.; Karsann, D.; Lau, M.L.; et al. A non-aqueous sodium hexafluorophosphate-based electrolyte degradation study: Formation and mitigation of hydrofluoric acid. *J. Power Sources* **2020**, *447*, 227363–227370. [CrossRef]
14. Mosallanejad, B.; Malek, S.S.; Ershadi, M.; Daryakenari, A.A.; Cao, Q.; Ajdari, F.B.; Ramakrishna, S. Cycling degradation and safety issues in sodium-ion batteries: Promises of electrolyte additives. *J. Electroanal. Chem.* **2021**, *895*, 115505–115522. [CrossRef]
15. Eom, J.Y.; Jung, I.H.; Lee, J.H. Effects of vinylene carbonate on high temperature storage of high voltage Li-ion batteries. *J. Power Sources* **2011**, *196*, 9810–9814. [CrossRef]
16. Genieser, R.; Loveridge, M.; Bhagat, R. Practical high temperature (80 °C) storage study of industrially manufactured Li-ion batteries with varying electrolytes. *J. Power Sources* **2018**, *386*, 85–95. [CrossRef]
17. Hu, D.; Zhang, Q.; Tian, J.; Chen, L.; Li, N.; Su, Y.; Bao, L.; Lu, Y.; Cao, D.; Yan, K.; et al. High-Temperature Storage Deterioration Mechanism of Cylindrical 21700-Type Batteries Using Ni-Rich Cathodes under Different SOCs. *ACS Appl. Mater. Interfaces* **2021**, *13*, 6286–6297. [CrossRef] [PubMed]
18. Kraft, V.; Weber, W.; Grutzke, M.; Winter, M.; Nowak, S. Study of decomposition products by gas chromatography-mass spectrometry and ion chromatography-electrospray ionization-mass spectrometry in thermally decomposed lithium hexafluorophosphate-based lithium ion battery electrolytes. *RSC Adv.* **2015**, *5*, 80150–80157. [CrossRef]
19. Gachot, G.; Ribiere, P.; Mathiron, D.; Grugeon, S.; Armand, M.; Leriche, J.B.; Pilard, S.; Laruelle, S. Gas chromatography/mass spectrometry as a suitable tool for the Li-ion battery electrolyte degradation mechanisms study. *Anal. Chem.* **2011**, *83*, 478–485. [CrossRef]
20. Horsthemke, F.; Friesen, A.; Monnighoff, X.; Stenzel, Y.P.; Grutzke, M.; Andersson, J.T.; Winter, M.; Nowak, S. Fast screening method to characterize lithium ion battery electrolytes by means of solid phase microextraction-gas chromatography-mass spectrometry. *RSC Adv.* **2017**, *7*, 46989–46998. [CrossRef]
21. Peschel, C.; Horsthemke, F.; Leissing, M.; Wiemers-Meyer, S.; Henschel, J.; Winter, M.; Nowak, S. Analysis of Carbonate Decomposition During Solid Electrolyte Interphase Formation in Isotope-Labeled Lithium Ion Battery Electrolytes: Extending the Knowledge about Electrolyte Soluble Species. *Batter. Supercaps* **2020**, *3*, 1183–1192. [CrossRef]
22. Campion, C.L.; Li, W.; Lucht, B.L. Thermal Decomposition of $LiPF_6$-Based Electrolytes for Lithium-Ion Batteries. *J. Electrochem. Soc.* **2005**, *152*, A2327–A2334. [CrossRef]
23. Gauthier, N.; Courrèges, C.; Demeaux, J.; Tessier, C.; Martinez, H. Impact of the cycling temperature on electrode/electrolyte interfaces within $Li_4Ti_5O_{12}$ vs $LiMn_2O_4$ cells. *J. Power Sources* **2020**, *448*, 227573–227583. [CrossRef]
24. Ravdel, B.; Abraham, K.M.; Gitzendanner, R.; DiCarlo, J.; Lucht, B.; Campion, C. Thermal stability of lithium-ion battery electrolytes. *J. Power Sources* **2003**, *119–121*, 805–810. [CrossRef]
25. Lee, H.H.; Wan, C.C.; Wang, Y.Y. Thermal Stability of the Solid Electrolyte Interface on Carbon Electrodes of Lithium Batteries. *J. Electrochem. Soc.* **2004**, *151*, A542–A547. [CrossRef]
26. Wang, Q.; Jiang, B.; Li, B.; Yan, Y. A critical review of thermal management models and solutions of lithium-ion batteries for the development of pure electric vehicles. *Renew. Sust. Energ. Rev.* **2016**, *64*, 106–128. [CrossRef]
27. Bandhauer, T.M.; Garimella, S.; Fuller, T.F. A Critical Review of Thermal Issues in Lithium-Ion Batteries. *J. Electrochem. Soc.* **2011**, *158*, R1–R25. [CrossRef]
28. Ma, S.; Jiang, M.; Tao, P.; Song, C.; Wu, J.; Wang, J.; Deng, T.; Shang, W. Temperature effect and thermal impact in lithium-ion batteries: A review. *Prog. Nat. Sci.* **2018**, *28*, 653–666. [CrossRef]
29. Hofmann, A.; Uhlmann, N.; Ziebert, C.; Wiegand, O.; Schmidt, A.; Hanemann, T. Preventing Li-ion cell explosion during thermal runaway with reduced pressure. *Appl. Therm. Eng.* **2017**, *124*, 539–544. [CrossRef]
30. Chen, J.; Huang, Z.; Wang, C.; Porter, S.; Wang, B.; Lie, W.; Liu, H.K. Sodium-difluoro(oxalato)borate (NaDFOB): A new electrolyte salt for Na-ion batteries. *Chem. Commun.* **2015**, *51*, 9809–9812. [CrossRef]
31. Purushotham, U.; Takenaka, N.; Nagaoka, M. Additive effect of fluoroethylene and difluoroethylene carbonates for the solid electrolyte interphase film formation in sodium-ion batteries: A quantum chemical study. *RSC Adv.* **2016**, *6*, 65232–65242. [CrossRef]
32. Zheng, X.; Huang, L.; Ye, X.; Zhang, J.; Min, F.; Luo, W.; Huang, Y. Critical effects of electrolyte recipes for Li and Na metal batteries. *Chem* **2021**, *7*, 2312–2346. [CrossRef]
33. Dahbi, M.; Nakano, T.; Yabuuchi, N.; Fujimura, S.; Chihara, K.; Kubota, K.; Son, J.-Y.; Cui, Y.-T.; Oji, H.; Komaba, S. Effect of Hexafluorophosphate and Fluoroethylene Carbonate on Electrochemical Performance and the Surface Layer of Hard Carbon for Sodium-Ion Batteries. *ChemElectroChem* **2016**, *3*, 1856–1867. [CrossRef]
34. Dugas, R.; Ponrouch, A.; Gachot, G.; David, R.; Palacin, M.R.; Tarascon, J.M. Na Reactivity toward Carbonate-Based Electrolytes: The Effect of FEC as Additive. *J. Electrochem. Soc.* **2016**, *163*, A2333–A2339. [CrossRef]
35. Cheng, Z.; Mao, Y.; Dong, Q.; Jin, F.; Shen, Y.; Chen, L. Fluoroethylene Carbonate as an Additive for Sodium-Ion Batteries: Effect on the Sodium Cathode. *Acta Phys. Chim. Sin.* **2019**, *35*, 868–875. [CrossRef]
36. Lu, H.; Wu, L.; Xiao, L.; Ai, X.; Yang, H.; Cao, Y. Investigation of the Effect of Fluoroethylene Carbonate Additive on Electrochemical Performance of Sb-Based Anode for Sodium-Ion Batteries. *Electrochim. Acta* **2016**, *190*, 402–408. [CrossRef]
37. Eshetu, G.G.; Martinez-Ibañez, M.; Sánchez-Diez, E.; Gracia, I.; Li, C.; Rodriguez-Martinez, L.M.; Rojo, T.; Zhang, H.; Armand, M. Electrolyte Additives for Room-Temperature, Sodium-Based, Rechargeable Batteries. *Chem. Asian J.* **2018**, *13*, 2770–2780. [CrossRef]

38. Bouibes, A.; Takenaka, N.; Kubota, K.; Komaba, S.; Nagaoka, M. Development of advanced electrolytes in Na-ion batteries: Application of the Red Moon method for molecular structure design of the SEI layer. *RSC Adv.* **2021**, *12*, 971–984. [CrossRef]
39. Hou, X.; Li, T.; Qiu, Y.; Jiang, M.; Zheng, Q.; Li, X. Weak coulomb interaction between anions and Na+ during solvation enabling desirable solid electrolyte interphase and superior kinetics for HC-based sodium ion batteries. *Chem. Eng. J.* **2023**, *453*, 139932. [CrossRef]
40. Shin, H.; Park, J.; Sastry, A.M.; Lu, W. Effects of Fluoroethylene Carbonate (FEC) on Anode and Cathode Interfaces at Elevated Temperatures. *J. Electrochem. Soc.* **2015**, *162*, A1683–A1692. [CrossRef]
41. Eshetu, G.G.; Elia, G.A.; Armand, M.; Forsyth, M.; Komaba, S.; Rojo, T.; Passerini, S. Electrolytes and Interphases in Sodium-Based Rechargeable Batteries: Recent Advances and Perspectives. *Adv. Energy Mater.* **2020**, *10*, 2000093–2000133. [CrossRef]
42. Gao, L.; Chen, J.; Liu, Y.; Yamauchi, Y.; Huang, Z.; Kong, X. Revealing the chemistry of an anode-passivating electrolyte salt for high rate and stable sodium metal batteries. *J. Mater. Chem. A* **2018**, *6*, 12012–12017. [CrossRef]
43. Gao, L.; Chen, J.; Chen, Q.; Kong, X. The chemical evolution of solid electrolyte interface in sodium metal batteries. *Sci. Adv.* **2022**, *8*, eabm4606. [CrossRef] [PubMed]
44. Xie, M.; Wu, F.; Huang, Y. *Sodium-Ion Batteries: Advanced Technology and Applications*; De Gruyter: Berlin, Germany, 2022.
45. Chen, X.; Shen, X.; Hou, T.-Z.; Zhang, R.; Peng, H.-J.; Zhang, Q. Ion-Solvent Chemistry-Inspired Cation-Additive Strategy to Stabilize Electrolytes for Sodium-Metal Batteries. *Chem* **2020**, *6*, 2242–2256. [CrossRef]
46. Sun, Z.; Fu, W.; Liu, M.Z.; Lu, P.; Zhao, E.; Magasinski, A.; Liu, M.; Luo, S.; McDaniel, J.; Yushin, G. A nanoconfined iron(iii) fluoride cathode in a NaDFOB electrolyte: Towards high-performance sodium-ion batteries. *J. Mater. Chem. A* **2020**, *8*, 4091–4098. [CrossRef]
47. Law, H.M.; Yu, J.; Kwok, S.C.T.; Zhou, G.; Robson, M.J.; Wu, J.; Ciucci, F. A hybrid dual-salt polymer electrolyte for sodium metal batteries with stable room temperature cycling performance. *Energy Stor. Mater.* **2022**, *46*, 182–191. [CrossRef]
48. Metzger, M.; Strehle, B.; Solchenbach, S.; Gasteiger, H.A. Hydrolysis of Ethylene Carbonate with Water and Hydroxide under Battery Operating Conditions. *J. Electrochem. Soc.* **2016**, *163*, A1219–A1225. [CrossRef]
49. Caracciolo, L.; Madec, L.; Gachot, G.; Martinez, H. Impact of the Salt Anion on K Metal Reactivity in EC/DEC Studied Using GC and XPS Analysis. *ACS Appl. Mater. Interfaces* **2021**, *13*, 57505–57513. [CrossRef]
50. Laruelle, S.; Pilard, S.; Guenot, P.; Grugeon, S.; Tarascon, J.M. Identification of Li-based electrolyte degradation products through DEI and ESI high-resolution mass spectrometry. *J. Electrochem. Soc.* **2004**, *151*, A1202–A1209. [CrossRef]
51. Grützke, M.; Weber, W.; Winter, M.; Nowak, S. Structure determination of organic aging products in lithium-ion battery electrolytes with gas chromatography chemical ionization mass spectrometry (GC-CI-MS). *RSC Adv.* **2016**, *6*, 57253–57260. [CrossRef]
52. Stenzel, Y.P.; Horsthemke, F.; Winter, M.; Nowak, S. Chromatographic Techniques in the Research Area of Lithium Ion Batteries: Current State-of-the-Art. *Separations* **2019**, *6*, 26. [CrossRef]
53. Fang, C.; Tran, T.N.; Zhao, Y.Z.; Liu, G. Electrolyte decomposition and solid electrolyte interphase revealed by mass spectrometry. *Electrochim Acta* **2021**, *399*, 139362–139371. [CrossRef]
54. Mogensen, R.; Brandell, D.; Younesi, R. Solubility of the Solid Electrolyte Interphase (SEI) in Sodium Ion Batteries. *ACS Energy Lett.* **2016**, *1*, 1173–1178. [CrossRef]
55. Ma, L.A.; Naylor, A.J.; Nyholm, L.; Younesi, R. Strategies for Mitigating Dissolution of Solid Electrolyte Interphases in Sodium-Ion Batteries. *Angew. Chem. Int. Ed.* **2021**, *60*, 4855–4863. [CrossRef]
56. Single, F.; Horstmann, B.; Latz, A. Dynamics and morphology of solid electrolyte interphase (SEI). *Phys. Chem. Chem. Phys.* **2016**, *18*, 17810–17814. [CrossRef] [PubMed]
57. Nayak, P.K.; Yang, L.; Brehm, W.; Adelhelm, P. From Lithium-Ion to Sodium-Ion Batteries: Advantages, Challenges, and Surprises. *Angew. Chem. Int. Ed.* **2018**, *57*, 102–120. [CrossRef]
58. Xu, G.-L.; Amine, R.; Abouimrane, A.; Che, H.; Dahbi, M.; Ma, Z.-F.; Saadoune, I.; Alami, J.; Mattis, W.L.; Pan, F.; et al. Challenges in Developing Electrodes, Electrolytes, and Diagnostics Tools to Understand and Advance Sodium-Ion Batteries. *Adv. Energy Mater.* **2018**, *8*, 1702403–1702465. [CrossRef]
59. Zhang, S.S. A review on electrolyte additives for lithium-ion batteries. *J. Power Sources* **2006**, *162*, 1379–1394. [CrossRef]
60. Chang, Z.; Qiao, Y.; Deng, H.; Yang, H.; He, P.; Zhou, H. A stable high-voltage lithium-ion battery realized by an in-built water scavenger. *Energy Environ. Sci.* **2020**, *13*, 1197–1204. [CrossRef]
61. Cho, I.H.; Kim, S.-S.; Shin, S.C.; Choi, N.-S. Effect of SEI on Capacity Losses of Spinel Lithium Manganese Oxide/Graphite Batteries Stored at 60 °C. *Electrochem. Solid-State Lett.* **2010**, *13*, A168–A172. [CrossRef]
62. Sinha, N.N.; Burns, J.C.; Dahn, J.R. Storage Studies on Li/Graphite Cells and the Impact of So-Called SEI-Forming Electrolyte Additives. *J. Electrochem. Soc.* **2013**, *160*, A709–A714. [CrossRef]
63. Müller, C.; Wang, Z.; Hofmann, A.; Stüble, P.; Liu-Théato, X.; Klemens, J.; Smith, A. Influences on Reliable Capacity Measurements of Hard Carbon in Highly Loaded Electrodes. *Batter. Supercaps* **2023**, e202300322. [CrossRef]
64. Hofmann, A.; Müller, F.; Schöner, S.; Jeschull, F. Revealing the Formation of Dialkyl Dioxahexane Dioate Products from Ethylene Carbonate based Electrolytes on Lithium and Potassium Surfaces. *Batter. Supercaps* **2023**, e202300325. [CrossRef]
65. Mrozik, W.; Rajaeifar, M.A.; Heidrich, O.; Christensen, P. Environmental impacts, pollution sources and pathways of spent lithium-ion batteries. *Energy Environ. Sci.* **2021**, *14*, 6099–6121. [CrossRef]

66. Zhao, Y.; Kang, Y.; Wozny, J.; Lu, J.; Du, H.; Li, C.; Li, T.; Kang, F.; Tavajohi, N.; Li, B. Recycling of sodium-ion batteries. *Nat. Rev. Mater.* **2023**, *8*, 623–634. [CrossRef]
67. Pfeiffer, L.F.; Li, Y.; Mundszinger, M.; Geisler, J.; Pfeifer, C.; Mikhailova, D.; Omar, A.; Baran, V.; Biskupek, J.; Kaiser, U.; et al. Origin of Aging of a P2-$Na_xMn_{3/4}Ni_{1/4}O_2$ Cathode Active Material for Sodium-Ion Batteries. *Chem. Mater.* **2023**, *35*, 8065–8080. [CrossRef]
68. Zhang, W.; Zhang, Y.; Zhou, J.; Li, X.; Zhou, W.; Zhang, D.; Mao, J.; Dai, K. Optimizing Electrochemical Performance in Sodium-Ion Batteries using O3-type $Na_{0.90}Cu_{0.22}Fe_{0.30}Mn_{0.48}O_2$ and Hard Carbon. *J. Electrochem. Soc.* **2023**, *170*, 70518–70533. [CrossRef]
69. Xia, X.; Lamanna, W.M.; Dahn, J.R. The Reactivity of Charged Electrode Materials with Sodium Bis(trifluoromethanesulfonyl)imide (NaTFSI) Based-Electrolyte at Elevated Temperatures. *J. Electrochem. Soc.* **2013**, *160*, A607–A609. [CrossRef]
70. Ding, M.S.; Xu, K.; Jow, T.R. Liquid-Solid Phase Diagrams of Binary Carbonates for Lithium Batteries. *J. Electrochem. Soc.* **2000**, *147*, 1688–1694. [CrossRef]
71. Nanbu, N.; Suzuki, K.; Yagi, N.; Sugahara, M.; Takehara, M.; Ue, M.; Sasaki, Y. Use of Fluoroethylene Carbonate as Solvent for Electric Double-Layer Capacitors. *Electrochemistry* **2007**, *75*, 607–610. [CrossRef]
72. Xu, K. *Electrolytes, Interfaces and Interphases: Fundamentals and Applications in Batteries*; Royal Society of Chemistry: London, UK, 2023.
73. Ponrouch, A.; Marchante, E.; Courty, M.; Tarascon, J.-M.; Palacín, M.R. In search of an optimized electrolyte for Na-ion batteries. *Energy Environ. Sci.* **2012**, *5*, 8572–8583. [CrossRef]
74. Hofmann, A.; Wang, Z.; Bautista, S.P.; Weil, M.; Müller, F.; Löwe, R.; Schneider, L.; Mohsin, I.U.; Hanemann, T. Comprehensive characterization of propylene carbonate based liquid electrolyte mixtures for sodium-ion cells. *Electrochim. Acta* **2022**, *403*, 139670–139687. [CrossRef]
75. Zhang, D.; Zhu, Q.; Wang, Y.; Zhao, C.L.; Liu, S.B.; Xu, S.D. Electrochemical Performance of Sodium Difluoro(oxalato)borate as the Additive of Non-Aqueous Electrolytes for Sodium-Ion Batteries. *J. Electrochem.* **2017**, *23*, 473–479.
76. Eshetu, G.G.; Grugeon, S.; Kim, H.; Jeong, S.; Wu, L.; Gachot, G.; Laruelle, S.; Armand, M.; Passerini, S. Comprehensive Insights into the Reactivity of Electrolytes Based on Sodium Ions. *ChemSusChem* **2016**, *9*, 462–471. [CrossRef]
77. Sun, Y.; Shi, P.; Xiang, H.; Liang, X.; Yu, Y. High-Safety Nonaqueous Electrolytes and Interphases for Sodium-Ion Batteries. *Small* **2019**, *15*, 1805479–1805494. [CrossRef]

Disclaimer/Publisher's Note: The statements, opinions and data contained in all publications are solely those of the individual author(s) and contributor(s) and not of MDPI and/or the editor(s). MDPI and/or the editor(s) disclaim responsibility for any injury to people or property resulting from any ideas, methods, instructions or products referred to in the content.

batteries

Article

Gravure-Printed Anodes Based on Hard Carbon for Sodium-Ion Batteries

Maria Montanino [1,*], Claudia Paoletti [2], Anna De Girolamo Del Mauro [1] and Giuliano Sico [1]

[1] ENEA—Italian National Agency for New Technologies, Energy and Sustainable Economic Development, Portici Research Centre, Piazzale E. Fermi 1, 80055 Portici, Italy; anna.degirolamo@enea.it (A.D.G.D.M.); giuliano.sico@enea.it (G.S.)

[2] ENEA—Italian National Agency for New Technologies, Energy and Sustainable Economic Development, Via Anguillarese 301, 00123 Roma, Italy; claudia.paoletti@enea.it

* Correspondence: maria.montanino@enea.it

Abstract: Printed batteries are increasingly being investigated for feeding small, wearable devices more and more involved in our daily lives, promoting the study of printing technologies. Among these, gravure is very attractive as a low-cost and low-waste production method for functional layers in different fields, such as energy, sensors, and biomedical, because it is easy to scale up industrially. Thanks to our research, the feasibility of gravure printing was recently proved for rechargeable lithium-ion batteries (LiBs) manufacturing. Such studies allowed the production of high-quality electrodes involving different active materials with high stability, reproducibility, and good performance. Going beyond lithium-based storage devices, our attention was devoted on the possibility of employing highly sustainable gravure printing for sodium-ion batteries (NaBs) manufacturing, following the trendy interest in sodium, which is more abundant, economical, and ecofriendly than lithium. Here a study on gravure printed anodes for sodium-ion batteries based on hard carbon as an active material is presented and discussed. Thanks to our methodology centered on the capillary number, a high printing quality anodic layer was produced providing typical electrochemical behavior and good performance. Such results are very innovative and relevant in the field of sodium-ion batteries and further demonstrate the high potential of gravure in printed battery manufacturing.

Keywords: printed batteries; sodium-ion batteries; hard carbon; gravure printing; capillary number; anodes; anodic active materials

1. Introduction

The necessity to feed small, portable, and wearable devices increasingly moved the research on printed batteries. Such storage devices, which have a volume below 10 mm^3 and capacity in the range of 5–10 mAh cm^{-2}, must be highly customizable to be perfectly integrated into the device needing the power supply [1]. For this reason, printing technologies were considered a useful manufacturing method, and different techniques were investigated. Printing technologies, in fact, offer the possibility to produce low-cost, thin, flexible, and arbitrary bi-dimensional shapes, different functional layers, and devices (even in the case of multifunctional layered structures), at high throughput, as well as in large areas, with high layer control in terms of homogeneity and thus reproducibility, especially if compared to conventional coating techniques [2]. In fact, coatings are limited in terms of customizability and their main drawback is the low homogeneity of the produced layers, which has consequences on their performances and reproducibility.

In addition, printing technologies are widely investigated for the production of electronic devices. In such cases, the industrial manufacturing of printed batteries could be in-line with the manufacturing of printed electronics in which they have to be inserted. This could result in a great advantage for the economy of the production process.

Printed batteries are already industrially produced, mainly as not-rechargeable primary systems. The most used printing technology, to date, is the screen printing [1].

Among the printing technologies, gravure is very appealing for the production of functional layers, as electrodes for batteries, thanks to its capability to couple high throughput (speed up to 600 m min^{-1}) and high printing quality (resolution < 2 μm) [3,4]. In this regard, such a technique can be useful in the study of new materials, when the deposition method is important as the investigated material to obtain its best possible performance. The use of high-performance materials would enable gravure manufacturing for different fields and applications in which large areas are required since it is roll-to-roll compatible. Gravure is widely used in industry for packaging, magazines, and currency; for this reason, each result obtained at a lab scale would be easy to scale up at the industrial level. Moreover, such a technique is highly sustainable, since it has low waste of materials and low energy consumption [5–7]. Despite its many advantages, to date, gravure is few investigated probably because of its main drawback: the use of diluted inks. In fact, an ink suitable for gravure printing has to have a viscosity below 100 mPa. The necessity of diluted ink makes it difficult to obtain a functional layer, especially in the case of composites, and to achieve mass loading useful for practical applications. During the last decade, our studies made an important contribution to the investigation of gravure as a manufacturing method for functional layers, also in the case of electrodes for lithium-ion batteries, improving the comprehension of the process and overcoming some of the most important issues related to the technology [8]. Aiming to rule the formulation of inks suitable for gravure printing, a methodology was recently investigated [9]. Considering the complexity of the process and simplifying its description through the dimensional analysis, the capillary number (Ca) represents the main forces involved during the process:

$$Ca = \eta U / \gamma \quad (1)$$

where η and γ are the viscosity and the surface tension of the ink, respectively, at a fixed printing speed U. Experimentally, it was proved that when the Ca value is approximatively 1, good printing quality of the layer is achieved and the printed layer is able to display the desired functionality. Each time changing involved materials, different formulations have to be investigated, thus viscosity and surface tension have to be studied to find the best possible ink and the process parameters through the Ca = 1 method. However, our study demonstrated that the printed layer performance also depends on specific issues related to the materials involved in the layer preparation. In the case of electrodes for batteries factors such as mass loading, component ratio, materials size, component distribution, and layer density play a fundamental role in the electrodic performance [8]. Using such methodology, gravure-printed electrodes for lithium-ion batteries (LiBs) were successfully obtained, and high performance was shown when material-specific challenges were addressed [8]. The important results obtained for LiBs motivated our investigation of different storage systems, beyond the lithium chemistry, such as sodium-ion batteries (NaBs). NaBs are widely investigated as alternatives to LiBs due to the abundance and low cost of sodium [10,11]. To date, LiBs are the most investigated and employed storage system for portable and wearable electronic devices, most recently also for electric vehicles and stationary storage systems [12]; however, lithium is expensive and strategic. Na has similar proprieties to Li and the metal ions batteries also operate in a similar way. The NaB mechanism can be described as a reversible migration of Na^+ between the anode and cathode during a charge/discharge process. Recently, sodiated transition metal oxides, phosphates, and organic compounds have been introduced as cathode materials for NaBs; at the same time, studies on anodic materials are carried out, involving carbonaceous materials, transition metal oxides or sulfides, and intermetallic and organic compounds [13].

Na^+ has a radius (1.03 Å) larger than Li^+ (0.71 Å), affecting phase stability, transport proprieties, and interphase formation [14]. Due to the different ion sizes, graphite, which is widely employed as an anodic material for LiBs, cannot be used as an anode for NaBs: it has a small interlayered space and tends to form NaC_x too [13]. Most of the investigated active materials for NaBs anodes are not suitable for practical applications due to complex preparation processes [13]. Therefore, advanced anode materials for NaBs, with adequate voltage storage, large reversible capacity, and structural stability, have to be developed to achieve high-performance storage systems. Hard carbon (HC) is a class of carbonaceous materials appealing as an anodic material for NaBs due to their relatively low cost, abundance, and performance [13]. HC structure is different from the ordered layered structure of graphite, consisting of randomly distributed turbostratic domains, containing curved graphene nano-sheets, with expanded interlayer spacing and micropores [14]. Such structure is usually obtained through the pyrolysis of a carbon precursor at 1000 °C, obtaining highly branched and cross-linked structure. The branches prevent the ordered graphitization, even pyrolyzing at higher temperatures (up to 3000 °C). The specific value of HC theoretical capacity depends on differences in structure resulting from the synthesis process [14]. Several issues have limited the practical use of HC as an active material for anode manufacturing, such as safety problems related to electroplating and poor cyclability [15]. In this regard, the HC irreversible capacity is the most critical aspect. Such feature depends on several characteristics such as particle sizes, electrolytes, vacancy defects, and porosity [13]. For this reason, further investigation are necessary for boosting the development of HC for NaBs and stabilizing the initial coulombic efficiency. In this paper, the study on a gravure-printed anode for NaBs based on a commercial HC as an active material is reported and discussed. An investigation of the mixing time during the ink preparation was also proposed to evaluate its effect on the electrochemical performance of the active material and of the printed anodes. Sodium-printed batteries were very recently investigated [16]. This is the first example of a gravure-printed anode for NaBs.

2. Materials and Methods

The used active material, hard carbon BHC-300, was provided by MTI Corporation (Richmond, CA, USA). According to the provider, the particle size is in the range of 3–7 μm and the specific surface area is 8 m^2 g^{-1}. The inks were prepared using active material (88 wt%), sodium salt of the carboxymethyl cellulose (CMC) (by Dow Wolff Cellulosics (Bomlitz, Germany)) (6 wt%) as a binder, and carbon Super P (by Imerys (Paris, France)) as an electric conductor (6 wt%). The used solvent was a mixture of water and 2-propanol (by Sigma–Aldrich (St. Louis, MO, USA)) (90/10 wt/wt%). Preliminary tests were carried out to determine the best solid content of the ink in terms of viscosity and thus printability. The inks were characterized from a rheological point of view at different temperatures and different shear rates using a Viscotester (by Thermofisher (Waltham, MA, USA)). The anodic layers were printed on Copper foils pre-treated by Corona (by Tantec (Roskilde, Denmark)). The used lab-scale gravure printer was an IGT-G1–5 (Almere, The Netherlands) equipped with a cylinder with a line density of 40 lines cm^{-1}, a stylus angle of 120°, a cell depth of 72 μm, and a screen angle of 53°. To increase the thickness and the mass loading of the printed anodes, a multilayer approach was adopted, printing 10 and 15 consecutive layers, one on each other, overlapping the same ink at the same printing conditions. Between each printed layer, a fast dryer was performed using a blow of air. The layers were printed at a speed of 36 m min^{-1} and a printing force of 700 N and finally dried at 100 °C for 1 h. Printed anodes were morphologically investigated using scanning electron microscopy (Zeiss (Jena, Germany); LEO 1530) and electrochemically characterized in half cells versus metallic Na. The cells were prepared using glass fiber separators (by Whatman (Maidstone, UK) and a 1 M solution of $NaClO_4$ (by Sigma–Aldrich) in propylene carbonate by Fluka (Charlotte, NC, USA) as electrolyte. Galvanostatic cycling measurements were performed on the cells using a Maccor 4000 (Tulsa, OK, USA) at 20 °C and fixed 0.1 C rate.

3. Results and Discussion

The gravure process consists of the direct transfer of ink from the engraved cells of a printing cylinder onto a flexible substrate thanks to the pressure of a counter cylinder, as depicted in Figure 1.

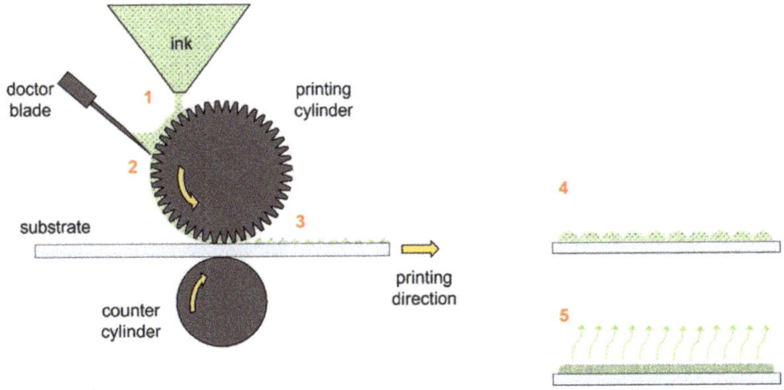

Figure 1. Operating roll-based gravure printing principle, showing inking (1), doctoring (2), transfer (3), spreading (4), and drying (5) sub-processes. Adapted with permission from Ref. [8].

Printing quality depends on several parameters related to the following: a. Gravure cylinder, such as cell geometry, cell density, surface energy, and diameter; b. Process, such as printing speed, printing force, and web tension; c. Ink, such as viscosity, surface tension, solvent evaporation rate, and solid content; d. Substrate, such as surface energy, roughness, and porosity. The process can be described as a series of sub-processes: 1. Inking, by filling the engravings of the printer cylinder with ink; 2. Doctoring, by removing the excess of ink from the printing cylinder; 3. Transfer of ink onto the flexible substrate; 4. Spreading, through the coalescence of ink droplets onto the substrate; 5. Drying, by completely removing the solvent. Each of such stages has its own ideal regime and all of them concur to determine the final printing quality [17]. The gravure process is very complex from the fluid-dynamic point of view and is not easy to model. The dimensionless analysis can help to simplify the process description and study through the capillary number. The ink formulation has a key role in the printing process; the Ca methodology was adopted to determine the best and most useful ink characteristics for the formulation and the process.

Experimentally, several inks containing different solid contents were prepared and rheologically tested to verify their Newtonian behavior. The solid content was chosen to give a viscosity below 100 mPa s. Among the inks respecting such requirements, we chose the one with the highest solid content. After determining the ink viscosity and its surface tension, Ca values were calculated at different printing speeds, selecting the practical one in a way to maximize the speed and contemporarily obtain a Ca value approaching the unity. The necessity to maximize the printing speed belongs to the idea of approaching the industrial conditions, minimizing time and cost. The research was also devoted to decreasing the amount of electric conductors and polymers in the final layer. The sustainability of the process and the products was enhanced through the use of water as a prevalent solvent since a water-soluble binder, carboxymethyl cellulose, was used in the ink composition. A small amount of 2-Propanol was used as a co-solvent to decrease the high ink surface tension due to the use of water (from 72 to 42 mN m^{-1}) and make the ink printable with a surface tension lower than the surface energy of the printing cylinder and of the substrate [8]. The ink was prepared by keeping the relative ratio of the solid components constant and changing the total solid amount.

The tested solid contents were 18, 15, and 14 wt%. The prepared inks were rheologically tested and all of them showed a Newtonian behavior (see Figure 2), namely, having a viscosity independent of the shear rate; the representative value of viscosity, at shear rate 100 s^{-1}, for the tested inks is reported in Table 1.

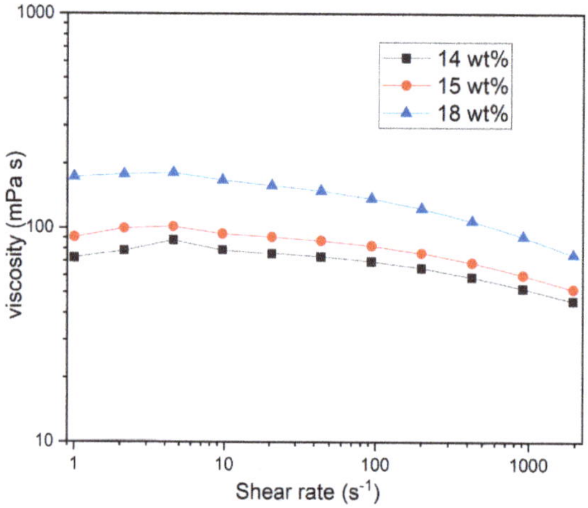

Figure 2. Viscosity versus shear rate for the prepared inks at different solid content (14, 15, 18 wt%).

Table 1. Viscosity and Ca for the inks at different solid content.

Solid Content (wt%)	Viscosity @ 20 °C (mPa s)	Printing Speed (m s^{-1})	Ca
18	138	0.6 0.4 0.2	1.97 1.31 0.66
15	83	0.6 0.4 0.2	1.18 0.80 0.40
14	70	0.6 0.4 0.2	1.00 0.67 0.34

To achieve high printing quality, we proposed a methodology mainly based on the capillary number approaching the unity [8]. In Table 1, the calculated values of Ca were reported for the inks at different solid contents.

The Ca was determined considering the surface tension of the solvent (42 mN m^{-1}) [18]. The choice of the printing speed follows the calculation of the Ca based on the measurements of ink viscosity and surface tension. The best possible ink is the one with suitable viscosity for gravure (below 100 mPa s) and the highest possible solid content, which combined with the printing speed brings a Ca = 1. In other words, the choice is based on a compromise to maximize both the solid content and the printing speed (for industrial reasons), achieving the Ca = 1 target. Higher speeds are possible; however, they require lower viscosity, namely, lower solid content of the ink, decreasing the mass loading of the printed layer, thus requiring a higher number of layers to be overlapped. The investigated inks, reported in Table 1, show that the 18 wt% ink had a viscosity out of the gravure printing range. Regarding the 15 wt% ink, the 0.4 m s^{-1} speed was not preferred in the view of speed maximization, while the 0.6 m s^{-1} speed gave a Ca higher than the one obtained at the same speed for the 14 wt% ink. Such excess on the Ca favors viscous forces over surface tension, causing possible pin-holes or void formation due to a lower ink spreading. For all such reasons, the 14 wt% ink was selected to realize the printed layer as the best compromise and was printed at 0.6 m s^{-1} and at a printing force of 700 N. Different ball milling time treatments were tested on the selected ink to investigate the homogenization of the particle sizes and distribution: 15 min, 60 min, and 3 h. All the inks were printed overlapping ten layers, while the ink ball milled for three hours was printed overlapping 15 layers too. All the printed samples showed high printing quality, and thus good coverage and homogeneity, without critical defects, such as voids and scratches, as shown by the SEM images (see Figures 3 and 4).

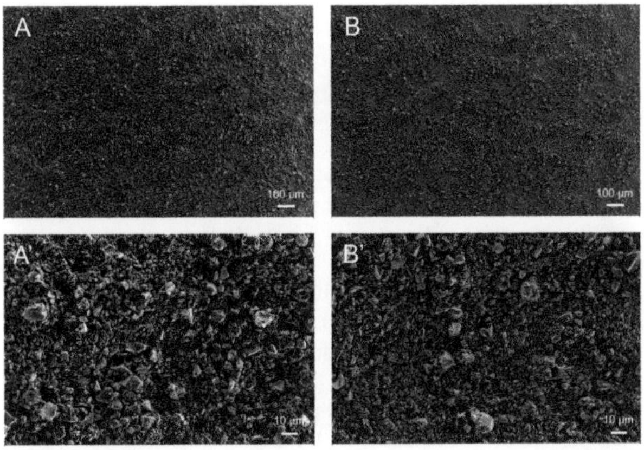

Figure 3. SEM images of the gravure printed layers: 15 min ball-milled ink (**A**,**A′**) and 60 min ball milled (**B**,**B′**) at different magnifications.

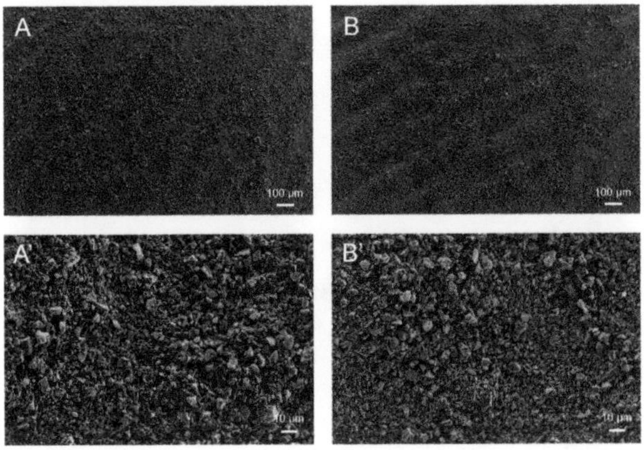

Figure 4. SEM images of the gravure printed layers: 3 h ball-milled ink overlapping 10 layers (**A**,**A′**) and 15 layers (**B**,**B′**) at different magnifications.

Sixty-minute ball milling increased the layer homogeneity decreasing the differences in particle size (see Figure 3). Further increase in ball milling time, up to 3 h, further increased the homogeneity of printed layers, as it is possible to notice by comparing the B, B′ images in Figure 3 and A, A′ images in Figure 4. When 15 layers of ink were overlapped instead of 10 layers, working at fixed printing parameters, a slight worsening in the printed layer quality can be observed (see Figure 4). The characteristics of the 10 printed layers are reported in Table 2.

Table 2. Characteristics of the printed layer, obtained by using different inks and calculated apparent density [19].

Ink/BM Time	Mass Loading (mg cm^{-2})	Active Material (mg cm^{-2})	Thickness (cm)	Apparent Layer Density (g cm^{-3})
15 min	0.80	0.70	2.6×10^{-3}	0.31
60 min	0.94	0.83	2.6×10^{-3}	0.36
3 h	0.92	0.81	2.0×10^{-3}	0.46

The increase in homogeneity of particle size with the increase in the ball milling time provoked a better packing of the materials in the printed layer, which is proved by the increase in apparent layer density from 0.31 to 0.46 g cm^{-3}.

The printed layers were characterized from the electrochemical point of view in half cell versus Na displaying anodic functionality. In Figure 5 voltage vs. specific capacity curves are reported for the discharge process of selected cycles of 10-layer anodes obtained by using 15 min and 3 h ball-milled inks. Such trends were typical of the sodiation-desodiation process occurring in the HC-based anode, showing adsorption/desorption of Na^+ at high potential and the reversible intercalation/deintercalation process, resulting in a plateau at low potential [20]. The 10-layer printed anode obtained using 15 min ball-milled ink showed an irreversible capacity in the range of 400–440 mAh g^{-1} for the first 20–30 cycles, following a drop in capacity to 130 mAh g^{-1}, as can be observed in Figure 6. The reversible capacity of 130 mAh g^{-1} was stable up to 100 cycles cycling at a fixed 0.1 C rate, without fading yet. The anodes printed using inks ball-milled for longer times (60 min and 3 h) did not show irreversible behavior, having from the beginning a reversible capacity of about 120 mAh g^{-1}, stable over 100 cycles at a fixed 0.1 C rate. In Figure 6, the specific capacity behavior of the printed layer obtained using 3 h ball-milled ink was reported as an example. The 15-layer gravure printed anode obtained with the 3 h ball-milled ink showed similar electrochemical behavior to the 10-printed-layer anode obtained with the same ink but displayed a lower specific capacity (90 mAh g^{-1}).

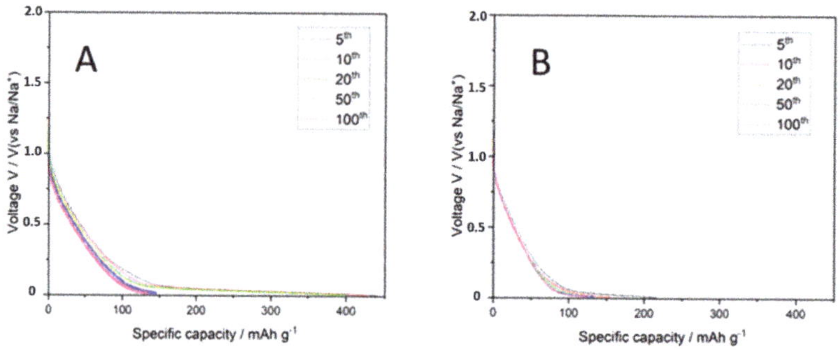

Figure 5. Voltage vs. specific capacity in discharge for gravure printed electrodes based on HC obtained using ball-milled inks at 15 min (**A**) and 3 h (**B**) for selected cycles (5th, 10th, 20th, 50th, 100th).

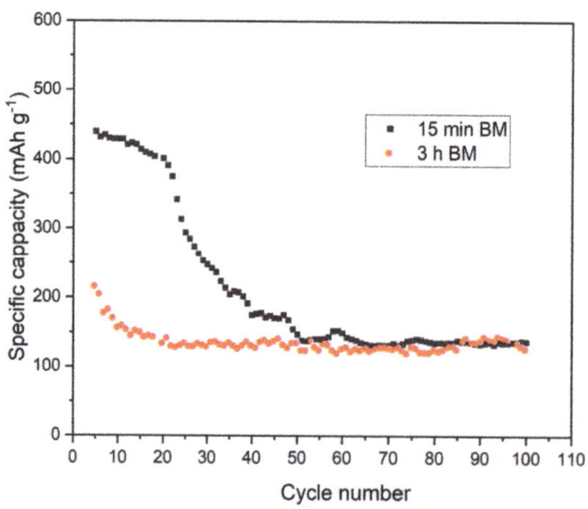

Figure 6. Discharge-specific capacity for gravure printed electrodes based on HC obtained using ball-milled inks at 15 min and 3 h.

The Na^+ storage mechanism for the HC is still under debate. According to the most recent results, this process can be described by a three-step insertion model: Na^+ ions are adsorbed on the defects of the HC-based anodic layer, then the Na^+ is intercalated in the HC structure and finally adsorbed onto the pore surface [21]. On such a basis, the obtained electrochemical results can be explained by considering the HC structure in the investigated gravure printed anodic layers. In fact, HC is composed of pseudo-graphitic and amorphous regions, which can be represented as a "house of cards" model, including a large number of defects, voids, and pores [11]. Due to such morphology a part of Na^+ forms clusters which are irreversibly trapped into the HC layers. The trapped sodium is not released in the subsequent de-sodiation process due to the strong interaction occurring between the graphene sheets, decreasing the sodium taking part in the redox process and causing the loss of Coulombic efficiency. This phenomenon, together with the Na^+ involved in the solid electrolyte interface (SEI) formation, causes a decrease in electrochemical performance leading to the irreversible fraction of capacity. Most likely, the prolonged ball-milling treatment is able to modify the HC structure affecting the charge/discharge process and dramatically cutting the irreversible part of the capacity. This could be the reason why the specific capacity results lower (from 400 to 120 mAh g^{-1}) since the first cycle. However, the capacity remains stable over a hundred cycles, demonstrating the high stability of the active material and the anode itself. Future works for enhancing the performances of the gravure printed anodes based on HC may aim for the stabilization of the HC structure within the printed layer, e.g., suitably tuning the porosity.

4. Conclusions

In this work, the feasibility of employing HC as an active material for Na-ion batteries in an anodic layer produced by gravure printing was investigated. The HC-based gravure printed anode is able to display up to 440 mAh g^{-1}. However, the intrinsic characteristic of the active material leads to a decrease in charge/discharge capacity, showing an irreversible part of capacity. The prolonged ball-milling of the ink probably affects the morphology of the printed anode, cutting out the irreversible part of the specific capacity, displaying about 120 mAh g^{-1}. Nevertheless, the displayed capacity remains unchanged over a hundred cycles, proving the stability of the active material and of the printed anode. On the base of such findings, further work can be employed to stabilize the HC structure in the printed layer, investigating the effect of porosity and/or reducing the defects.

Author Contributions: Conceptualization, M.M. and G.S.; methodology, M.M., A.D.G.D.M., C.P. and G.S.; validation, M.M. and G.S.; formal analysis, M.M. and G.S.; investigation, M.M., A.D.G.D.M., C.P. and G.S.; resources, M.M.; data curation, M.M. and G.S.; writing—original draft preparation, M.M.; writing—review and editing, M.M. and G.S.; visualization, M.M., A.D.G.D.M. and G.S.; supervision, M.M. and G.S.; funding acquisition, M.M. All authors have read and agreed to the published version of the manuscript.

Funding: This research was funded by Ministry of the Environment, grant number I53C22003080001, by Accordo di Programma 2022–2024 Progetto di ricerca integrato: Tecnologie di accumulo elettrochimico e termico; WP1-Accumulo elettrochimico: materiali avanzati.

Data Availability Statement: The raw data supporting the conclusions of this article will be made available by the authors upon request.

Conflicts of Interest: The authors declare no conflicts of interest.

References

1. Oliveira, J.; Costa, C.M.; Lanceros-Méndez, S. *Printed Batteries Materials, Technologies and Applications*; John Wiley & Sons Ltd.: Chichester, UK, 2018; pp. 1–14.
2. Krebs, F.C. Fabrication and processing of polymer solar cells: A review of printing and coating techniques. *Sol. Energy Mater. Sol. Cells* **2009**, *93*, 394–412. [CrossRef]
3. Grau, G.; Kitsomboonloha, R.; Subramanian, V. Fabrication of a high-resolution roll for gravure printing of 2 µm features. In Proceedings of the SPIE Organic Photonics + Electronics, San Diego, CA, USA, 9–13 August 2015.

4. Tiara, A.M.; Moon, H.; Cho, G.; Lee, J. Fully roll-to-roll gravure printed electronics: Challenges and the way to integrating logic gates. *Jpn. J. Appl. Phys.* **2022**, *61*, SE0802.
5. Søndergaard, R.R.; Hosel, M.; Krebs, F.C. Roll-to-Roll Fabrication of Large Area Functional Organic Materials. *J. Polym. Sci. Part B Polym. Phys.* **2013**, *51*, 16–34. [CrossRef]
6. Khan, S.; Lorenzelli, L.; Dahiya, R. Technologies for printing sensors and electronics over large flexible substrates: A review. *IEEE Sens. J.* **2015**, *15*, 3164–3185. [CrossRef]
7. Huang, Q.; Zhu, Y. Printing conductive nanomaterials for flexible and stretchable electronics: A review of materials, processes, and applications. *Adv. Mater. Technol.* **2019**, *4*, 1800546. [CrossRef]
8. Montanino, M.; Sico, G. Gravure Printing for Lithium-Ion Batteries Manufacturing: A Review. *Batteries* **2023**, *9*, 535. [CrossRef]
9. Montanino, M.; Paoletti, C.; De Girolamo Del Mauro, A.; Sico, G. The Influence of the Gravure Printing Quality on the Layer Functionality: The Study Case of LFP Cathode for Li-Ion Batteries. *Coatings* **2023**, *13*, 1214. [CrossRef]
10. Yabuuchi, N.; Kubota, K.; Dahbi, M.; Komada, S. Research development on sodium-ion batteries. *Chem. Rev.* **2014**, *114*, 11636. [CrossRef] [PubMed]
11. Slater, M.D.; Kim, D.; Lee, E.; Johnson, C.S. Sodium-ion batteries. *Adv. Funct. Mater.* **2013**, *23*, 947. [CrossRef]
12. Scrosati, B.; Garche, J. Lithium batteries: Status, prospects and future. *J. Power Sources* **2010**, *195*, 2419–2430. [CrossRef]
13. Yang, Y.; Wu, C.; He, X.-X.; Zhao, J.; Yang, Z.; Li, L.; Wu, X.; Li, L.; Chou, S.-L. Boosting the development of hard carbon for sodium-ion batteries: Strategies to optimize the initial coulombic efficiency. *Adv. Funct. Mater.* **2024**, *34*, 2302277. [CrossRef]
14. Xiao, B.; Rojo, T.; Li, X. Hard carbon as Sodium-ion battery anodes: Progress and challenges. *ChemSusChem* **2019**, *12*, 133–144. [CrossRef] [PubMed]
15. Hasegawa, G.; Kanamori, K.; Kannari, N.; Ozaki, J.-I.; Nakanishi, K.; Abe, T. Hard carbon anodes for Na-ion batteries: Toward a practical use. *ChemElectroChem* **2015**, *2*, 1917–1920. [CrossRef]
16. Ren, H.; Zhang, X.; Liu, Q.; Tang, W.; Liang, J.; Wu, W. Fully-printed flexible aqueous rechargable soldium-ion batteries. *Small* **2024**, *20*, 2312207. [CrossRef] [PubMed]
17. Sico, G.; Montanino, M.; Prontera, C.T.; De Girolamo Del Mauro, A.; Minarini, C. Gravure printing for thin film ceramics manufacturing from nanoparticles. *Ceram. Int.* **2018**, *44*, 19526–19534. [CrossRef]
18. Vázquez, G.; Alvarez, E.; Navaza, J.M. Surface Tension of Alcohol + Water from 20 to 50 °C. *J. Chem. Eng. Data* **1995**, *40*, 611–614. [CrossRef]
19. Montanino, M.; De Girolamo Del Mauro, A.; Paoletti, C.; Sico, G. Gravure Printing of Graphite-Based Anodes for Lithium-Ion Printed Batteries. *Membranes* **2022**, *12*, 999. [CrossRef] [PubMed]
20. Morikawa, Y.; Nishimura, S.-I.; Hashimoto, R.-I.; Ohnuma, M.; Yamada, A. Mechanism of sodium storage in hard carbon: An X-ray scattering analysis. *Adv. Energy Mater.* **2020**, *10*, 1903176. [CrossRef]
21. Hwang, J.-Y.; Myung, S.-T.; Sun, Y.-K. Sodium-ion batteries: Present and future. *Chem. Soc. Rev.* **2017**, *46*, 3529. [CrossRef] [PubMed]

Disclaimer/Publisher's Note: The statements, opinions and data contained in all publications are solely those of the individual author(s) and contributor(s) and not of MDPI and/or the editor(s). MDPI and/or the editor(s) disclaim responsibility for any injury to people or property resulting from any ideas, methods, instructions or products referred to in the content.

batteries

MDPI

Article

Dendrite-Free Zn Anode Modified by Organic Coating for Stable Aqueous Zinc Ion Batteries

Fujie Li, Hongfei Zhang, Xuehua Liu, Binghui Xu and Chao Wang *

Institute of Materials for Energy and Environment, College of Materials Science and Engineering, Qingdao University, Qingdao 266071, China
* Correspondence: wangc@qdu.edu.cn

Abstract: Aqueous zinc-ion batteries (AZIBs) have emerged as highly promising options for large-scale energy storage systems due to their cost-effectiveness, substantial energy capacity, and improved safety features. However, the Zn anode faces challenges such as self-corrosion and dendrite formation, which limit its practical use in AZIB applications. In this work, a simple blade-coating method was used to successfully coat poly (vinylidene fluoride–hexafluoro propylene) (PVDF-HFP) on the Zn anode. The coated Zn anode (P-Zn) displayed a stable cycling performance (700 h) at 1 mA cm^{-2} current density in the symmetric cell. In addition, the full cell using MnO_2 as the cathode and P-Zn as the anode retained almost full capacity even after 1400 cycles at 2C, far outperforming the full cell using the unmodified Zn anode with only 50% capacity retention after 600 cycles. In situ optical observations of Zn deposition demonstrate that the special organic coating significantly enhances the uniform deposition of Zn^{2+}, thus effectively mitigating corrosion and hydrogen evolution. Density Functional Theory (DFT) calculations show that the PVDF-HFP coating effectively narrows the adsorption energy gap between the P-Zn (002) and (101) planes, leading to the homogeneous deposition of Zn^{2+} with fewer Zn dendrites. A simple and feasible strategy for designing ultra-stable AZIBs by coating an organic protective layer on the Zn surface is provided by this work.

Keywords: aqueous zinc ion battery; organic coating; zinc dendrites; long cycling stability

1. Introduction

Rechargeable aqueous zinc ion batteries (AZIBs) have become a popular research target in recent years as a candidate for large-scale energy storage devices with low cost, high safety, and environmental friendliness [1–4]. Owing to the advantages of abundant reserves, low redox potential (−0.76 V vs. SHE), and ultra-high theoretical volumetric energy density (5851 mAh cm^3), AZIBs usually directly use pure metallic Zn foil as the anode [5,6].

In AZIBs, the cycling stability of the Zn anode is a significant challenge due to the interplay of complex electrochemical and chemical processes [7,8]. One of the primary issues is the growth of Zn dendrites, which occurs during the deposition of Zn^{2+} on the anode surface. These dendrites not only increase the electrode surface area but also penetrate the separator, leading to the potential short-circuiting of the battery [9]. Additionally, due to the low standard electrode potential of Zn, the electrode may reach the threshold for the hydrogen evolution reaction (HER), resulting in hydrogen gas generation [10]. The evolution of hydrogen gas further accelerates electrolyte consumption. Moreover, the Zn dissolution and redeposition processes are often non-uniform due to uneven current density distribution, which leads to unstable electrode structures and deteriorated electrochemical performance [11]. Beyond these electrochemical reactions, the Zn anode is also susceptible to chemical corrosion, which is exacerbated by the increased contact area between the electrode and the electrolyte, particularly after dendrite growth [6,12]. Furthermore, the formation of insulating by-products like $Zn_4SO_4(OH)_6{\cdot}5H_2O$ (ZSH) on the anode surface contributes to a passivation layer that impedes uniform zinc deposition, ultimately impairing the redox kinetics and cycling stability [13,14].

The aforementioned issues pertaining to the Zn anode have been addressed through the implementation of various strategies, including structural construction [15,16], electrolyte optimization [17–19], surface modification [20,21], and Zn alloy preparation [22,23]. Various surface modification techniques have been extensively investigated to enhance the electrode–electrolyte interface using different coatings to improve the performance [24–27]. For example, the Zn anode modified with a metal–organic framework coating showed an 88.9% high-capacity retention after 600 cycles. This coating helps to form a supersaturated surface with low interfacial resistance [28]. Hu et al. utilized a simple drop-coating approach to cover the Zn anode with polyacrylonitrile (PAN), which led to a more uniform Zn^{2+} deposition and enhanced cycling durability [29]. Likewise, Cui et al. demonstrated that polyamide (PA) coatings, known for their robust coordination with Zn^{2+} and unique hydrogen bonding networks, contribute positively to the performance of Zn electrodes [30]. Notably, a PA-coated Zn anode exhibited a stable current over 8000 h of cycling without the formation of Zn dendrites, highlighting its potential to improve electrode stability.

In this study, we successfully applied a poly(vinylidene fluoride–hexafluoropropylene) (PVDF-HFP) coating on the Zn anode using a simple blade-coating method. The objective was to enhance the long-term cycling stability of the Zn anode by reducing Zn dendrite formation. The modified Zn anode (P-Zn) symmetric cell demonstrated stable Coulombic efficiency and an impressive cycling life of 700 h at 1 mA cm^{-2} current density. Moreover, the full cell configured with MnO_2 as the cathode and P-Zn as the anode demonstrated remarkable stability, maintaining its full capacity even after 1400 cycles at 2C. In comparison, the full cell utilizing an uncoated Zn anode maintained only 50% of its original capacity after just 600 cycles, highlighting the significant improvement in performance due to the P-Zn modification. The PVDF-HFP coating effectively reduces the adsorption energy difference between the P-Zn (002) and (101) planes, as predicted by the Density Functional Theory (DFT), leading to uniform Zn^{2+} deposition and fewer Zn dendrites. This results in reduced corrosion and fewer by-products, improving the reversibility of the Zn^{2+}/Zn stripping/plating process. This straightforward approach of applying an organic coating to the Zn anode presents a promising solution for developing AZIBs with enhanced stability and dendrite-free performance.

2. Materials and Synthesis

Materials. All reagents were utilized as provided without further purification. This included Zn foil sourced from Aladdin (Aladdin Biochemical Technology Co., Ltd., Shanghai, China), concentrated H_2SO_4 (95–98%) from Sinopharm Chemical Reagent (Sinopharm Chemical Reagent Co., Ltd., Shanghai, China), as well as $MnSO_4{\cdot}H_2O$, $ZnSO_4$, and MnO_2, all obtained from Aladdin (Aladdin Biochemical Technology Co., Ltd., Shanghai, China). Additional chemicals like poly(vinylidene fluoride–hexafluoropropylene) (PVDF-HFP) from Solvay (Solvay (Shanghai) Co., Ltd., Shanghai, China), conductive carbon (Super P, Timcal Super C65), and N-methyl-2-pyrrolidone (NMP) from Aladdin (Aladdin Biochemical Technology Co., Ltd., Shanghai, China) were also employed.

Preparation of P-Zn. The PVDF-HFP film was applied onto the Zn surface using a straightforward blade-coating technique. Initially, 50 mg of PVDF-HFP was dissolved in 4 mL of NMP to create a uniform solution. This solution was then spread over the Zn foil using a scraper blade with a 100 μm thickness. The Zn sample with the coating was then vacuum-dried at 60 °C for a duration of 12 h, yielding the modified version referred to as P-Zn. For the comparison, two additional samples were prepared using the same method but with 25 mg and 75 mg of PVDF-HFP, referred to as P-Zn25 and P-Zn75, respectively.

3. Results and Discussion

Figure 1a presents a comparative analysis of the X-ray diffraction (XRD) patterns of the bare Zn and P-Zn samples. The Zn and P-Zn diffraction peaks are in accordance with the standard card (PDF # 04-0831), and no significant variations were observed in the peak positions and intensities for both samples [31,32]. This was due to the thin PVDF-HFP layer, which did not affect the crystal structure of zinc metal. The high-resolution F 1s X-ray photoelectron spectroscopy (XPS) spectrum of P-Zn is shown in Figure 1b. There is a characteristic peak at 686.2 eV, corresponding to the F 1s of the PVDF-HFP coating layer [31,33].

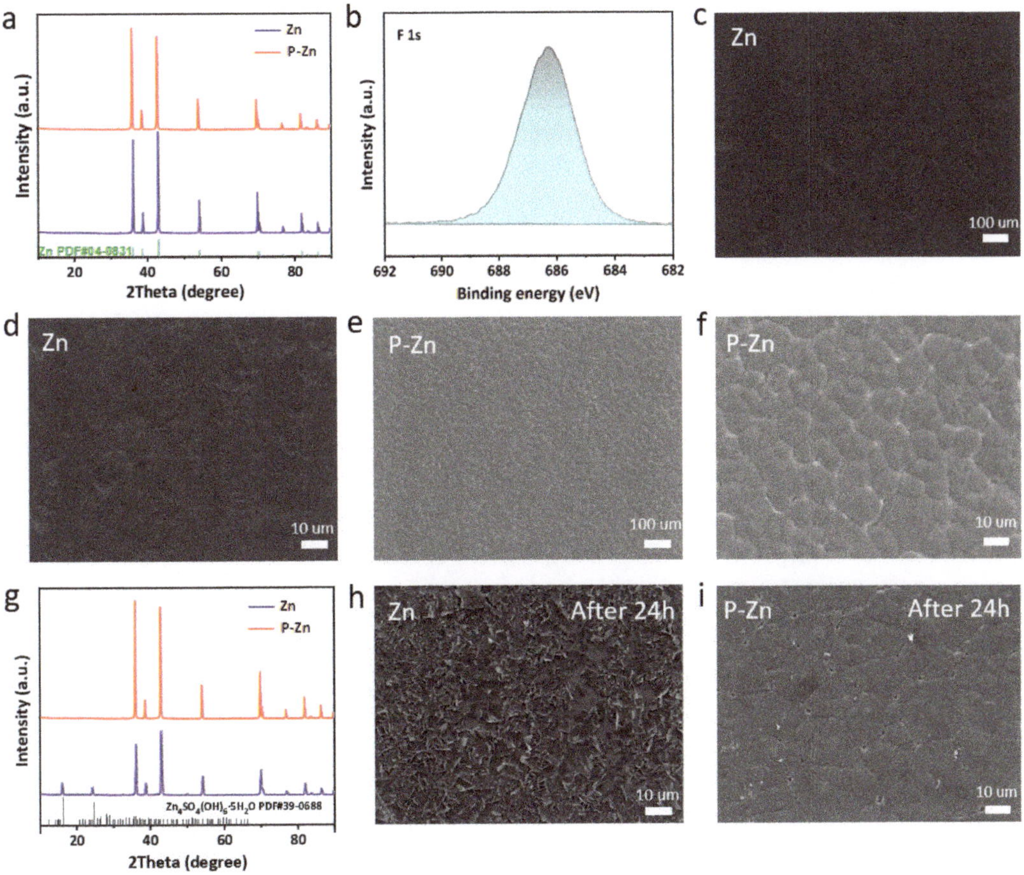

Figure 1. Characterization results: (**a**) XRD patterns comparing Zn and P-Zn. (**b**) High-resolution XPS spectrum for the F 1s region of P-Zn. (**c**,**d**) FESEM images showing the surface morphology of bare Zn and (**e**,**f**) FESEM images of P-Zn. (**g**) XRD patterns of both bare Zn and P-Zn after 24 h of immersion. (**h**,**i**) FESEM images depicting surface changes in bare Zn and P-Zn post 24 h of immersion.

Figure 1c–f present field-emission scanning electron microscopy (FESEM) images for uncoated Zn and P-Zn, respectively. In contrast to the bare Zn, which shows a smooth surface, the P-Zn sample reveals a dense PVDF-HFP coating layer. The dense surface structure effectively isolates the Zn anode from the aqueous electrolyte, thereby reducing the risk of self-corrosion and hydrogen evolution [20]. Figure S1 shows the cross-sectional FESEM image of P-Zn. It can be observed that the thickness of the coating layer is about 10 μm. Figures S2 and S3 present the FESEM images of P-Zn25 and P-Zn75 samples, respectively. It was found that a low concentration of PVDF-HFP resulted in the non-uniform distribution of the coating (Figure S2), while a too-high concentration resulted in an excessively thick coating, which is unfavorable for the transport of Zn^{2+} (Figure S3).

In order to more clearly assess the corrosion resistance of the Zn and P-Zn samples, they were placed in a concentrated $ZnSO_4$ solution for 24 h. Figure 1g presents the XRD patterns of the two samples following 24 h of immersion. The diffraction peaks corresponding to the by-products ZSH (PDF # 39-0688) were observed in the bare Zn sample, with a noticeable decrease in peak intensity [31]. This indicates that an active interfacial reaction occurred, and the corrosion by-product covered the surface of the Zn anode. In contrast, the XRD pattern of P-Zn remained unaltered, thereby corroborating the positive impact of the PVDF-HFP coating in impeding the generation of by-products. Figure 1h,i show the FESEM images of bare Zn and P-Zn after being immersed in a $ZnSO_4$ solution for 24 h, respectively. The bare Zn surface is filled with corrosion by-product deposits, which are usually responsible for low redox kinetics and the poor cycling stability of the Zn anode. However, the surface of P-Zn still remains intact with a dense morphology without any noticeable changes. This proves that the PVDF-HFP coating is beneficial in reducing the corrosion and the generation of the ZSH by-product, thus improving the reversibility of Zn^{2+}/Zn stripping/plating [21]. The FESEM image of P-Zn25 after immersion in $ZnSO_4$ solution for 24 h is shown in Figure S4. It can be observed that the areas covered by the PVDF-HFP coating in P-Zn25 remain intact, while the uncovered areas corrode completely, forming corrosion by-products. This further confirms the positive role of the PVDF-HFP coating in inhibiting the by-product's generation.

Figure 2a shows the Tafel curves of Zn and P-Zn electrodes in a 2 M Zn_2SO_4 aqueous electrolyte. It can be observed that the corrosion potential of the P-Zn electrode is −0.97 V, which is larger than that of the Zn electrode (−0.99 V). This indicates that the PVDF-HFP coating of the P-Zn electrode leads to the alleviation of the corrosion reaction [34]. In the Linear sweep voltammetry (LSV) test (Figure 2b), the P-Zn electrode exhibits a lower potential of −1.51 V than that of the Zn electrode (−1.36 V) at the same current density. This demonstrates that the PVDF coating layer of P-Zn is beneficial for the reduced HER, which inhibits the ZSH by-products [34].

Coulombic efficiency is a key factor in order to evaluate the stability and lifespan of the battery [35]. As shown in Figure 2c, half cells were assembled using bare Zn and P-Zn electrodes with pure Cu foil and were tested for long-cycle performance at a current density of 1 mA cm^{-2}. The Zn | | Cu half-cell demonstrated irregular Coulombic efficiency after reaching 300 cycles. This deterioration can be attributed to the short circuit caused by the inhomogeneous deposition of Zn^{2+} on the surface of the bare Zn electrode. In contrast, the P-Zn | | Cu half-cell can be stably cycled for about 1500 cycles with the Coulombic efficiency maintained at 100%, which highlights the commendable stability possessed by the PVDF-HFP coating. Figure 2d illustrates the Galvanostatic cycling profiles for both Zn | | Cu and P-Zn | | Cu half-cells. The data indicate that the P-Zn electrode shows reduced polarization voltage within the same voltage window in comparison to the bare Zn electrode, which contributes to its improved stability [18].

Figure 2e shows the long-term stability of Zn | | Zn and P-Zn | | P-Zn symmetric cells at a current density of 1 mA cm^{-2} to evaluate the effect of the PVDF-HFP coating on cell performance. The results show that the bare Zn electrode has a large polarization voltage and suddenly fails after about 100 h of cycling, which is caused by the formation of Zn dendrites during the Zn^{2+} stripping/plating process, leading to a short circuit [36]. On the other hand, the P-Zn | | P-Zn symmetric cell maintains a steady polarization voltage with only a slight increase over 700 h, indicating a highly reversible Zn^{2+}/Zn stripping and plating process. Figure 2f compares the rate of performance of the symmetric cells. The polarization voltage of the bare Zn electrode is consistently larger as the current increases, and the bare Zn electrode short circuits and fails when the current reaches 5 mA cm^{-2}. However, the P-Zn electrode shows an excellent rate of performance and a relatively stable voltage without significant polarization. These tests consistently emphasize the excellent and stable performance of the PVDF-HFP coating in terms of Zn^{2+} stripping/plating, which greatly extends the lifespan of the cell.

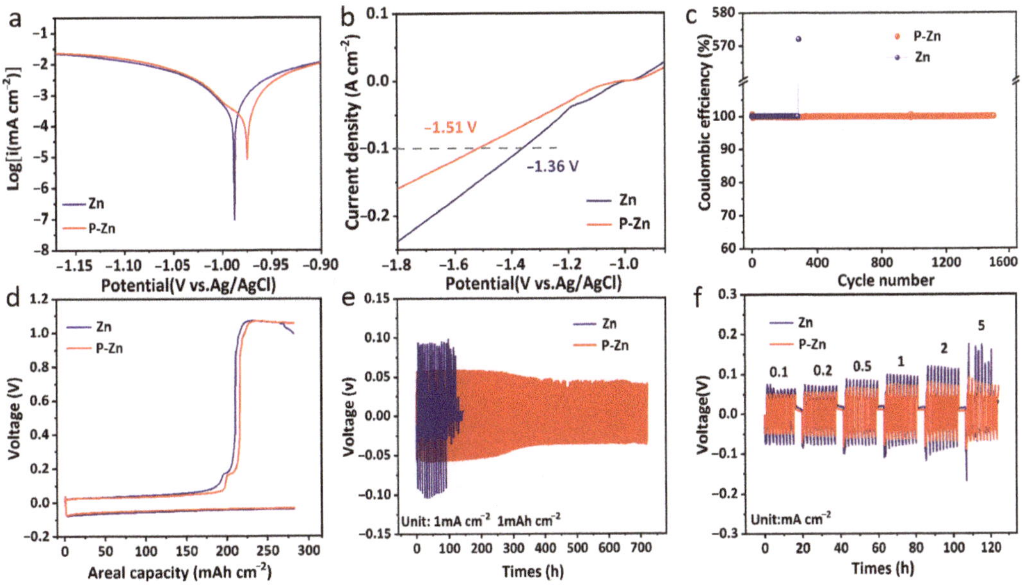

Figure 2. (**a**) Tafel corrosion curves and (**b**) LSV curves of bare Zn and P-Zn electrodes in the 2 M Zn_2SO_4 aqueous electrolyte at the scan rate of 5 mV s^{-1}. (**c**) Coulombic efficiencies of bare Zn and P-Zn stripping/plating on Cu foil at 1 mA cm^{-2}. (**d**) Galvano static cycling curves on Cu foil at 1 mA cm^{-2}. (**e**) Long-term stability of Zn I I Zn and P-Zn I I P-Zn symmetric cells at 1 mA cm^{-2}. (**f**) The rate of performance of Zn I I Zn and P-Zn I I P-Zn symmetric cells evaluated across various current densities ranging from 0.1 to 5 mA cm^{-2}.

The dendrite growth behavior of bare Zn and P-Zn electrodes was carefully investigated using in situ optical microscopy, as shown in Figure 3. It was noticeable that irregular dendrites started to appear on the surface of the bare Zn electrode after just 20 min at a current density of 10 mA cm^{-2} (Figure 3a). By 40 min, these dendrites rapidly expanded to form a distinct network. By 60 min, the surface of the bare Zn electrode was almost completely covered by a dense network of irregular dendrites, which could easily lead to a short circuit of the battery, seriously affecting its stability and lifespan. In contrast, the P-Zn electrode was very stable throughout the process, with no significant dendrite growth on the surface (Figure 3b). This observation convincingly demonstrates the key role of the PVDF-HFP coating in minimizing dendrite formation, thereby improving the stability and safety of AZIBs. This further confirms the excellent performance and potential value of the P-Zn anode for AZIB applications.

To further illustrate the excellent performance of this coating, the bare Zn or P-Zn electrode was used as the anode and matched with the MnO_2 cathode to assemble the full cells (Zn I I MnO_2 or P-Zn I I MnO_2) for electrochemical testing. Figure 4a presents the Cyclic Voltammetry (CV) curves for the Zn I I MnO_2 and P-Zn I I MnO_2 full cells, recorded at a scan rate of 0.1 mV s^{-1} over a voltage window of 0.8 to 1.8 V. Both cells display two sets of redox peaks, which correspond to the sequential redox reactions involving the Mn^{4+}/Mn^{3+} transition [37]. Figure 4b compares the impedance of Zn I I MnO_2 and P-Zn I I MnO_2 full cells. The resistance of the P-Zn electrode is reduced by almost 50%. There are two reasons for this: one is the inevitable formation of ZnO on the bare Zn anode surface, which increases the resistance; the other is the 3D structure of the PVDF-HFP coating, which provides more channels for the migration of Zn^{2+}, promoting homogeneous deposition, and increasing the nucleation sites, thus reducing the resistance [38].

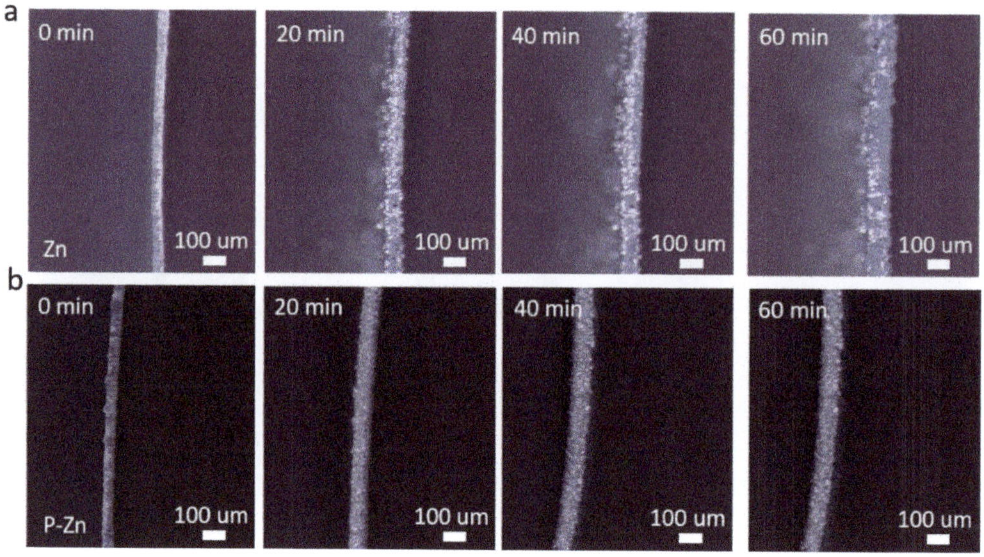

Figure 3. The dendrite growth behavior on (**a**) bare Zn and (**b**) P-Zn electrodes operating at 10 mA cm^{-2} for 60 min were observed using in situ optical microscopy.

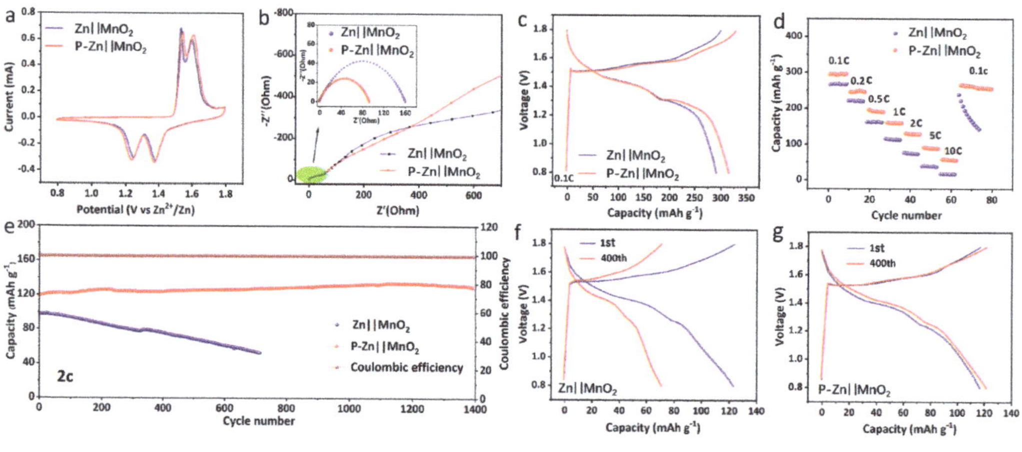

Figure 4. Electrochemical performance analysis of full cells utilizing either bare Zn or P-Zn electrodes as anodes, with MnO_2 serving as the cathode. (**a**) CV curves. (**b**) Nyquist impedance plots obtained prior to cycling. (**c**) Galvanostatic charge–discharge profiles at a current rate of 0.1C. (**d**) Rate capability assessment. (**e**) Long-term cycling stability evaluated at 2C. Galvanostatic charge–discharge profiles of (**f**) Zn||MnO_2 and (**g**) P-Zn||MnO_2 at 2C for the 1st and 400th cycle.

Figure 4c presents a comparison of the initial charge–discharge profiles for the Zn||MnO_2 and P-Zn||MnO_2 cells at a rate of 0.1C (where 1C equals 308 mAh g^{-1}). The specific capacity of the P-Zn||MnO_2 cell is about 320 mAh g^{-1}. On the other hand, although the Zn||MnO_2 cell shows comparable charge–discharge behavior, its specific capacity is reduced, achieving only 287 mAh g^{-1}. The rate of performance of Zn||MnO_2 and P-Zn||MnO_2 full cells are compared in Figure 4d. At 0.1, 0.2, 0.5, 1, 2, 5, and 10C, the

P-Zn||MnO_2 cell achieved specific capacities of 300, 253, 203, 167, 134, 95, and 82 mAh g^{-1}, respectively, which were superior to those of the Zn||MnO_2 cell in each case, demonstrating an excellent rate capability.

Figure 4e compares the long-cycle stability of Zn||MnO_2 and P-Zn||MnO_2 full cells at 2 C. The capacity of the Zn||MnO_2 cells gradually decreased during cycling, and the capacity of retention was 50% after 700 cycles. This degradation can be attributed to the self-corrosion and dendrite generation of the bare Zn electrode during the cycling process, which ultimately led to the short circuit. In contrast, the capacity retention of the P-Zn||MnO_2 cell was 100% after 1400 cycles. The PVDF-HFP coating minimizes electrode–electrolyte interactions, thereby reducing side reactions and enhancing the reversibility of Zn^{2+}/Zn stripping/plating [26,34]. This improvement in interface stability contributes to the enhanced cycling performance of the P-Zn||MnO_2 cell. The increased capacity during cycling can be attributed to the activation of MnO_2 in the electrolyte and the nanoscale transformation of manganese ions, which continuously generate nanoscale MnO_2 at the cathode [39]. In addition, the charge–discharge profiles of two full cells after 400 cycles were selected, as shown in Figure 4f,g. It is clear that the Zn||MnO_2 cell rapidly lost 50% of its capacity after 400 cycles, while the P-Zn||MnO_2 cell maintained a stable charge–discharge profile. All these test results demonstrate the potential of PVDF-HFP coatings to enhance the performance of AZIBs. Figure S5 shows the long-term cycling performance of the P-Zn25||MnO_2 and P-Zn75||MnO_2 cells at 2C. The P-Zn25||MnO_2 cell demonstrated capacity retention of 37% after 600 cycles (Figure S5a), while the P-Zn75||MnO_2 cell maintained only 9% capacity retention after 1400 cycles (Figure S5b). In the comparison of the long cycling performance of PVDF coating samples in Figures 4e and S5, P-Zn||MnO_2 exhibits the best stability. The low content of PVDF-HFP in P-Zn25 does not encapsulate Zn well to inhibit the occurrence of the side reaction, while the high content of PVDF-HFP in P-Zn75 is too dense for the transport of Zn^{2+}.

In order to reveal the growth of dendrites during the cycling process, the surface morphology of bare Zn and P-Zn electrodes was observed using FESEM after 400 cycles at 2C. The surface of the bare Zn electrode displayed noticeable irregularities in the deposition layers, as shown in Figure 5a,b. Lamellar dendritic growth resulted in poor cycling stability and slow redox kinetics. In contrast, on the surface of the P-Zn electrode, the deposited layer was more uniform (Figure 5c,d) and remained smooth and dense without dendritic growth, which effectively improved the cycling stability of P-Zn||MnO_2 cells. The cycling stability of the P-Zn||MnO_2 cell was effectively improved.

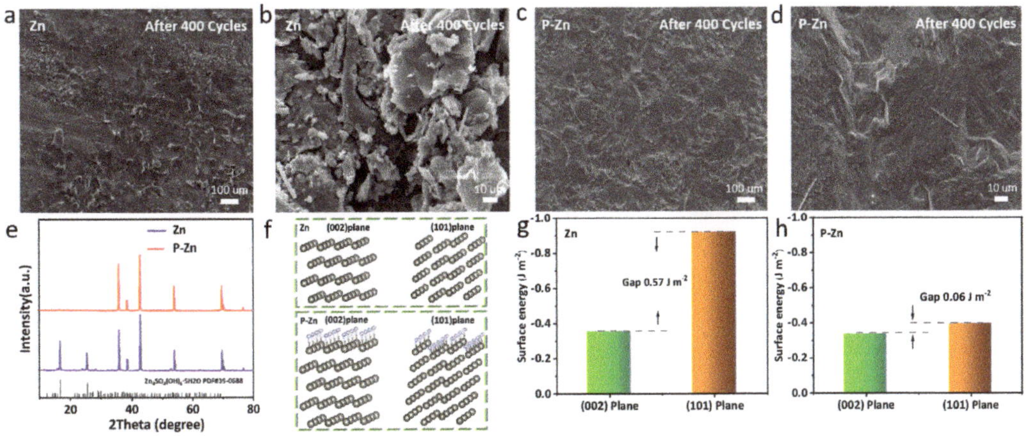

Figure 5. FESEM images of Zn (**a**,**b**) and P-Zn (**c**,**d**) electrodes; (**e**) XRD patterns of Zn and P-Zn electrodes. (**f**) The crystal models of DFT calculation. Surface energies of Zn atoms adsorbed on bare Zn (**g**), P-Zn (**h**) (002), and (101) planes.

The XRD patterns of the two electrodes after 400 cycles are shown in Figure 5e. It was clear that the bare Zn electrode showed obvious sulfate corrosion peaks, which was the consequence of destroying the stability of the anode [31]. In contrast, the XRD pattern of the P-Zn electrode was basically unchanged, further proving its corrosion resistance during the cycling process. This can primarily be attributed to the strong hydrophobic nature of the PVDF-HFP coating, which effectively minimizes direct interactions between the P-Zn anode and the electrolyte. As a result, this reduces both hydrogen evolution and the risk of corrosion. The contact angle tests of bare Zn and P-Zn electrodes are shown in Figures S6 and S7. The results show that the contact angle of the PVDF- HFP coated surface was reduced by 20 degrees compared with that of the bare Zn surface, indicating that the coating had a favorable hydrophobicity performance.

In order to deeply understand the effect of the PVDF-HFP coating on Zn^{2+} stripping/plating, using the Vienna Abinitio Simulation Package (VASP) to carry out the surface energy of the Zn and P-Zn (002) and (101) planes by DFT calculations, the crystal models of Zn and P-Zn are shown in Figure 5f [11,40]. For the bare Zn anode (Figure 5g), the surface energy on the (002) plane was -0.36 J m^{-2}, which is significantly larger than that on the (101) plane (-0.92 J m^{-2}, the gap of plans: 0.57 J m^{-2}), indicating the preferential deposition of Zn atoms on the (101) plane [28]. This preferential deposition is the main reason for the inhomogeneous deposition of Zn^{2+} and the generation of dendrites. On the contrary, after coating the PVDF-HFP coating, the difference in surface energies between the P-Zn anode (002) and (101) planes was not significant (the gap of plans: 0.06 J m^{-2}, Figure 5h), which effectively suppressed the inhomogeneous deposition of Zn^{2+}, and thus, favored the reduction in Zn dendrite growth.

Figure 6a vividly illustrates the side reactions and dendrite growth that occur in bare Zn during the cell operation. Firstly, Zn^{2+} was unevenly deposited on the bare Zn surface during plating. Due to the concentration of the electric field and concentration polarization, Zn^{2+} tends to preferentially accumulate on these protrusions. This phenomenon leads to the uneven nucleation of Zn^{2+}, fostering dendrite growth [41]. Secondly, the contact of bare Zn with the aqueous solution during the stripping/plating process led to hydrogen evolution. The precipitation of H_2 results in localized pH increases and the accumulation of by-products such as ZSH [42], which contribute to the degradation of the cell's performance over extended cycling and leads to the reduced lifespan of the cells over prolonged usage [43]. In the case of Zn^{2+} stripping and plating on the P-Zn electrode, as shown in Figure 6b, it exhibits a more stable performance over cycles. This PVDF-HFP coating has a dense structure, which reduces the surface energy gap between the (002) and (101) surfaces (Figure 5g,h) and guides the homogeneous Zn^{2+} deposition. After plating, the P-Zn electrode remained smooth, and no dendrites were detected (Figure 3b), indicating that the PVDF-HFP coating favored the homogenization of the nuclear sites.

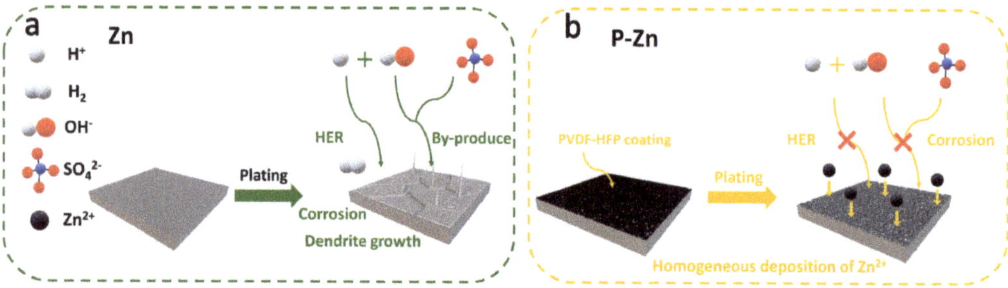

Figure 6. The behavior of Zn^{2+} stripping and plating on bare Zn (**a**) and P-Zn (**b**) electrodes in the $ZnSO_4$ aqueous electrolyte.

4. Conclusions

In this study, we used an economical and simple blade-coating technique to successfully apply a uniform and dense PVDF-HFP coating on the Zn metal surface. This organic coating, with its compact microstructure, effectively isolates the Zn metal from the aqueous electrolyte, significantly reducing self-corrosion and hydrogen evolution. Notably, the unique porous structure of the coating not only improves the transport efficiency of Zn^{2+} but also provides a large number of active sites. DFT calculations show that the PVDF-HFP coating facilitates the homogeneous deposition of Zn^{2+} and effectively inhibits the growth of dendrites, which significantly improves the overall stability and lifespan of the AZIBs. P-Zn as an anode and MnO_2 as an anode are assembled as full cells, and their capacity retention is 100% after 1400 cycles at 2C. The P-Zn materials reported in this study are promising for the preparation of highly stable AZIBs.

Supplementary Materials: The following supporting information can be downloaded at https://www.mdpi.com/article/10.3390/batteries10120420/s1. Figure S1: FESEM image of the thickness of P-Zn; Figure S2: FESEM image of P-Zn25; Figure S3: FESEM image of P-Zn75; Figure S4: FESEM image of P-Zn25 after 24 h of immersion; Figure S5: Long-term cycling performance of full cells using the bare P-Zn25 (a) or P-Zn75 (b) electrode as the anode and MnO_2 as the cathode at 2 C; Figure S6: Contact angle of P-Zn anode electrode material; Figure S7: Contact angle of bare Zn anode electrode material. References [44–47] are cited in the supplementary materials.

Author Contributions: Conceptualization, F.L. and C.W.; methodology, F.L. and H.Z.; software, F.L.; validation, F.L. and C.W.; formal analysis, F.L.; investigation, X.L. and B.X.; resources, X.L. and C.W.; data curation, F.L.; writing—original draft preparation, F.L.; writing—review and editing, B.X. and C.W.; visualization, F.L.; supervision, C.W.; project administration, C.W.; funding acquisition, C.W. All authors have read and agreed to the published version of the manuscript.

Funding: We are grateful for the financial support from the Natural Science Foundation of Shandong Province (No. ZR2022QB025) and the Start-up Foundation of Qingdao University (No. DC2000005025).

Data Availability Statement: The original contributions presented in the study are included in the article/Supplementary Materials, further inquiries can be directed to the corresponding author.

Conflicts of Interest: The authors declare no conflicts of interest.

References

1. Jia, X.; Liu, C.; Neale, Z.G.; Yang, J.; Cao, G. Active Materials for Aqueous Zinc Ion Batteries: Synthesis, Crystal Structure, Morphology, and Electrochemistry. *Chem. Rev.* **2020**, *120*, 7795–7866. [CrossRef] [PubMed]
2. Chen, H.; Wang, C.; Wu, H.; Li, L.; Xing, Y.; Zhang, C.; Long, X. Host-Guest-Induced Electronic State Triggers Two-Electron Oxygen Reduction Electrocatalysis. *Nat. Commun.* **2024**, *15*, 9222. [CrossRef] [PubMed]
3. Gao, L.; Li, F.; Jin, X.; Li, Z.; Wang, C. Stabilizing 4.6 V $LiCoO_2$ via Surface-to-Bulk Titanium Modification. *Adv. Funct. Mater.* **2024**, 2416338. [CrossRef]
4. Wang, B.; Xu, H.; Hao, J.; Du, J.; Wu, C.; Ma, Z.; Qin, W. Mini-Review on the Regulation of Electrolyte Solvation Structure for Aqueous Zinc Ion Batteries. *Batteries* **2023**, *9*, 73. [CrossRef]
5. Wang, T.; Li, C.; Xie, X.; Lu, B.; He, Z.; Liang, S.; Zhou, J. Anode Materials for Aqueous Zinc Ion Batteries: Mechanisms, Properties, and Perspectives. *ACS Nano* **2020**, *14*, 16321–16347. [CrossRef] [PubMed]
6. Xie, C.; Li, Y.; Wang, Q.; Sun, D.; Tang, Y.; Wang, H. Issues and Solutions Toward Zinc Anode in Aqueous Zinc-Ion Batteries: A Mini Review. *Carbon Energy* **2020**, *2*, 540–560. [CrossRef]
7. Zheng, J.; Zhao, Q.; Tang, T.; Yin, J.; Quilty, C.D.; Renderos, G.D.; Liu, X.; Deng, Y.; Wang, L.; Bock, D.C.; et al. Reversible Epitaxial Electrodeposition of Metals in Battery Anodes. *Science* **2019**, *366*, 645–648. [CrossRef]
8. Shen, Q.; Wang, Y.; Han, G.; Li, X.; Yuan, T.; Sun, H.; Gong, Y.; Chen, T. Recent Progress in Electrolyte Additives for Highly Reversible Zinc Anodes in Aqueous Zinc Batteries. *Batteries* **2023**, *9*, 284. [CrossRef]
9. Zhang, Y.; Wei, C.; Wu, M.-X.; Wang, Y.; Jiang, H.; Zhou, G.; Tang, X.; Liu, X. A High-Performance COF-Based Aqueous Zinc-Bromine Battery. *Chem. Eng. J.* **2023**, *451*, 138915. [CrossRef]
10. Huang, J.; He, B.; Zhi, J.; Chen, P.; Wang, C.; Chen, H.; Zhao, X.S. Formation and Conversion of Hydrate Zinc Sulfate Hydroxide in Aqueous Zn/MnO_2 Batteries. *J. Power Sources* **2024**, *623*, 235409. [CrossRef]

11. Wang, R.; Xin, S.; Chao, D.; Liu, Z.; Wan, J.; Xiong, P.; Luo, Q.; Hua, K.; Hao, J.; Zhang, C. Fast and Regulated Zinc Deposition in a Semiconductor Substrate toward High-Performance Aqueous Rechargeable Batteries. *Adv. Funct. Mater.* **2022**, *32*, 2207751. [CrossRef]
12. Liang, P.; Yi, J.; Liu, X.; Wu, K.; Wang, Z.; Cui, J.; Liu, Y.; Wang, Y.; Xia, Y.; Zhang, J. Highly Reversible Zn Anode Enabled by Controllable Formation of Nucleation Sites for Zn-Based Batteries. *Adv. Funct. Mater.* **2020**, *30*, 1908528. [CrossRef]
13. He, P.; Huang, J. Chemical Passivation Stabilizes Zn Anode. *Adv. Mater.* **2022**, *34*, 2109872. [CrossRef] [PubMed]
14. Thieu, N.A.; Li, W.; Chen, X.; Hu, S.; Tian, H.; Tran, H.N.N.; Li, W.; Reed, D.M.; Li, X.; Liu, X. An Overview of Challenges and Strategies for Stabilizing Zinc Anodes in Aqueous Rechargeable Zn-Ion Batteries. *Batteries* **2023**, *9*, 41. [CrossRef]
15. Zeng, Y.; Zhang, X.; Qin, R.; Liu, X.; Fang, P.; Zheng, D.; Tong, Y.; Lu, X. Dendrite-Free Zinc Deposition Induced by Multifunctional CNT Frameworks for Stable Flexible Zn-Ion Batteries. *Adv. Mater.* **2019**, *31*, 1903675. [CrossRef]
16. Li, C.; Shi, X.; Liang, S.; Ma, X.; Han, M.; Wu, X.; Zhou, J. Spatially Homogeneous Copper Foam as Surface Dendrite-Free Host for Zinc Metal Anode. *Chem. Eng. J.* **2020**, *379*, 122248. [CrossRef]
17. Guo, S.; Qin, L.; Zhang, T.; Zhou, M.; Zhou, J.; Fang, G.; Liang, S. Fundamentals and Perspectives of Electrolyte Additives for Aqueous Zinc-Ion Batteries. *Energy Storage Mater.* **2021**, *34*, 545–562. [CrossRef]
18. Blanc, L.E.; Kundu, D.; Nazar, L.F. Scientific Challenges for the Implementation of Zn-Ion Batteries. *Joule* **2020**, *4*, 771–799. [CrossRef]
19. De Cachinho Cordeiro, I.; Li, A.; Lin, B.; Ma, D.; Xu, L.; Eh, A.; Wang, W. Solid Polymer Electrolytes for Zinc-Ion Batteries. *Batteries* **2023**, *9*, 343. [CrossRef]
20. Xia, A.; Pu, X.; Tao, Y.; Liu, H.; Wang, Y. Graphene Oxide Spontaneous Reduction and Self-Assembly on the Zinc Metal Surface Enabling a Dendrite-Free Anode for Long-Life Zinc Rechargeable Aqueous Batteries. *Appl. Surf. Sci.* **2019**, *481*, 852–859. [CrossRef]
21. Dong, L.; Yang, W.; Yang, W.; Tian, H.; Huang, Y.; Wang, X.; Xu, C.; Wang, C.; Kang, F.; Wang, G. Flexible and Conductive Scaffold-Stabilized Zinc Metal Anodes for Ultralong-Life Zinc-Ion Batteries and Zinc-Ion Hybrid Capacitors. *Chem. Eng. J.* **2020**, *384*, 123355. [CrossRef]
22. Wang, S.-B.; Ran, Q.; Yao, R.-Q.; Shi, H.; Wen, Z.; Zhao, M.; Lang, X.-Y.; Jiang, Q. Lamella-Nanostructured Eutectic Zinc–Aluminum Alloys as Reversible and Dendrite-Free Anodes for Aqueous Rechargeable Batteries. *Nat. Commun.* **2020**, *11*, 1634. [CrossRef] [PubMed]
23. Liu, B.; Wang, S.; Wang, Z.; Lei, H.; Chen, Z.; Mai, W. Novel 3D Nanoporous Zn-Cu Alloy as Long-Life Anode toward High-Voltage Double Electrolyte Aqueous Zinc-Ion Batteries. *Small* **2020**, *16*, 2001323. [CrossRef] [PubMed]
24. Cui, Y.; Zhao, Q.; Wu, X.; Chen, X.; Yang, J.; Wang, Y.; Qin, R.; Ding, S.; Song, Y.; Wu, J.; et al. An Interface-Bridged Organic-Inorganic Layer that Suppresses Dendrite Formation and Side Reactions for Ultra-Long-Life Aqueous Zinc Metal Anodes. *Angew. Chem. Int. Ed.* **2020**, *59*, 16594–16601. [CrossRef]
25. Zhou, Y.; Xie, S.; Li, Y.; Zheng, Z.; Dong, L. Sieve-Like Interface Built by ZnO Porous Sheets Towards Stable Zinc Anodes. *J. Colloid. Interf. Sci.* **2023**, *630*, 676–684. [CrossRef]
26. Lu, H.; Jin, Q.; Jiang, X.; Dang, Z.-M.; Zhang, D.; Jin, Y. Vertical Crystal Plane Matching Between $AgZn_3$ (002) and Zn (002) Achieving a Dendrite-Free Zinc Anode. *Small* **2022**, *18*, 2200131. [CrossRef]
27. He, H.; Tong, H.; Song, X.; Song, X.; Liu, J. Highly Stable Zn Metal Anodes Enabled by Atomic Layer Deposited Al_2O_3 Coating for Aqueous Zinc-Ion Batteries. *J. Mater. Chem. A* **2020**, *8*, 7836–7846. [CrossRef]
28. Zhao, Z.; Wang, R.; Peng, C.; Chen, W.; Wu, T.; Hu, B.; Weng, W.; Yao, Y.; Zeng, J.; Chen, Z.; et al. Horizontally Arranged Zinc Platelet Electrodeposits Modulated by Fluorinated Covalent Organic Framework Film for High-Rate and Durable Aqueous Zinc Ion Batteries. *Nat. Commun.* **2021**, *12*, 6606. [CrossRef]
29. Chen, P.; Yuan, X.; Xia, Y.; Zhang, Y.; Fu, L.; Liu, L.; Yu, N.; Huang, Q.; Wang, B.; Hu, X.; et al. An Artificial Polyacrylonitrile Coating Layer Confining Zinc Dendrite Growth for Highly Reversible Aqueous Zinc-Based Batteries. *Adv. Sci.* **2021**, *8*, 2100309. [CrossRef]
30. Zhao, Z.; Zhao, J.; Hu, Z.; Li, J.; Li, J.; Zhang, Y.; Wang, C.; Cui, G. Long-Life and Deeply Rechargeable Aqueous Zn Anodes Enabled by a Multifunctional Brightener-Inspired Interphase. *Energy Environ. Sci.* **2019**, *12*, 1938–1949. [CrossRef]
31. Deng, D.; Fu, K.; Yu, R.; Zhu, J.; Cai, H.; Zhang, X.; Wu, J.; Luo, W.; Mai, L. Ion Tunnel Matrix Initiated Oriented Attachment for Highly Utilized Zn Anodes. *Adv. Mater.* **2023**, *35*, 2302353. [CrossRef] [PubMed]
32. Baghodrat, M.; Zampardi, G.; Glenneberg, J.; La Mantia, F. Influence of the Thermal Treatment on the Structure and Cycle Life of Copper Hexacyanoferrate for Aqueous Zinc-Ion Batteries. *Batteries* **2023**, *9*, 170. [CrossRef]
33. Yin, H.; Liu, Y.; Zhu, Y.; Ye, F.; Xu, G.; Lin, M.; Kang, W. Bimetal-Initiated Concerted Zn Regulation Enabling Highly Stable Aqueous Zn-Ion Batteries. *Batteries* **2024**, *10*, 70. [CrossRef]
34. Feng, J.; Li, X.; Cui, X.; Zhao, H.; Xi, K.; Ding, S. Periodically Alternating Electric Field Layers Induces the Preferential Growth of Zn (002) Plane for Ultralow Overpotential Zinc-Ion Batteries. *Adv. Energy Mater.* **2023**, *13*, 2204092. [CrossRef]
35. Bao, S.; Tu, M.; Huang, H.; Wang, C.; Chen, Y.; Sun, B.; Xu, B. Heterogeneous Iron Oxide Nanoparticles Anchored on Carbon Nanotubes for High-Performance Lithium-Ion Storage and Fenton-Like Oxidation. *J. Colloid. Interf. Sci.* **2021**, *601*, 283–293. [CrossRef]
36. Liu, Y.; Chen, S.; Yuan, H.; Xiong, F.; Liu, Q.; An, Y.; Zhang, J.; Wu, L.; Sun, J.; Zhang, Y.-W.; et al. Achieving Highly Reversible Zinc Metal Anode via Surface Termination Chemistry. *Sci. Bull.* **2023**, *68*, 2993–3002. [CrossRef]

37. Chen, M.; Yang, M.; Zhou, W.; Tian, Q.; Han, X.; Chen, J.; Zhang, P. Oriented Zn Plating Guided by Aligned ZnO Hexagonal Columns Realizing Dendrite-Free Zn Metal Electrodes. *J. Colloid. Interf. Sci.* **2023**, *644*, 368–377. [CrossRef]
38. Jia, R.; Zhang, R.; Yu, L.; Kong, X.; Bao, S.; Tu, M.; Liu, X.; Xu, B. Engineering a Hierarchical Carbon Supported Magnetite Nanoparticles Composite from Metal Organic Framework and Graphene Oxide for Lithium-Ion Storage. *J. Colloid. Interf. Sci.* **2023**, *630*, 86–98. [CrossRef]
39. Liang, R.; Fu, J.; Deng, Y.-P.; Pei, Y.; Zhang, M.; Yu, A.; Chen, Z. Parasitic Electrodeposition in Zn-MnO_2 Batteries and Its Suppression for Prolonged Cyclability. *Energy Storage Mater.* **2021**, *36*, 478–484. [CrossRef]
40. Zhao, R.; Dong, X.; Liang, P.; Li, H.; Zhang, T.; Zhou, W.; Wang, B.; Yang, Z.; Wang, X.; Wang, L.; et al. Prioritizing Hetero-Metallic Interfaces via Thermodynamics Inertia and Kinetics Zincophilia Metrics for Tough Zn-Based Aqueous Batteries. *Adv. Mater.* **2023**, *35*, 2209288. [CrossRef]
41. Yang, Y.; Liu, C.; Lv, Z.; Yang, H.; Zhang, Y.; Ye, M.; Chen, L.; Zhao, J.; Li, C.C. Synergistic Manipulation of Zn^{2+} Ion Flux and Desolvation Effect Enabled by Anodic Growth of a 3D ZnF_2 Matrix for Long-Lifespan and Dendrite-Free Zn Metal Anodes. *Adv. Mater.* **2021**, *33*, 2007388. [CrossRef] [PubMed]
42. Hao, J.; Li, X.; Zhang, S.; Yang, F.; Zeng, X.; Zhang, S.; Bo, G.; Wang, C.; Guo, Z. Designing Dendrite-Free Zinc Anodes for Advanced Aqueous Zinc Batteries. *Adv. Funct. Mater.* **2020**, *30*, 2001263. [CrossRef]
43. Li, Y.; Wu, P.; Zhong, W.; Xie, C.; Xie, Y.; Zhang, Q.; Sun, D.; Tang, Y.; Wang, H. A Progressive Nucleation Mechanism Enables Stable Zinc Stripping-Plating Behavior. *Energy Environ. Sci.* **2021**, *14*, 5563–5571. [CrossRef]
44. Hafner, J. Ab-initio simulations of materials using VASP: Density-functional theory and beyond. *J. Comput. Chem.* **2008**, *29*, 2044. [CrossRef]
45. Blochl, P.E. Projector augmented-wave method. *Phys. Rev. B* **1994**, *50*, 17953. [CrossRef]
46. Perdew, J.P.; Burke, K.; Ernzerhof, M. Generalized Gradient Approximation Made Simple. *Phys. Rev. Lett.* **1996**, *77*, 3865. [CrossRef]
47. Yang, H.G.; Sun, C.H.; Qiao, S.Z.; Zou, J.; Liu, G.; Smith, S.C.; Cheng, H.M.; Lu, G.Q. Anatase TiO_2 Single Crystals with a Large Percentage of Reactive Facets. *Nature* **2008**, *453*, 638. [CrossRef]

Disclaimer/Publisher's Note: The statements, opinions and data contained in all publications are solely those of the individual author(s) and contributor(s) and not of MDPI and/or the editor(s). MDPI and/or the editor(s) disclaim responsibility for any injury to people or property resulting from any ideas, methods, instructions or products referred to in the content.

batteries

MDPI

Article

Synthesis and Characterization of Lithium Phosphate (Li_3PO_4) as a Solid Electrolyte

Seybou Yacouba Zakariyaou, Hua Ye * and Chongwen Jiang

Department of Chemical Engineering and Technology, School of Chemistry and Chemical Engineering, Central South University, Changsha 410083, China; 212318002@csu.edu.cn (S.Y.Z.); jcwcsu@csu.edu.cn (C.J.)
* Correspondence: yehua308@csu.edu.cn

Abstract: Due to its high thermal stability, environmental friendliness, and safety, lithium phosphate (Li_3PO_4) is used as a solid electrolyte in battery applications, but it is usually used with dopants due to its lower ionic conductivity, which is required for ion transport. However, due to its stability and environmentally friendly aspect, lithium phosphate is still a hot topic among suitable energy materials that need further research to improve its electrochemical properties. In the current work, a novel synthesis of lithium phosphate was proposed from the raw materials lithium carbonate (Li_2CO_3) and trisodium phosphate dodecahydrate (Na_3PO_4*$12H_2O$) under suitable stoichiometric conditions using the co-precipitation method. In the set of synthesized samples, a single-phase β-Li_3PO_4 (named LPO-4) with 99.7% purity and 93.49% yield was successfully prepared under appropriate stoichiometric conditions and pH 13 at 90 °C. The average particle size was 10 nm with a large surface area of 9.02 m^2g^{-1}. Electrochemical impedance spectroscopy (EIS) of LPO-4 revealed a conductivity of 7.1×10^{-6} $S.cm^{-1}$ at room temperature and 2.7×10^{-5} $S.cm^{-1}$ at 80 °C with a low activation energy of 0.38 eV. This performance is attributed to the morphology of the nanotubes and the smaller particle size, which enlarge the reaction interfaces and shorten the diffusion distance of lithium ions. The kinetic and thermodynamic key parameters showed that the β-Li_3PO_4 exhibits thermal stability in the room temperature range up to 208.8 °C. All these property values indicate a promising application of lithium phosphate as a solid electrolyte in solid-state batteries and a new route for further investigation.

Keywords: lithium phosphate; synthesis; characterization; lithium carbonate; trisodium phosphate dodecahydrate; solid electrolyte; lithium-ion battery

Highlights

1. The single-phase and high-purity β-Li_3PO_4 was synthesized by a facile co-precipitation method.
2. A particle nanoscale sample with a high surface area was successfully prepared.
3. A solid electrolyte, lithium phosphate, with high ionic conductivity was achieved.
4. Kinetic and thermodynamic studies established lithium phosphate behavior as a function of the temperature.

1. Introduction

Lithium element has become an essential component of the new generation of sustainable and green energy technologies. Lithium phosphate produced from lithium plays a crucial role in modern battery technology and biomass conversion as a catalyst [1] due to its excellent electrochemical and catalytic performance in alkylene oxide isomerization reactions. In lithium-ion batteries, it is usually used as a solid electrolyte or as a protective coating rather than directly as a cathode material [2,3]. Due to the safety issues associated with liquid electrolytes, such as leakage, explosion, and fire risks, a solid electrolyte has been developed as an alternative solution to overcome these problems and improve the energy density of lithium-ion batteries [4,5].

Therefore, researchers have applied various techniques to improve conductivity, including particle size modulation and doping with some elements (iron, cobalt, and nickel) [6], using different synthesis methods, including the plasma spray [2,7], the co-precipitation [8], the pulsed laser decomposition [9], wet chemical (conductivity of 4.65×10^{-8} S.cm^{-1}) [10] and the solid-state methods [11]. These approaches for producing lithium phosphate include a co-precipitation method, which is the best way to obtain a homogeneous lithium phosphate product under suitable stoichiometric conditions and pH values due to the high ion exchange. In the co-precipitation method, several routes were explored by a combination of various sources of lithium (hydroxide, carbonate, etc.) and phosphate ions (phosphoric acid, ammonium dihydrogen phosphate, sodium phosphate, etc.), as shown in Table 1.

Table 1. Basic conditions for the synthesis of lithium phosphate.

Lithium Source	Phosphate Source	pH Control Reagent	Reaction Temperature °C	Reaction Time/min	pH	Ref.
Li_2CO_3	$Na_3PO_4 \bullet 10H_2O$	HCl; NaOH	70–78	20–25	12.5–13	[12]
$LiOH \bullet H_2O$	$NH_4H_2PO_4$	-	200	60	-	[13]
LiOH	H_3PO_4	-	>80	-	7.5	[14]
Li_2CO_3	$Na_3PO_4 \bullet 10H_2O$	HCl	>90	60	>12	[15]
Spent battery	H_3PO_4	-	70–90	60–120	7–8	[16]

He Guoduan et al. [12] extracted the lithium from the mother liquor of lithium carbonate precipitation to obtain high purity of lithium phosphate in the temperature range of 70–78 °C at a pH of 12.5–13. Norikazu Ishigaki and Junji Akimoto [17] synthesized lithium phosphate as a solid electrolyte for batteries at room temperature and investigated its phase transformation from β-Li_3PO_4 to γ-Li_3PO_4 type. The challenge is to propose the most cost-effective high-purity lithium phosphate and improve its ionic conductivity (10^{-10} S.cm^{-1} at room temperature [8]) with excellent electrochemical performance. Some researchers propose to reduce the particle size to the nanoscale due to the increased surface area, which could establish a balance between the ion transport pathways and structural integrity [2].

However, a simple method to prepare Li_3PO_4 by considering the proper stoichiometry and studying the key kinetic and thermodynamic parameters may be crucial to understanding the properties, behavior, and potential of lithium phosphate (Figure 1).

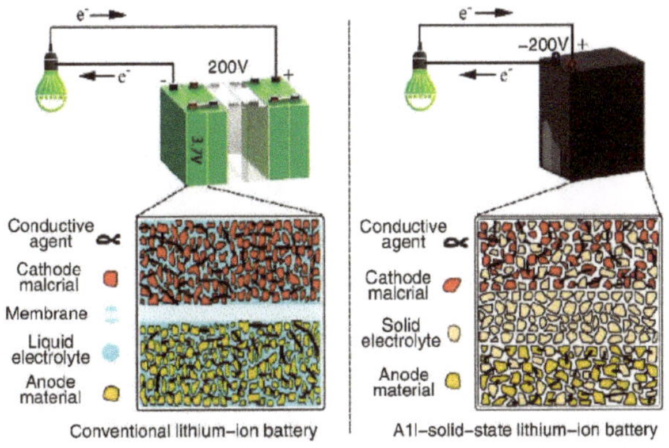

Figure 1. A typical solid-state battery: A solid electrolyte, anode material, and conductive agent. Copyright permission obtained from the author [18].

Herein, we proposed a new preparation of lithium phosphate (Li_3PO_4) as a solid electrolyte from lithium mother liquor (Li_2CO_3) and the phosphate source trisodium phosphate dodecahydrate (Na_3PO_4*$12H_2O$) for solid-state batteries. To improve the lower conductivity of lithium phosphate, the average particle size was set to a nanoscale of 10 nm with a micropore volume close to zero. This resulted in nanostructured Li_3PO_4, which improves its electrochemical performance by reducing the diffusion pathways for lithium ions. The kinetic and thermodynamic models of Coats and Redfern were developed using the constants of Plank and Boltzmann to determine the sample thermal behavior, including mass loss phases, activation energy (Ea), entropy changes (ΔS), enthalpy changes (ΔH), Gibbs free energy changes (ΔG), and pre-exponential factor (A).

2. Experimental Procedures and Characterization

The raw materials of lithium mother liquor (Li_2CO_3) and the phosphate source trisodium phosphate dodecahydrate (Na_3PO_4*$12H_2O$) were purchased from the National Pharmaceutical Group Chemical Reagent Co., Ltd. (Shanghai, China).

2.1. Titration

The titration procedure (acid mixture): 0.25 g of the lithium carbonate sample was dissolved in 50 mL of deionized solution and stirred well until the solid was completely dissolved. Eight drops of the indicator bromocresol green-methyl red solution (3:1) were added to this solution, which turned green. The titration acid solution was prepared from 50 mL chloric acid (HCl, 0.4996 mol/L) and distilled water (1:1). Using the titrator, the acid solution was first gradually added to the sample solution in the round bottom flask and stirred until the solution turned red. This solution was then heated for 2 min (the color must turn green again) and allowed to cool naturally. The titration was continued the second time; generally, one or two drops of acid were sufficient for the sample solution to turn red. The titration was stopped when the solution turned red, and the acid solution used was measured, as presented in Figure 2. Then, the purity was calculated through the following equation: $W_{L_iCO_3} = \frac{C_B*V*M}{m*1000}*100$. C_B: HCl concentration in the solution (1:1) (C_B = 0.2498 mol/L); V: measured volume (mL) of HCl injected in the tested solution during titration; M: ½ of the mass molar (g/mol) of lithium Li_2CO_3; m: mass (g) of Li_2CO_3 tested sample. For example: $W_{L_iCO_3} = \frac{0.2498\text{mol/L}*28.80\text{mL}*36.94\text{g/mol}}{0.2720\text{g}*1000}*100 = 97.704\%$.

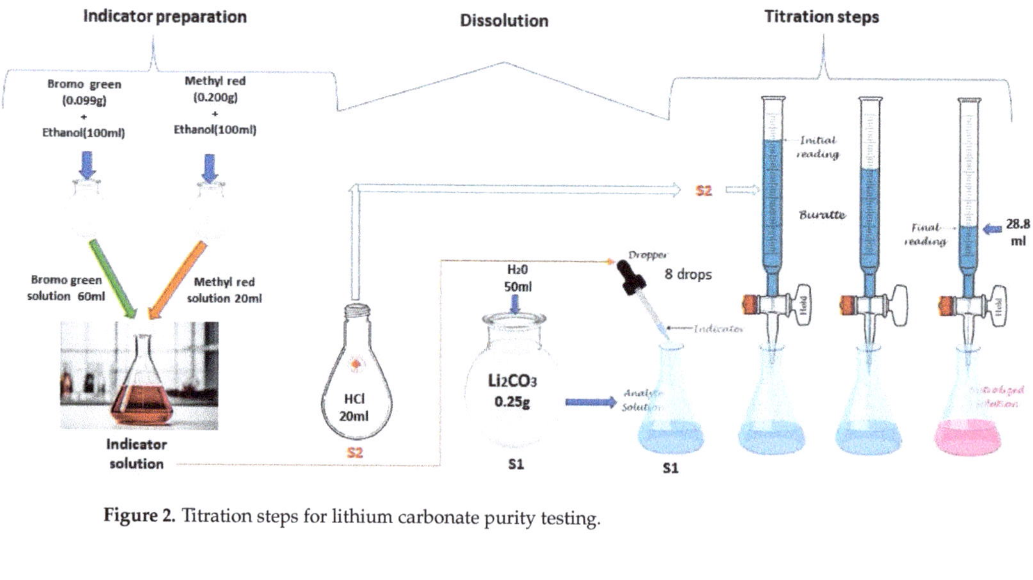

Figure 2. Titration steps for lithium carbonate purity testing.

2.2. Preparation

Figure 3 describes the Lithium phosphate preparation procedure and the main steps for removal of impurities. The Li_3PO_4 samples were prepared using co-precipitation. First, 3 g of lithium carbonate (Li_2CO_3) sample was dissolved in deionized water at room temperature to obtain a lithium concentration of 0.21 mol/L (Li^+) in the solution. The initial pH of the sample solution was adjusted to 5–5.5 by adding chloric acid (HCl, 36–38%). After stirring at 70 °C for 20–25 min to remove the carbonate ions in the solution, the pH was readjusted to 12–13 by adding NaOH (1%) and 0.025–0.5 g.L^{-1} and EDTA (ethylenediaminetetraacetic acid).

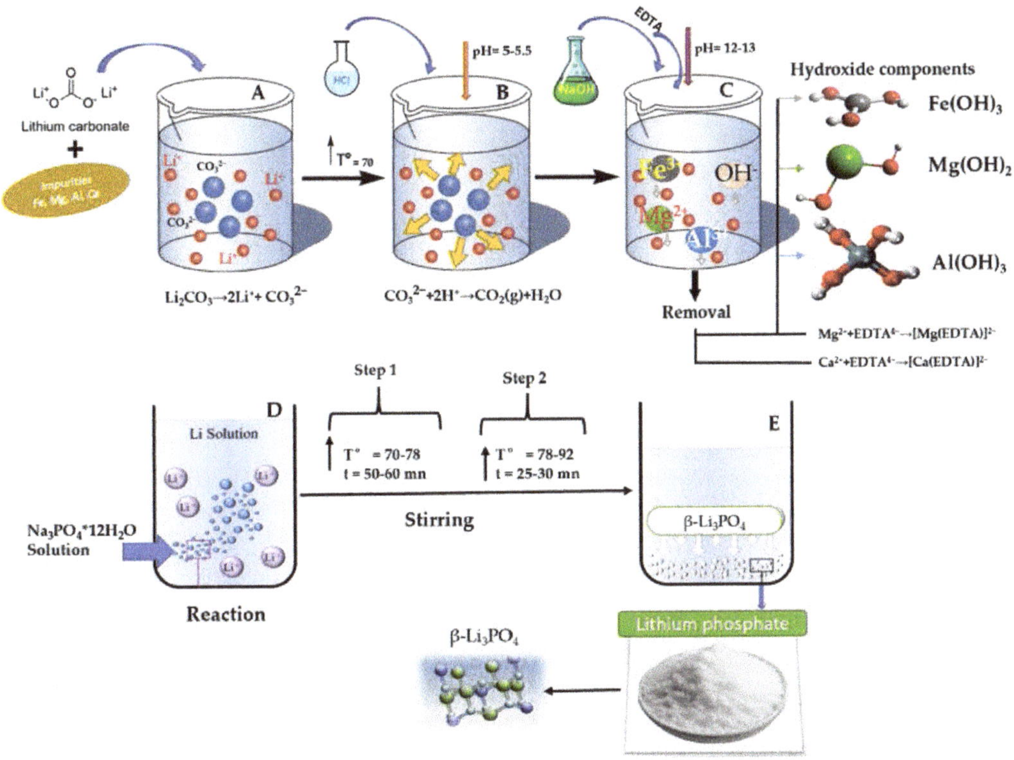

Figure 3. Lithium phosphate synthesis and raw materials impurities removal process.

During the increase in the lithium carbon solution's pH, some impurities such as Al, Fe, and Mg formed insoluble hydroxides in basic conditions and were removed from the solution. The added EDTA formed soluble complexes with Mg^{2+} and Ca^{2+} under vigorous stirring for 25–30 min and supported the precipitation of the remaining impurities.

The obtained solution was then filtered using a multifunction vacuum pump with water circulation and filter paper.

Secondly, an equivalent molarity of sodium phosphate source was dissolved in deionized water according to the following relationship: n[$Na_3PO_4.12H_2O$] = n[Li^+]/3. The sodium phosphate solution was gradually added to the lithium source solution at 70–78 °C with vigorous stirring for 50–60 min. Additional heating and stirring for 25–30 min at 78–92 °C were necessary to complete the reaction, and a white precipitate (Li_3PO_4) began to form immediately. The final white solution was filtered and washed; the solid phase was collected and dried at 105 °C for 2 h in an electrothermal fan dryer (Shanghai Lichen

Bangxi Instrument Technology Co., Ltd., Shanghai, China, No. 101-185). Table 2 presents the basic conditions of the process and statistics of the different samples.

Table 2. Basic conditions and statistics.

Samples	Li_2CO_3	$Na_3PO_4.12H_2O$	pH Control Reagent	Reaction Temperature/°C	Reaction Time/min	pH	Li_3PO_4 (g)	Yield (%)
LPO-1	3 g	9.942 g	HCl-NaOH	78	40 min	12	2.40	79.19
LPO-2		9.942 g		78	50 min	12.5	2.37	78.20
LPO-3		9.942 g		78–92	1 h 30 min	13	2.50	82.49
LPO-4		10 g		78–90	1 h 30 min	12.7	2.85	93.49

$\text{yield}(\%) = \frac{Experimental\ mass}{Theorical\ mass} \cdot 100$; where Theorical mass = 3.04 to 3.05 g.

2.3. *Sample Characterization Methods*

Various techniques were used to characterize the different samples. The purity of the Li_2CO_3 source and the content of impurities were determined using the standard titration method and inductively coupled plasma optical emission spectroscopy (ICP-OES) (Optima 5300 DV model, S09022867, manufacturing unit: PerkinElmer, USA).

The crystal phase studies for all samples were performed using the Empyrean Malvinpanakry X-ray diffractometer (Goniometer reproducibility: 0.0001 degrees; controllable step: 0.0001 degrees; test voltage 45 kv 40 mA). All reflections were assigned using the International Center for Diffraction Data (ICDD).

The surface morphology of the particles and the distribution of the samples were characterized using scanning electron microscopy (SEM) (Company: Zeiss, Germany, Model: Sigma300) in combination with energy dispersive X-ray spectroscopy (EDS) (Oxford Energy Spectrograph 30 xplore).

X-ray photoelectron spectroscopy (XPS) was performed using ThermoFisher Scientific (America) with X-ray radiation of 10–400 μm (5 μm step size setting). The binding energies of the chemical elements were corrected with the internal standard C 1s (284.5 eV). The peaks were fitted with a Gaussian line shape in Origin 2024b software.

The electrochemical properties of the sample were analyzed by electrochemical impedance spectroscopy (EIS) using the DH7000 Electrochemical Workstation. The test was performed at room temperature, 40, 60, and 80 °C. The frequency was set to 1 MHz–0.01 Hz, and the amplitude to 5 mV. After the test, the EIS card was adjusted.

The atmospheric plasma spraying (APS, Jiangsu Donghua Analytical instrument Co., Ltd., Jiangyan, China) technique was used to deposit materials as coatings on surfaces with a surface area of 0.785 cm^2 and a thickness of 0.0781 cm measured with a spiral micrometer (model IP64). It used a symmetric cell with stainless steel blocking electrodes, with an amplitude of 10 mV and a frequency range of 0.1 to 1 MHz. This method consisted of measuring the conductivity from high temperature to room temperature to ensure the thermal stability of the sample. The ionic conductivity (σ) of LPO-4 was calculated from the impedance data at RT, 40, 80 °C.

The thermal properties of the tested solid electrolyte were investigated using the TG/DTA7300 system (Seiko, Japon: Hitachi Corporation). The LPO-4 powder was heated to 820 °C under nitrogen supply at a rate of 10 °C min^{-1}. Subsequently, the TG data were used to develop Coats and Redfern models using Plank and Boltzmann constants and to calculate the key kinetic and thermodynamic parameters for each thermal phase of the LPO-4 sample.

2.4. *Modeling*

We assumed that all conversions were first-order reactions. Coats and Redfern models were developed to calculate the kinetic and thermodynamic key parameters for each thermal phase, such as pre-exponential factor (A), Gibbs free energy changes (ΔG), entropy changes (ΔS), and enthalpy changes (ΔH) [19–23]. The following equations and constants were used in this procedure:

$$\ln[-\ln(1-x)] = ln\frac{ART^2}{\beta E_a} - \frac{E_a}{RT},$$

where: $x = \frac{wi-wt}{wi-wf}$; wi: initial sample weight; wt: weight of sample at particular temperature T; wf: final sample weight; A: pre-exponential factor; β: heating rate (10 °C/min); R: gas constant (8.3143 $Jmol^{-1}K^{-1}$); Ea: activation energy, T: temperature (K) at the peak of the DTG curves.

$$Ea = slope \times 8.3143$$

$$A = \frac{\beta E_a e^{\frac{Ea}{RT^2}}}{RT^2}$$

$$\Delta H = E_a - RT$$

$$\Delta G = E_a + T.\ln\left(\frac{KT}{hA}\right)$$

$$\Delta S = \frac{\Delta H - \Delta G}{T}$$

where K is Boltzmann's constant (1.381×10^{-23} $J.K^{-1}$), and h is Planck's constant (6.626×10^{-34} J s).

The slope and intercept values of the plot of $\ln[-\ln(1-x)]$ versus 1000/T were used to determine the activation energies and other thermodynamic parameters.

3. Results and Discussion

3.1. Lithium Mother Source Study

The lithium mother liquor (Li_2CO_3) was first analyzed to understand its composition, purity, and water content. Titration is still one of the most popular and oldest methods for determining the purity of chemical components. The titration results of the different samples showed a purity of 96–97%.

Inductively coupled plasma mass spectroscopy indicates the presence of lithium and other impurities that were further removed during the treatment, as shown in Figure 3. The percentage and the concentration of each element were calculated using Equations (1) and (2), respectively, as presented in Table 3. The water content of lithium phosphate was calculated using Equation (3), as shown in Table 4.

$$\%\text{element} = C\cdot\frac{V_1}{m_{L2CO3}}\cdot\frac{V_2}{V}\cdot 0.0001 \quad (1)$$

$$\text{oncentraton element (g/L)} = (C\cdot\frac{V_1}{m_{L2CO3}}\cdot\frac{V_2}{V})/(1000) \quad (2)$$

where, V_1 = 100 mL, V_2 = 100 mL, V = 2 mL, $m_{Li_2CO_3}$ = 0.507 g, and C (mg/L): concentration from ICP.

$$\%\text{ Water content} = [(A + B) - (C)]/B \quad (3)$$

Table 3. Composition of lithium carbonate.

		Li_2CO_3		
Element	**Intensity**	**Concentration (C) (mg/L)**	**with Equation (1) (%)**	**with Equation (2) (g/L)**
Al	396.163	0.242	0.239	2.390
As	188.984	−0.320	−0.316	−3.163

Table 3. *Cont.*

		Li_2CO_3		
Element	Intensity	Concentration (C) (mg/L)	with Equation (1) (%)	with Equation (2) (g/L)
B	249.778	−0.082	−0.081	−0.816
Ba	233.533	0.002	0.002	0.020
Bi	223.065	−0.009	−0.009	−0.096
Ca	317.941	0.358	0.353	3.531
Cd	214.443	−0.003	−0.003	−0.039
Cr	283.568	0.001	0.0015	0.015
Co	228.618	0.001	0.001	0.018
Cu	324.762	0.001	0.001	0.016
Fe	238.205	0.125	0.123	1.236
K	766.51	0.124	0.123	1.230
Li	670.803	18.460	18.205	182.059
Mg	285.217	0.134	0.132	1.323
Mn	257.615	0.011	0.011	0.110
Na	589.606	1.150	1.134	11.344
Ni	231.605	0.003	0.003	0.036
P	213.621	0.101	0.099	0.999
Pb	220.36	−0.009	−0.009	−0.097
Se	196.032	0.059	0.058	0.582
Sr	407.78	0.003	0.003	0.034
V	290.886	0.0006	0.0006	0.006
Zn	213.861	0.047	0.047	0.473

Table 4. Water content.

Sample	Weigh of Empty Glass (A)	Weigh of Sample (B)	(A + B)	Weigh After Drying (C)	Water Content (%)
Li_2CO_3	42.10	16	58.10	57.10	6.25

3.2. *Impurities Removal Process and Their Distribution in Synthesized Samples*

The preparation of lithium phosphate by co-precipitation of lithium carbonate (Li_2CO_3) and trisodium phosphate dodecahydrate ($Na_3PO_4*12H_2O$) has attracted great interest (simple and inexpensive). However, the challenge is the presence of some impurities that need to be removed or reduced to minimize their effects in the synthesized sample. As described in the previous section, these different steps cannot definitively remove some impurities. As can be seen in Figure 4, there were still traces of impurities in varying amounts in the samples due to the preparation conditions and severity. Since sodium is difficult to control, it was reduced from 1.26% in the first sample (LPO-1) to 0.28% in LPO-4 (only this sample was used to study ionic conductivity). Other impurities such as Al (0.03–0%), Ca (0.02–0%), Fe (0.01%), K (0.08–0.02) and Mg (0.02–0.01) were significantly reduced or eliminated. This performance in removing impurities was related to various factors, such as stirring of the solution, pH, amount of HCl, NaOH, and EDTA, as described in the "Experiment" section.

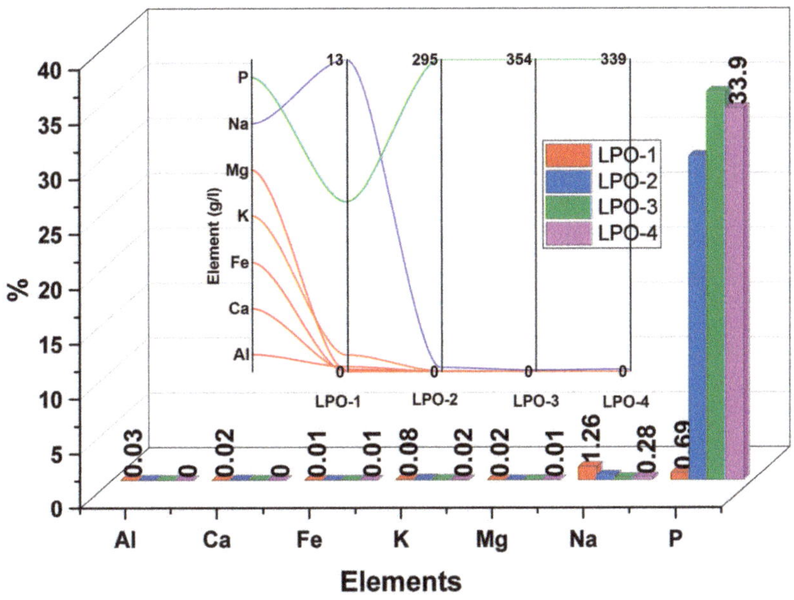

Figure 4. ICP analysis of element distribution in different samples.

3.3. *Surfaces, Pore Distribution, and Isotherm (N_2) Analysis*

As can be seen from Table 5, the BET surface area (9.0293 m^2g^{-1}) is significantly lower than the Langmuir surface area (28.1896 m^2g^{-1}). This is because the BET method considers multilayer adsorption and provides a more conservative estimate compared to the Langmuir method, which assumes single-layer adsorption. The higher Langmuir value indicates that the sample has a significant amount of fine surface features that provide additional adsorption sites. On the other hand, the BET surface area (9.0293 m^2g^{-1}) is lower than the external t-plot surface area (19.6989 m^2g^{-1}). This is because BET measures the total accessible surface area, including contributions from micropores and mesopores, while the external t-plot surface area focuses on the surface area without micropores, suggesting that the sample has significant mesoporous and microporous structure. However, when the Langmuir and t-plot surface areas are compared, it is found that the Langmuir surface area is larger than the t-plot surface area. The Langmuir method tends to overestimate the surface area as it only takes into account the adsorption of single layers at all available sites, whereas the t-plot method provides a more realistic measure of the surface area accessible to larger molecules, excluding micropores.

Table 5. LPO-4 properties.

Samples	S_{BET} (m^2g^{-1})	$S_{t\text{-}Plot\text{-}Ext}$ (m^2g^{-1})	$S_{Langmuir}$ (m^2g^{-1})	V_{total} (cm^3g^{-1})	V_{micro} (cm^3g^{-1})	Size
LPO-4	9.029	19.69	28.189	0.022	-	10.08 nm

The differences in these surface area measurements indicate the complex pore structure of the lithium phosphate sample. A higher Langmuir surface area indicates many fine pores, while a significant external t-plot surface area indicates a substantial contribution from larger mesopores and macropores. This balance is critical to understanding the material's potential applications in solid electrolytes, where both large surface area and appropriate pore sizes are important for ionic conductivity and structural integrity. The average BET adsorption pore width (size) indicates a predominance of mesopores with an average width of 10.08 nm.

Figure 5a shows the isothermal N_2 adsorption–desorption curve of the LPO-4 sample. The adsorption–desorption curve has characteristics similar to those of type IV, according to IUPAC rules [23]. The hysteresis loop appears above p/p0 = 0.057, which represents a low-pressure point in the adsorption isotherm where mesoporous materials exhibit initial multilayer adsorption. The virtual absence of micropores is also observed.

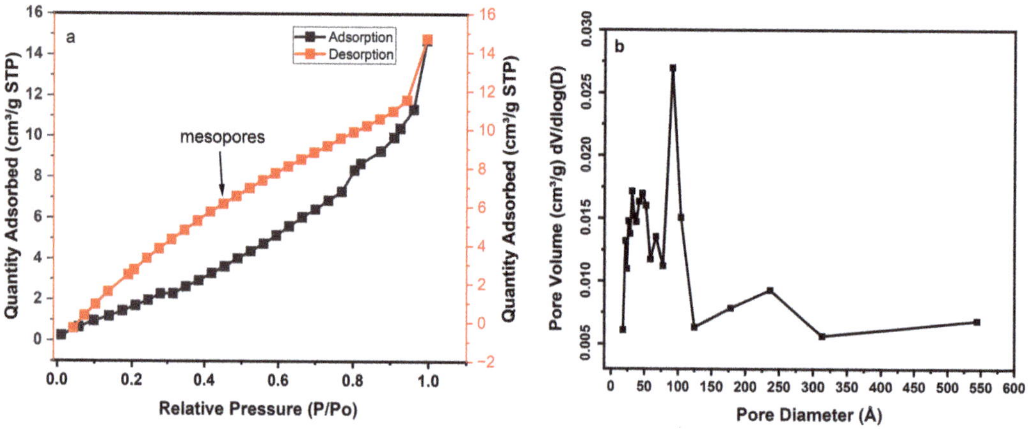

Figure 5. Isothermal adsorption–desorption curves (**a**) and the pore size distribution as the function of pore volume (**b**).

Figure 5b shows the particle size distributions as a function of the pore volume of LPO-4. From the observations in the figure and the table, the density of the small particles below 5 nm is important at a moderate pore volume compared to the particles above 10 nm. The 10 nm particles have a large pore volume, which increases the free ion transport and improves the sample's electrochemical performance.

3.4. Crystallography Analysis

The precipitation method is still available and simple to form a wide range of lithium phosphate structures and well-crystalline materials. In addition, the stochiometric conditions, solution dissolution, stirring, time, and temperature of the preparation must be strict to ensure a good particle size distribution. Figure 6 shows the XRD patterns of the synthesized samples performed in the MDI Jade V9.1 software using WPF and the Rietveld refinement method. LPO-2, LPO-3, and LPO-4 spectrums reveal a single phase of Li_3PO_4, which agrees well with the patterns of lithium phosphate structure corresponding to the orthorhombic crystal system with space group Pmn21 (31) (PDF of 04-027-6504 and 01-086-3942) [10,24]. In all samples, the diffraction peaks appeared at close positions of 16.9, 22.32, 23.37, 24.98, 28.99, 36.99, 38.78, 43.64, 51.06, 60.08, 60.51, 73.58°, corresponding to the reflections (010), (110), (101), (011), (111), (002), (211), (112), (212), (222), (400), and (232) planes. However, the spectrum of LPO-1 shows lithium phosphate as the main phase and two other phases of gallium phosphate ($Ga(PO_4)$) with a weak intensity peak of $2\theta \approx 31.77°$ and $Li_4(SO_4)_2$ with a peak of $2\theta \approx 25.86°$, corresponding to the reflections (112) and (111) planes, (PDF: 04-014-2299 and 01-085-5310) [25,26]. The formation of the $Ga(PO_4)$ phase of 0.1% probably comes from other reagents used in the process, while the formation of 0.2% $Li_4(SO_4)_2$ *phase* may attributed to the lithium reach and the synthesis conditions as presented in the experience and XPS analysis sections. As shown in Table 6, the calculations of the XRD data revealed the amount and the lattice constants of the individual phases formed in the different samples. The LPO-2, LPO-3, and LPO-4 samples showed a single phase (100%) of β-Li_3PO_4, as shown in Figure 7 (left).

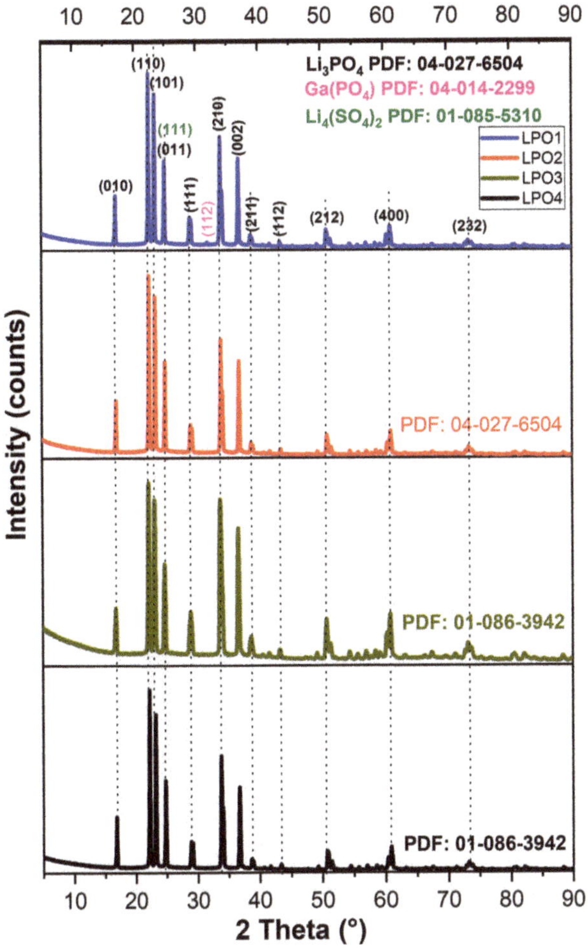

Figure 6. XRD patterns of samples LPO1, LPO2, LPO3, and LPO4.

Table 6. Calculated lattice Constants of samples.

Samples	Phase	a (Å)	b (Å)	C (Å)	Cell Volume (Å^3)	Density (g/cm^3)
LPO-1	β-Li_3PO_4	6.115	5.2394	4.855	155.56	2.472
LPO-2	β-Li_3PO_4	6.115	5.2394	4.854	155.56	2.471
LPO-3	β-Li_3PO_4	6.115	5.2394	4.854	155.56	2.471
LPO-4	β-Li_3PO_4	6.115	5.2394	4.854	155.56	2.471

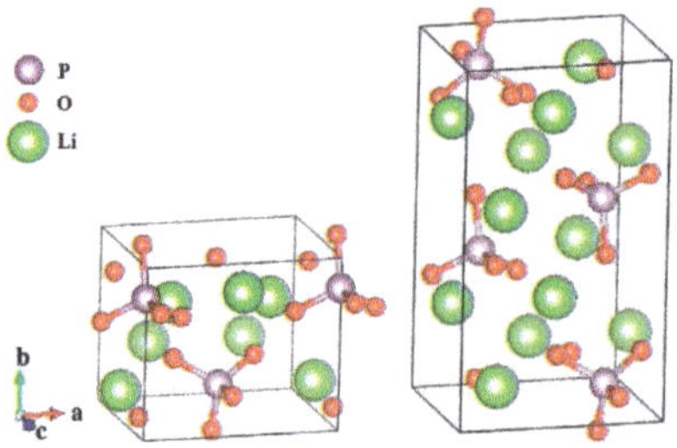

Figure 7. Crystal structure of β-Li_3PO_4 (**left**) and γ-Li_3PO_4 (**right**) [10].

3.5. Morphology and Impurities Distribution Analysis (SEM-EDS and ICP-OES)

Figure 8 shows the SEM images of samples LPO-1, LPO-2, LPO_3, and LPO-4, which were prepared from lithium carbonate and trisodium phosphate dodecahydrate at different pH values and temperatures. All samples exhibited irregular (rectangular) nanotube shapes. A comparison of the observations revealed that the particle shapes of LPO-1 (pH = 12) may be due to an incomplete reaction during production. LPO-2 (pH = 12.5), LPO-3 (pH = 13), and LPO-4 (pH = 13) exhibited complete particle shapes with spherical, cross-linked small particles, with average small particle sizes increasing between 2–100 nm. In the bulk of the samples, LPO-4 showed high purity with a trace of the element sodium (0.28–0.22%, 2.88 g/L), confirmed by coupled plasma mass spectroscopy (ICP) and energy dispersive X-ray spectroscopy (EDS) analysis, as shown in Figures 4 and 9, respectively.

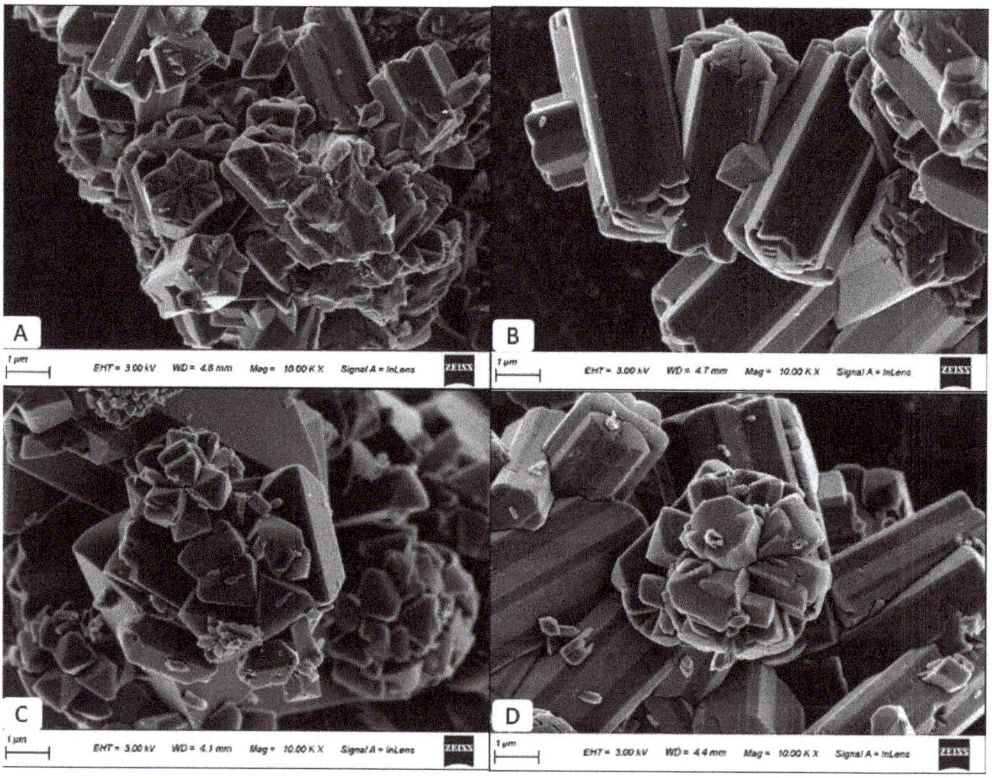

Figure 8. SEM images of (**A**) (LPO-1), (**B**) (LPO-2), (**C**) (LPO-3), and (**D**) (LPO-4).

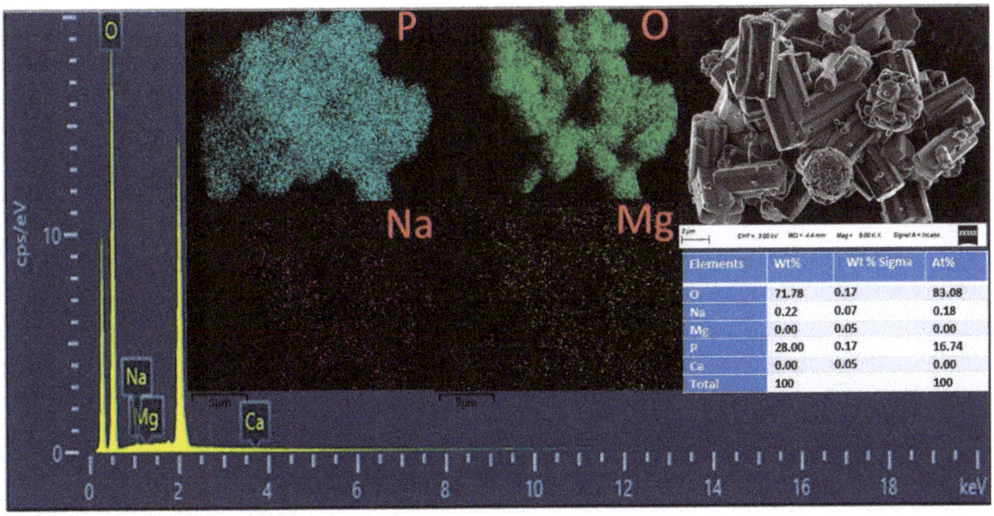

Figure 9. Energy Dispersive X-ray Spectroscopy (EDS) of LPO-4 sample—purities distribution.

3.6. X-Ray Photoelectron Analysis

The X-ray photoelectron (XPS) analysis was investigated to determine the main elements bending energies, their atomic concentration, and ratio in the material. XPS is known to be a powerful technique that can provide a surface analysis with specific information about bonding between atoms and their oxidation in the material.

Table 7 shows the elemental states of lithium, phosphate, and oxygen and their atomic compositions, which were analyzed using X-ray photoelectron spectroscopy (XPS). In both samples, the atomic ratios of Li/P and O/P were close to the stoichiometric values of 3 and 4, respectively. These slight deviations from the stoichiometric values could be due to the slight excess of lithium and oxygen atoms compared to the ideal ratios. The bending energy values of Li 1s, P 2p, and O 1s of all samples were similar and close, as shown in Table 7. On the other hand, these values are in good agreement with those reported by Tae Yong Kim [1]. The atomic ratio of Na/P of all samples was also calculated to determine the degree of sample purities and understand their impact on the chemical environment as the main impurity. The LPO-4 presents the lowest Na/P ratio of 0.06 against 0.84, 0.66, and 0.52 for LPO-1, LPO-2, and LPO-3, respectively. Because of its lowest Na atomic concentrations (0.67% atomic), LPO-4 was considered as a reference to compare the local electronic environment of C 1s, Li 1s, P 2p, O 1s, and Na 1s species in different samples.

Table 7. Sample elements' binding energies and atomic ratios.

		LPO-1	LPO-2	LPO-3	LPO-4
Binding Energy (eV)	Li 1s	54.8	54.8	54.7	54.8
	P 2p	133.0	133.05	133.0	133.1
	O 1s	530.9	530.87	530.9	531.0
	C 1s	284.8	284.8	284.8	284.7
	Na 1s	1071.4	1071.4	1071.6	1071.8
Atomic ratio	Li/P	3.1	3.2	3.0	3.0
	O/P	4.3	4.3	4.2	4.1
	Na/P	0.84	0.66	0.52	0.06

Figure 10 shows C 1s, Li 1s, P 2p, O 1s, and Na 1s spectra, elements, and bending energies (as shown in Table 8) of various samples. For all samples, Li 1s spectra exhibited a major peak at 54.84 eV attributed to Li_2CO_3, which may originate from the lithium mother liquor source. LPO-1 showed an additional smaller Li_2CO_3 peak in the Li-1s spectra at the same position as the first peak. Observation shows that the intensity of the main peak of Li_2CO_3 decreases from LPO_1 to LPO-4, indicating the decrease in its concentration due to the stoichiometric approach to the ideal values (Li/P).

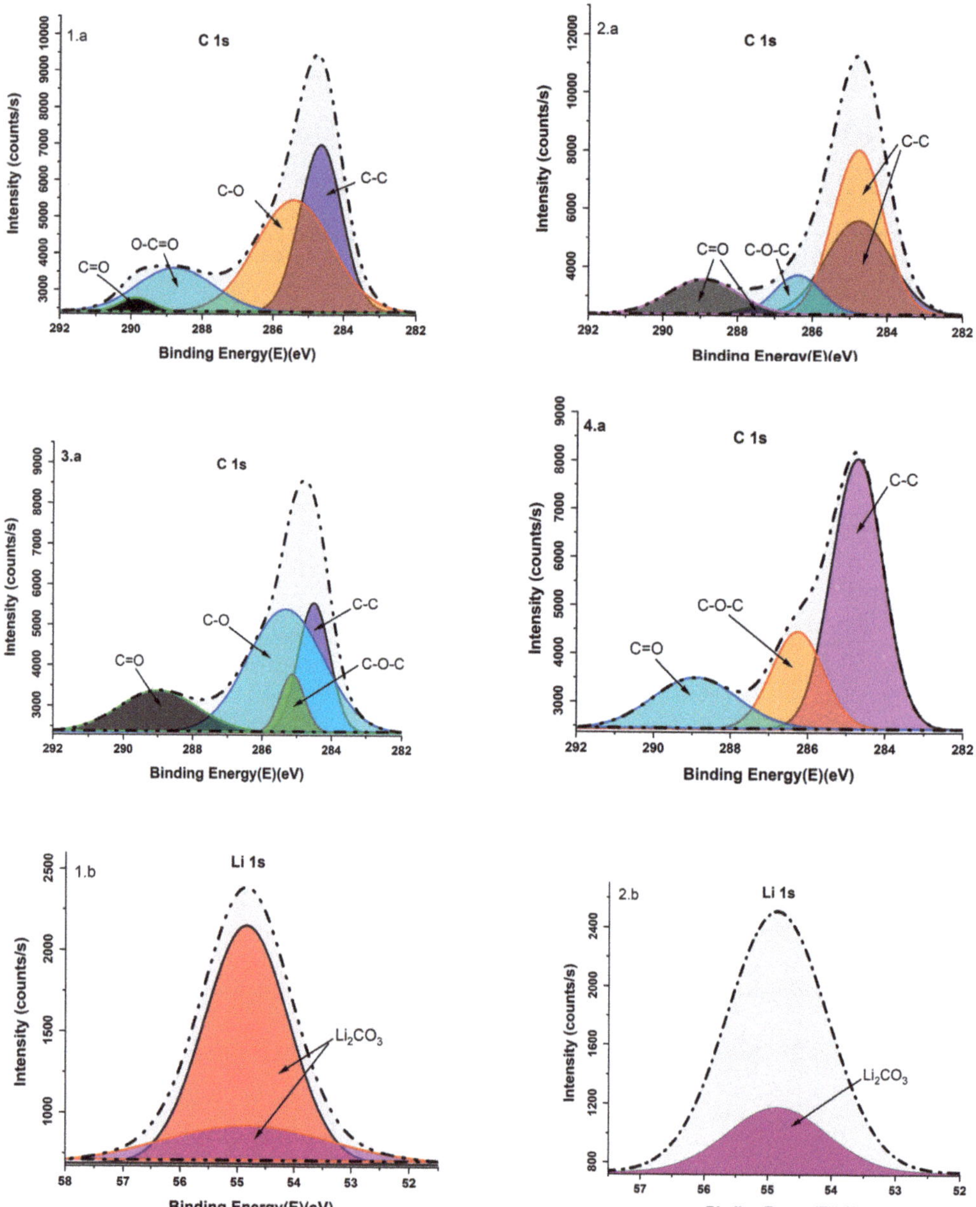

Figure 10. *Cont.*

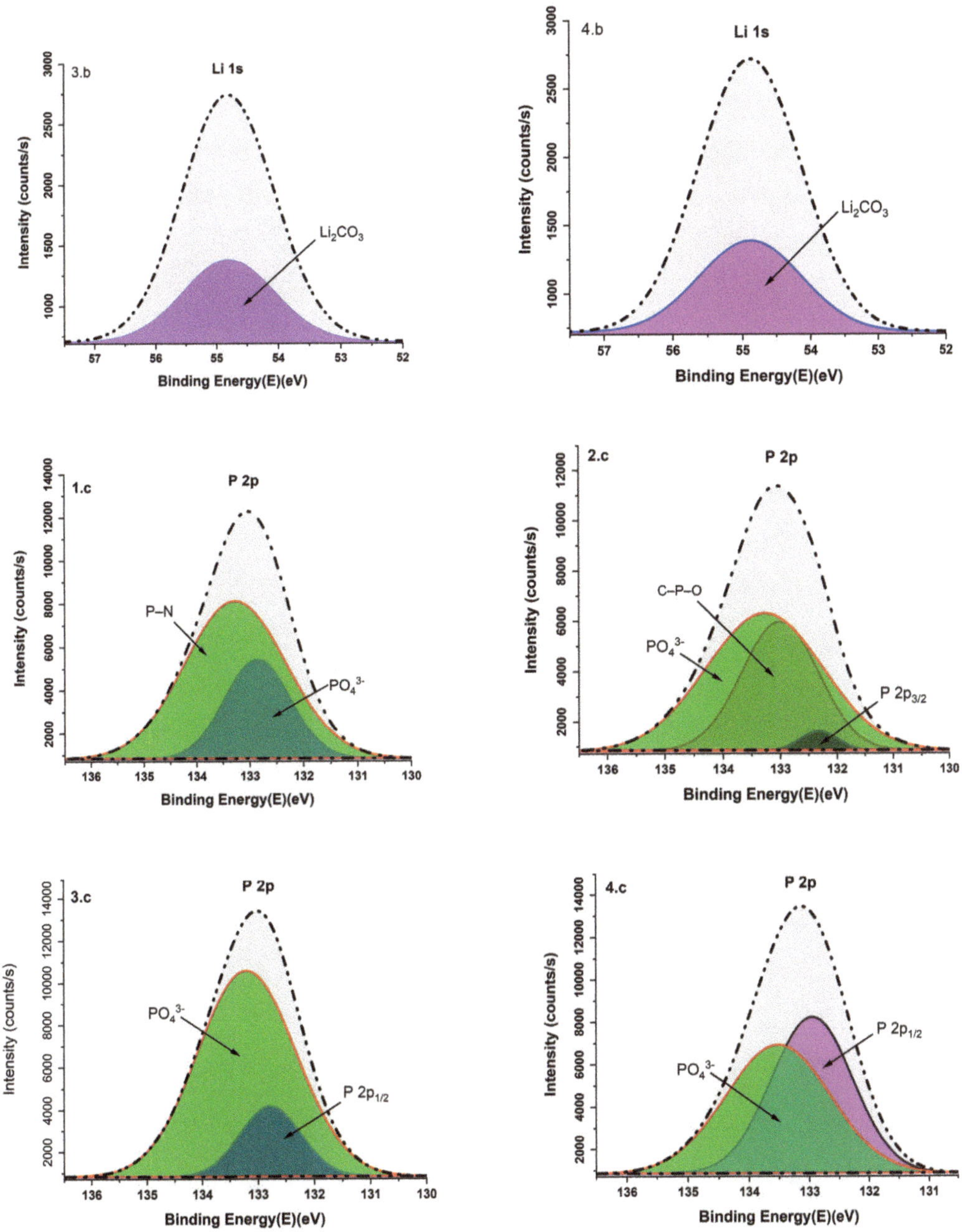

Figure 10. *Cont.*

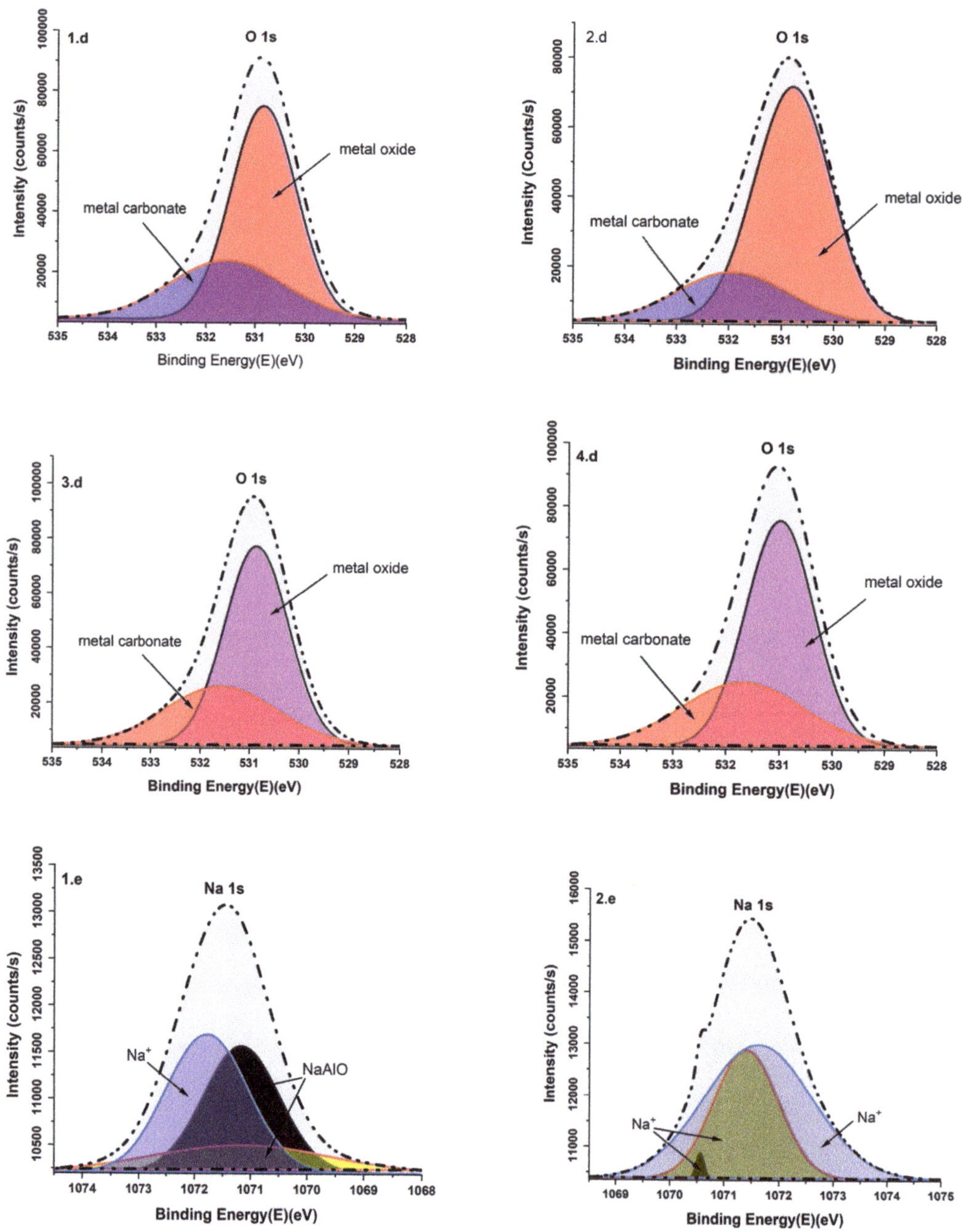

Figure 10. *Cont.*

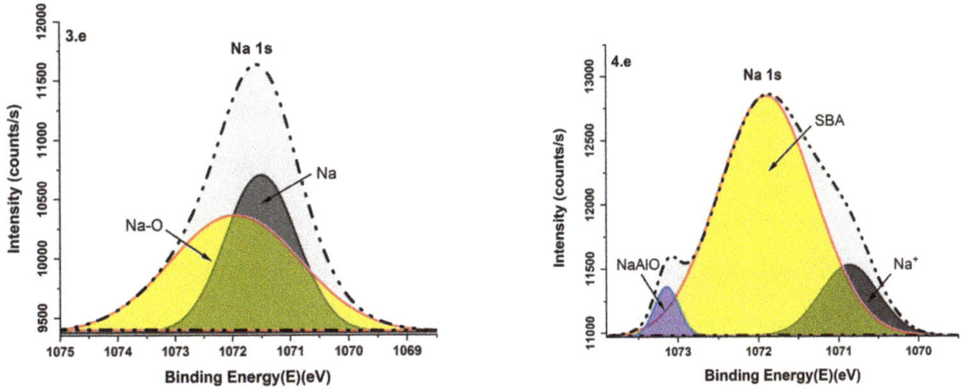

Figure 10. XPS C 1s, Li 1s, P 2p, O 1s, and Na 1s core spectra for LPO-1 (1.a, 1.b, 1.c, 1.d, and 1.e), LPO-2 (2.a, 2.b, 2.c, 2.d, and 2.e), LPO-3 (3.a, 3.b, 3.c, 3.d, and 3.e), and LPO-4 (4.a, 4.b, 4.c, 4.d, and 4.e).

Table 8. XPS peak Blinding Energies and Assignments.

Spectra	Binding Energy (eV)	Assignment	Refs.
C 1s	284.65/284.50/284.72	C-C	[27,28]
	285.41/285.32	C-O	[29,30]
	288.75	OC=O	[27]
	289.00/289.85	C=O	[27,31]
	285.12/286.26	C-O-C	[27]
Li 1s	54.84	Li_2CO_3	[32]
P 2p	133.27	P–N	[33]
	132.77/132.95	P $2p_{1/2}$	[34,35]
	133.27/133.52/133.20/132.85	PO_4^{3-}	[36–38]
	132.32	P $2p_{3/2}$	[39]
	133.00	C–P–O	[40]
O 1s	531.56	metal carbonate	[41]
	530.86	metal oxide	[41]
Na 1s	1071.7/1070.86	Na^+	[42]
	1071.16/1073.13	Sodium oxides (NaAlO)	[43]
	1071.89	SBA	[44]
	1071.94	Na-O	[45]
	1071.50	Na	[46]

The P 2p peaks of PO43- were identified in all samples with slight shifts at 133.2 eV for LPO-4/LPO-2 and at 132.85 eV and 133.5 eV for LPO-1 and 3, respectively. The P 2p spectra of LPO-1 showed a peak at 133.27 eV, which was assigned to the P-N species. The P 2p spectra showed a peak of the P 2p1/2 species at 132.77 and 132.95 eV for LPO-4 and 3, respectively. The additional LPO-2 P 2p peaks identified at 133.00 eV and 132.32 eV were assigned to C–P–O and P 2p3/2, respectively. However, the surface areas and intensities of the O 1s peaks were similar for all samples, with a slight shift. The peaks at 530.86 and 531.56 eV correspond to metal oxide (possibly Li_2O) and metal carbonate, respectively.

The C 1s XPS core spectra showed peaks with slight shifts attributable to C-C bonds at 284.65, 284.77, 284.50, and 284.72 eV in LPO-1, 2, 3, and 4, respectively. Additional peaks of 286.39, 285.12, and 286.26 eV were found in LPO-2, 3, and 4, respectively. These C 1s peaks corresponded to the C-0-C species. The C=O binding energy at 289.85 eV for LPO-1, 287.53–288.91 eV for LPO-2, 289.00 for LPO-3 and 288.91 eV for LPO-4. The O-C=0 binding energy at 288.75 eV was found only in the LPO-1 C1s spectra, while two other binding energies at 285.41 and 285.12 eV were found in LPO-1 and 3, which correspond to the C-O environment due to the carbon layer on most of the sample surfaces exposed to air. The carbon peaks of the samples originate from surface impurities.

The XPS spectrum of Na 1s in LPO4 is decomposed into two peaks at 1070.77 and 1073.13 eV, indicating the presence of Na^+. The Na 1s peaks in LPO3 with binding energies of 1071.50 and 1071.94 eV indicate the presence of Na and Na-O, respectively.

The same Na 1s peak of Na^+ was found at 1071.16, 1071.16, and 1071.77 eV for LPO-2 and 1071.63 eV. An additional Na 1s peak of sodium oxides (NaAlO) was found at 1070.55 and 1071.41 eV in LPO-1 and at 1073.131 eV in LPO-4. The XPS spectrum of LPO-4 for Na 1s showed another peak at 1071.89 eV, indicating the presence of SBA species in the sample structure.

As can be seen in Figure 10, LPO-4 exhibited clear and strong peaks of C 1s, Li 1s, P 2p, and O 1s, indicating that this sample may have more stable solid electrolytes. Other slight changes in peak positions or intensities could be related to the slight variations in particle structure (e.g., slight disorder or variations in crystal quality) of the different samples (as shown in Figure 8), which may affect the electronic structure of the elements, resulting in shifts in the observed binding energies. The change may also originate from the local chemical environment around the lithium, phosphorus, and oxygen atoms, which is essentially due to residual impurities in LPO-1, LPO-2, and LPO-3 samples.

3.7. Electrical Properties

Figure 11a shows the Nyquist plot curves of lithium phosphate (LPO-4, pH = 13) at different temperatures (room temperature, 40, 60, and 80 °C). The spectra were obtained in a frequency range of f = 0.01 Hz to 1 MHz. A clear semicircle was identified in the high-frequency range, and the Warburg impedance line was identified in the low-frequency range.

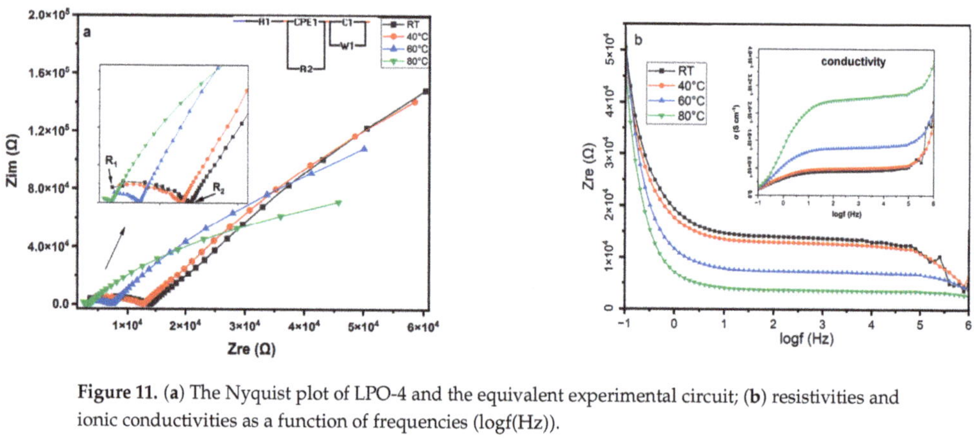

Figure 11. (**a**) The Nyquist plot of LPO-4 and the equivalent experimental circuit; (**b**) resistivities and ionic conductivities as a function of frequencies (logf(Hz)).

The first resistance (R1) was interpreted as the bulk's response to the electric field. The second resistance (R2) could indicate additional resistances, such as grain boundary resistance or secondary interfacial processes of the sample.

An electrical circuit containing three resistors, a constant phase (CPE1), a capacitor, and an open Warburg, which was associated with lithium-ion diffusion, was proposed for the electrical analysis. The following formula was used to determine the ionic conductivity of the sample: σ = L/(R.A)

Where: L is the thickness of the pellets in cm, R is the bulk resistance in Ω, and A is the paste area of the sample in cm^2. The intersection of the semicircle with the high-frequency region was used to determine the total resistance, which corresponds to the R2 of the circle. The ionic conductivities obtained were 7.1×10^{-6} $S.cm^{-1}$ at room temperature, 7.6×10^{-6} $S.cm^{-1}$, 1.3×10^{-5} $S.cm^{-1}$, and 2.7×10^{-5} $S.cm^{-1}$ at 40, 60 and 80 °C, respectively. As presented in Figure 11b, these values were higher than Li_3PO_4 conductivity values of $1.8.10^{-10}$ $s.cm^{-1}$ and 4.65×10^{-8} $S.cm^{-1}$, 3.10^{-7} $S.cm^{-1}$, and 4.10^{-8} $S.cm^{-1}$ reported by Lugas Dwi Prayogi et al. [3], Yin-Qiu Sun et al. [2], and Nur I.P. Ayu et al. [10], respectively. This performance in conductivity may be attributed to the short diffusion distance of lithium ions due to the particle nanoscales and nanotube morphologies with high surface area (9.029 m^2g^{-1}).

3.8. Kinetic and Thermodynamic Analysis

Figure 12 shows the thermogravimetric (TG) profile and a derived thermogravimetric (DTG) profile as a function of temperature. The TG curve shows the mass loss profile with increasing temperature, while the DTG curve shows two small and strong endothermic peaks at 222.7 and 450 °C. These peaks may correspond to decomposition processes. These peaks can be assigned to the decomposition phases of the sample from its original structure.

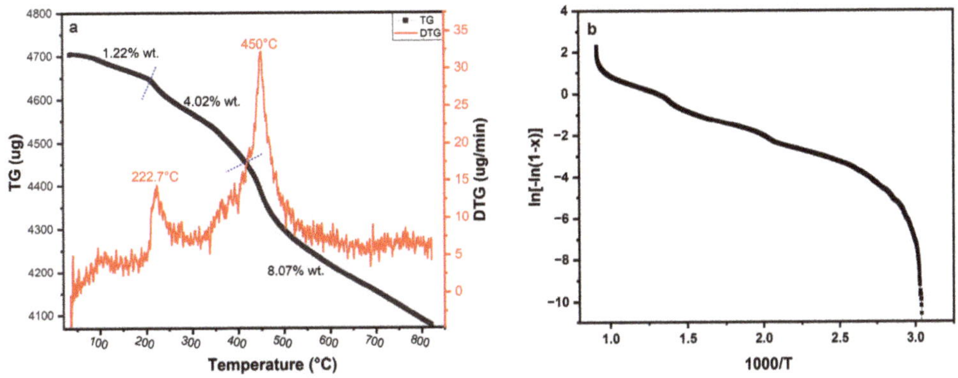

Figure 12. (**a**): Thermal phases and (**b**): plot of ln[−ln(1 − x)] over 1000/*T*.

The sample, with its key kinetic and thermodynamic parameters, provided important insights into the thermal stability and decomposition behavior of LPO-4 (β-Li_3PO_4). The data were divided into three different phases, each characterized by specific temperature ranges, weight loss percentages, activation energies (Ea), and thermodynamic parameters (ΔS, ΔH, and ΔG), as shown in Table 9. These parameters are crucial for understanding the stability of the material and its potential performance as a solid electrolyte in lithium batteries, pointing the way for further investigations.

Table 9. Kinetic and thermodynamic parameters of LPO-4.

Phases	T°C Range	Loss (Wt.%)	Ea (eV)	ΔS	Δ*H*	ΔG	Fitting Equations	A ($M.s^{-1}$)
1	RT–208.7 °C	1.22	0.38	−206.502	$1.49 \times 10^{+8}$	$1.49 \times 10^{+8}$	Y = −4.4647x + 7.498 R^2 = 80.56	163.89
2	208.7–414.8 °C	4.02	0.19	−261.54	$1.1 \times 10^{+8}$	$1.1 \times 10^{+8}$	Y = −2.3067x + 2.608 R^2 = 98.92	0.31
3	414.8–820.89 °C	8.07	0.29	−255.374	$2.58 \times 10^{+8}$	$2.58 \times 10^{+8}$	Y = −3.4057x + 4.352 R^2 = 95.09	1.04

The first phase was measured at room temperature up to 208.7 °C. This phase exhibits relatively low weight loss and moderate activation energy (0.38 eV). This loss can be related to the removed moisture in the sample (RT–208.7 °C). The negative entropy change (ΔS) indicates a decrease in disorder in this phase. The higher pre-exponential factors in the first phase lead to more frequent effective collisions and reactions, reflecting the efficiency of LPO-4 in facilitating ion movement in this region.

The second phase was defined in the temperature range of 208.7–414.8 °C. In this phase, a higher weight loss indicates a stronger decomposition, and the activation energy decreases to 0.19 eV. The more negative ΔS value indicates a greater decrease in disorder.

The third phase shows the highest weight loss, indicating a significant decomposition of the material. The activation energy ($Ea = slope \times 8.3143$) increases slightly to 0.29 eV compared to the second phase, indicating a certain resistance of the bonds and a degree of thermal stability, which is still relatively low. The entropy change (ΔS) remains negative. The high values for enthalpy change (ΔH) in the first and third phases indicate that the material is relatively stable and requires significant energy to decompose; thus, it may be suitable for applications requiring high thermal stability. The activation energies for all phases are below 5 ev and relate to the values for superionic conductors [47,48]. The R^2 values for all phases show a strong correlation between the experimental data and the fitting model.

4. Conclusions

Solid lithium phosphate electrolyte samples with the formula (Li_3PO_4) were synthesized from lithium carbonate (Li_2CO_3) and trisodium phosphate dodecahydrate (Na_3PO_4*$12H_2O$) by co-precipitation at different pHs, temperatures, and stoichiometric values. A single-phase, high-purity and ionically conductive β-Li_3PO_4 with nanotube morphologies was successfully prepared in the set of samples at 90 °C and pH 13. The ionic conductivities obtained were 7.1×10^{-6} S.cm^{-1} at room temperature, 7.6×10^{-6} S.cm^{-1}, 1.3×10^{-5} S.cm^{-1}, and 2.7×10^{-5} S.cm^{-1} at 40, 60, and 80 °C, respectively.

The LPO-4 sample exhibited excellent electrochemical properties and represents a promising route for the further industrial application of lithium phosphate as a sustainable and environmentally friendly solid electrolyte. The actual achievement compared with some previous work on lithium phosphate, including its conductivity, high surface area, and lower activation energy, may attributed to nanotubes and nanoscale size of the synthesized LPO-4, which enlarge the reaction interfaces and shorten the diffusion distance of lithium ions.

The kinetic and thermodynamic studies confirmed the excellent stability in the temperature range RT–208 °C and some endothermic reactions that can occur in high-temperature ranges.

Author Contributions: Conceptualization, S.Y.Z.; Methodology, S.Y.Z.; Software, S.Y.Z.; Validation, S.Y.Z., H.Y. and C.J.; Formal analysis, S.Y.Z.; Investigation, S.Y.Z.; Resources, S.Y.Z. and H.Y.; Data curation, S.Y.Z.; Writing—original draft, S.Y.Z.; Writing—review & editing, S.Y.Z. and C.J.; Visualization, S.Y.Z.; Supervision, H.Y.; Project administration, S.Y.Z., H.Y. and C.J.; Funding acquisition, H.Y. All authors have read and agreed to the published version of the manuscript.

Funding: This research was funded by National Natural Science Foundation project 21975288.

Data Availability Statement: The data presented in this study are available on request from the authors.

Conflicts of Interest: The authors declare no conflict of interest.

References

1. Kim, T.Y.; Song, C.K.; Yun, Y.S.; Yun, D.; Han, J.W.; Yi, J. Active site structure of a lithium phosphate catalyst for the isomerization of 2,3-epoxybutane to 3-buten-2-ol. *Mol. Catal.* **2018**, 445, 133–141. [CrossRef]
2. Sun, Y.Q.; Luo, X.T.; Zhu, Y.S.; Liao, X.J.; Li, C.J. Li_3PO_4 electrolyte of high conductivity for all-solid-state lithium battery prepared by plasma spray. *J. Eur. Ceram. Soc.* **2022**, 42, 4239–4247. [CrossRef]

3. Prayogi, L.D.; Faisal, M.; Kartini, E.; Honggowiranto, W. Supardi, Morphology and conductivity study of solid electrolyte Li_3PO_4. *AIP Conf. Proc.* **2016**, *1710*, 030047. [CrossRef]
4. Fan, L.; Wei, S.; Li, S.; Li, Q.; Lu, Y. Recent Progress of the Solid-State Electrolytes for High-Energy Metal-Based Batteries. *Adv. Energy Mater.* **2018**, *8*, 1702657. [CrossRef]
5. Rosen, M.; Hecker, P.; Mann, M.; Ma, Q.; Gross, J.P.; Schwaiger, R.; Guillon, O.; Fattakhova-Rohlfing, D.; Finsterbusch, M. Reducing the environmental footprint of solid-electrolytes—A green synthesis route for LATP. *Green Chem.* **2024**, *26*, 2712–2720. [CrossRef]
6. Saran, S.; Eker, Y.R. Synthesis, structural and conductive properties of Nd doped garnet-type $Li_7La_3Zr_2O_{12}$ Li-ion conductor. *Curr. Appl. Phys.* **2022**, *41*, 1–6. [CrossRef]
7. Xiao, B.; Li, D.; Dai, X.; Wei, Y.; Liao, Y.; Wang, C.; Ji, F.; Wu, F. Construction of a robust lithium cobalt phosphate layer for enhancement of the electrochemical performance of $LiNi_{0.8}Co_{0.1}Mn_{0.1}O_2$ at high voltage. *Electrochim. Acta* **2024**, *475*, 143648. [CrossRef]
8. Zeng, S.; Ding, X.; He, L.; Li, H.W.; Zhang, Q.; Li, Y. Realizing fast Li-ion conduction of Li_3PO_4 solid electrolyte at low temperature by mechanochemical formation of lithium-containing dual-shells. *Mater. Adv.* **2023**, *4*, 2780–2784. [CrossRef]
9. Kuwata, N.; Iwagami, N.; Matsuda, Y.; Tanji, Y.; Kawamura, J. Thin Film Batteries with Li_3PO_4 Solid Electrolyte Fabricated by Pulsed Laser Deposition. *ECS Trans.* **2009**, *16*, 53–60. [CrossRef]
10. Ayu, N.I.P.; Kartini, E.; Prayogi, L.D.; Faisal, M. Supardi, Crystal structure analysis of Li_3PO_4 powder prepared by wet chemical reaction and solid-state reaction by using X-ray diffraction (XRD). *Ionics* **2016**, *22*, 1051–1057. [CrossRef]
11. Zhao, S.; Jiang, W.; Zhu, X.; Ling, M.; Liang, C. Understanding the synthesis of inorganic solid-state electrolytes for Li ion batteries: Features and progress. *Sustain. Mater. Technol.* **2022**, *33*, e00491. [CrossRef]
12. Method for Recovering Lithium from Lithium Carbonate Precipitation Mother Liquor. CN111533146A. Available online: https://patents.google.com/patent/CN111533146A/zh (accessed on 25 October 2024).
13. Liu, W.Y.; Li, C.L.; Fu, Z.W. Stability of Nitrogen-Containing Lithium Phosphate Films in Air. *Acta Phys.-Chim. Sin.* **2006**, *11*, 1413–1418. Available online: https://qikan.cqvip.com/Qikan/Article/Detail?id=23426027&from=Qikan_Article_Detail (accessed on 2 October 2024).
14. Wu, J. Discussion on the process of preparing lithium dihydrogen phosphate from lithium phosphate tower skin. *XinJiang Youse Jinshu* **2012**, *S1*, 124–125. Available online: https://qikan.cqvip.com/Qikan/Article/Detail?id=1003444489&from=Qikan_Article_Detail (accessed on 2 October 2024).
15. Wu, S.; Zhang, M.; Zan, C.; Zhou, H. Process of Deep Recovery of Lithium from the Mother Liquor of Lithium Carbonate by Phosphate Precipitation. *J. Tianjin Univ. Sci. Technol.* **2023**, *38*, 35–41. [CrossRef]
16. Yao, X.; Wang, Y.; Xu, M.; Shuying, S.U.N. Preparation of lithium phosphate from waste liquid of retired lithium batteries. *Chem. Ind. Eng.* **2024**, *41*, 179–190. [CrossRef]
17. Ishigaki, N.; Akimoto, J. Room temperature synthesis and phase transformation of lithium phosphate Li3PO4 as solid electrolyte. *J. Asian Ceram. Soc.* **2021**, *9*, 452–458. [CrossRef]
18. Puente, P.M.G.; Song, S.; Cao, S.; Rannalter, L.Z.; Pan, Z.; Xiang, X.; Shen, Q.; Chen, F. Garnet-type solid electrolyte: Advances of ionic transport performance and its application in all-solid-state batteries. *J. Adv. Ceram.* **2021**, *10*, 933–972. [CrossRef]
19. Farrukh, M.A.; Butt, K.M.; Chong, K.K.; Chang, W.S. Photoluminescence emission behavior on the reduced band gap of Fe doping in CeO2-SiO2 nanocomposite and photophysical properties. *J. Saudi Chem. Soc.* **2019**, *23*, 561–575. [CrossRef]
20. Del Mar Graciani, M.; Rodríguez, A.; Muñoz, M.; Moyá, M.L. Micellar solutions of sulfobetaine surfactants in water-ethylene glycol mixtures: Surface tension, fluorescence, spectroscopic, conductometric, kinetic studies. *Langmuir* **2005**, *21*, 7161–7169. [CrossRef]
21. Silva, J.M.R.; de Morais Araújo, A.M.; da Costa Evangelista, J.P.; da Silva, D.R.; Gondim, A.D.; de Araujo, A.S. Evaluation of the kinetic and thermodynamic parameters in catalytic pyrolysis process of sunflower oil using Al-MCM-41 and zeolite H-ZSM-5. *Fuel* **2023**, *333*, 126225. [CrossRef]
22. Bhardwaj, G.; Kumar, M.; Mishra, P.K.; Upadhyay, S.N. Kinetic analysis of the slow pyrolysis of paper wastes. *Biomass Convers. Biorefinery* **2023**, *13*, 3087–3100. [CrossRef]
23. Zakariyaou, S.Y.; Ye, H.; Oumarou, A.D.M.; Aziz, M.S.A.; Ke, S. Characterization of Equilibrium Catalysts from the Fluid Catalytic Cracking Process of Atmospheric Residue. *Catalysts* **2023**, *13*, 1483. [CrossRef]
24. Keffer, C.; Mighell, A.; Mauer, F.; Swanson, H.; Block, S. The Crystal Structure of Twinned Low-Temperature Lithium Phosphate. *Inorg. Chem.* **1967**, *6*, 119–125. [CrossRef]
25. Lander, L.; Reynaud, M.; Carrasco, J.; Katcho, N.A.; Bellin, C.; Polian, A.; Baptiste, B.; Rousse, G.; Tarascon, J.M. Unveiling the electrochemical mechanisms of $Li_2Fe(SO_4)_2$ polymorphs by neutron diffraction and density functional theory calculations. *Phys. Chem. Chem. Phys.* **2016**, *18*, 14509–14519. [CrossRef]
26. Huang, Y.X.; Liu, J.Y.; Mi, J.X.; Zhao, J.T. $(Ga_{0.71}B_{0.29})PO_4$ with a high-cristobalite-type structure refined from powder data. *Acta Crystallogr. Sect. E Struct. Rep. Online* **2010**, *66*, i4. [CrossRef]
27. Carbon | XPS Periodic Table | Thermo Fisher Scientific | Thermo Fisher Scientific—CN. Available online: https://www.thermofisher.cn/cn/zh/home/materials-science/learning-center/periodic-table/non-metal/carbon.html (accessed on 18 November 2024).

28. Carbon Spectra—Li2CO3—Lithium Carbonate. Available online: https://xpsdatabase.net/carbon-spectra-li2co3-lithium-carbonate/ (accessed on 18 November 2024).
29. Al-Kadhi, N.S.; Hefnawy, M.A.; Alamro, F.S.; Pashameah, R.A.; Ahmed, H.A.; Medany, S.S. Polyaniline-Supported Nickel Oxide Flower for Efficient Nitrite Electrochemical Detection in Water. *Polymers* **2023**, *15*, 1804. [CrossRef]
30. Alamro, F.S.; Medany, S.S.; Al-Kadhi, N.S.; Ahmed, H.A.; Hefnawy, M.A. Modified $NiFe_2O_4$-Supported Graphene Oxide for Effective Urea Electrochemical Oxidation and Water Splitting Applications. *Molecules* **2024**, *29*, 1215. [CrossRef]
31. Shulga, Y.M.; Baskakov, S.A.; Knerelman, E.I.; Davidova, G.I.; Badamshina, E.R.; Shulga, N.Y.; Skryleva, E.A.; Agapov, A.L.; Voylov, D.N.; Sokolov, A.P.; et al. Carbon nanomaterial produced by microwave exfoliation of graphite oxide: New insights. *RSC Adv.* **2014**, *4*, 587–592. [CrossRef]
32. Lu, Y.C.; Crumlin, E.J.; Veith, G.M.; Harding, J.R.; Mutoro, E.; Baggetto, L.; Dudney, N.J.; Liu, Z.; Shao-Horn, Y. In Situ Ambient Pressure X-ray Photoelectron Spectroscopy Studies of Lithium-Oxygen Redox Reactions. *Sci. Rep.* **2012**, *2*, 715. [CrossRef]
33. Yin, X.; Sun, X.; Li, D.; Xie, W.; Mao, Y.; Liu, Z.; Liu, Z. 2D/2D Phosphorus-Doped g-C_3N_4/Bi_2WO_6 Direct Z-Scheme Heterojunction Photocatalytic System for Tetracycline Hydrochloride (TC-HCl) Degradation. *Int. J. Environ. Res. Public. Health* **2022**, *19*, 14935. [CrossRef]
34. Liu, X.; Yu, M.; Wu, S.; Gong, J. Composite nanoarchitectonics for efficient lithium storage by encapsulating black phosphorus quantum dots in cobalt/iron based Prussian blue analogues. *J. Alloys Compd.* **2023**, *969*, 172291. [CrossRef]
35. Chatterjee, T.; Raul, C.K.; Mandal, S.; Pradhan, S.K.; Meikap, A.K. Effect of surfactant-assisted hierarchical growth of cupric oxide-hydroxyapatite nanocomposite on the dielectric and electrical transport behavior. *Phys. B Condens. Matter* **2023**, *650*, 414560. [CrossRef]
36. Jung, W.D.; Jeon, M.; Shin, S.S.; Kim, J.S.; Jung, H.G.; Kim, B.K.; Lee, J.H.; Chung, Y.C.; Kim, H. Functionalized Sulfide Solid Electrolyte with Air-Stable and Chemical-Resistant Oxysulfide Nanolayer for All-Solid-State Batteries. *ACS Omega* **2020**, *5*, 26015–26022. [CrossRef]
37. Chen, K.; Tang, Y.; Zhang, S.; Hao, X.; Zhao, X.; Cheng, L.Q.; Xiao, Y.; Wen, Z. Promoted Stability and Reaction Kinetics in Ni-Rich Cathodes via Mechanical Fusing Multifunctional $LiZr_2(PO_4)_3$ Nanocrystals for High Mass Loading All-Solid-State Lithium Batteries. *ACS Appl. Mater. Interfaces* **2024**, *16*, 45459–45472. [CrossRef]
38. Fontecha, D.; Nuwayhid, R.B.; Kozen, A.C.; Stewart, D.M.; Rubloff, G.W.; Gregorczyk, K.E. Low temperature plasma-enhanced atomic layer deposition of sodium phosphorus oxynitride with tunable nitrogen content. *J. Vac. Sci. Technol. A* **2022**, *40*, 032403. [CrossRef]
39. López, E.O.; Bernardo, P.L.; Checca, N.R.; Rossi, A.L.; Mello, A.; Ellis, D.E.; Rossi, A.M.; Terra, J. Hydroxyapatite and lead-substituted hydroxyapatite near-surface structures: Novel modelling of photoemission lines from X-ray photoelectron spectra. *Appl. Surf. Sci.* **2022**, *571*, 151310. [CrossRef]
40. Wang, Y.; Zheng, X.; Qu, Q.; Liu, G.; Battglia, V.S.; Zheng, H. A novel maleic acid/graphite composite anode for lithium ion batteries with high energy and power density. *Carbon.* **2018**, *132*, 420–429. [CrossRef]
41. Oxygen | Thermo Fisher Scientific XPS Periodic Table | Thermo Fisher Scientific | Thermo Fisher Scientific—CN. Available online: https://www.thermofisher.cn/cn/zh/home/materials-science/learning-center/periodic-table/non-metal/oxygen.html (accessed on 18 November 2024).
42. Khan, Z.; Park, S.O.; Yang, J.; Park, S.; Shanker, R.; Song, H.K.; Kim, Y.; Kwak, S.K.; Ko, H. Binary N,S-doped carbon nanospheres from bio-inspired artificial melanosomes: A route to efficient air electrodes for seawater batteries. *J. Mater. Chem. A Mater.* **2018**, *6*, 24459–24467. [CrossRef]
43. García-Bordejé, E.; Dongil, A.B.; Conesa, J.M.; Guerrero-Ruiz, A.; Rodríguez-Ramos, I. Dual functional materials based on Ni and different alkaline metals on alumina for the cyclic stepwise CO_2 capture and methanation. *Chem. Eng. J.* **2023**, *472*, 144953. [CrossRef]
44. Fang, R.; Li, Y.; Wu, N.; Xu, B.; Liu, Y.; Manthiram, A.; Goodenough, J.B. Ultra-Thin Single-Particle-Layer Sodium Beta-Alumina-Based Composite Polymer Electrolyte Membrane for Sodium-Metal Batteries. *Adv. Funct. Mater.* **2022**, *33*, 2211229. [CrossRef]
45. Foucaud, Y.; Badawi, M.; Filippov, L.O.; Barres, O.; Filippova, I.V.; Lebègue, S. Synergistic adsorptions of Na_2CO_3 and Na_2SiO_3 on calcium minerals revealed by spectroscopic and ab initio molecular dynamics studies. *Chem. Sci.* **2019**, *10*, 9928–9940. [CrossRef] [PubMed]
46. Savinova, E.R.; Zemlyanov, D.Y.; Scheybal, A.; Schlögl, R.; Doblhofer, K. Ex Situ X-ray Photoelectron Spectroscopy Study of the Interface between a Ag(111) Electrode and an Alkaline Electrolyte. 2. Structure of the Double Layer. *Langmuir* **1999**, *15*, 6552–6556. [CrossRef]
47. He, X.; Zhu, Y.; Mo, Y. Origin of fast ion diffusion in super-ionic conductors. *Nat. Commun.* **2017**, *8*, 15893. [CrossRef] [PubMed]
48. Du, Y.A.; Holzwarth, N.A.W. Li Ion Diffusion Mechanisms in the Crystalline Electrolyte γ-Li_3PO_4. *J. Electrochem. Soc.* **2007**, *154*, A999–A1004. [CrossRef]

Disclaimer/Publisher's Note: The statements, opinions and data contained in all publications are solely those of the individual author(s) and contributor(s) and not of MDPI and/or the editor(s). MDPI and/or the editor(s) disclaim responsibility for any injury to people or property resulting from any ideas, methods, instructions or products referred to in the content.

batteries

MDPI

Article

Study of Cathode Materials for Na-Ion Batteries: Comparison Between Machine Learning Predictions and Density Functional Theory Calculations

Claudio Ronchetti [1], Sara Marchio [2], Francesco Buonocore [2,*], Simone Giusepponi [2], Sergio Ferlito [3] and Massimo Celino [2]

[1] Telespazio S.p.A., Via Tiburtina 965, 00156 Rome, Italy; claudio.ronchetti@telespazio.com
[2] Italian National Agency for New Technologies, Energy and Sustainable Economic Development (ENEA)—C. R. Casaccia, Via Anguillarese 301, 00123 Rome, Italy; sara.marchio@enea.it (S.M.); simone.giusepponi@enea.it (S.G.); massimo.celino@enea.it (M.C.)
[3] Italian National Agency for New Technologies, Energy and Sustainable Economic Development (ENEA)—C. R. Portici, Piazzale Enrico Fermi 1, 80055 Portici, Italy; sergio.ferlito@enea.it
* Correspondence: francesco.buonocore@enea.it

Abstract: Energy storage technologies have experienced significant advancements in recent decades, driven by the growing demand for efficient and sustainable energy solutions. The limitations associated with lithium's supply chain, cost, and safety concerns have prompted the exploration of alternative battery chemistries. For this reason, research to replace widespread lithium batteries with sodium-ion batteries has received more and more attention. In the present work, we report cutting-edge research, where we explored a wide range of compositions of cathode materials for Na-ion batteries by first-principles calculations using workflow chains developed within the AiiDA framework. We trained crystal graph convolutional neural networks and geometric crystal graph neural networks, and we demonstrate the ability of the machine learning algorithms to predict the formation energy of the candidate materials as calculated by the density functional theory. This materials discovery approach is disruptive and significantly faster than traditional physics-based computational methods.

Keywords: DFT calculations; neural networks; machine learning; electrochemical energy storage; Na-ion; high-throughput calculations

1. Introduction

As global technology advances, humanity faces significant challenges in terms of pollution and climate impact. The current trajectory is unsustainable; while technological progress improves our quality of life, it simultaneously endangers the planet. A critical shift is necessary, from relying on fossil fuels like oil and coal for energy production to adopting renewable energy sources [1]. Renewable energy primarily involves converting power from the Sun, wind, and oceans into electricity. However, these energy sources do not incessantly produce power, creating a need for effective energy storage solutions.

Lithium-ion batteries (LIBs) have been highly successful in meeting energy storage demands in recent years [2]. Yet, with the growing energy needs, particularly in sectors like transportation where the shift from fossil fuels to electricity is accelerating, LIBs face two major challenges: an increase in the cost of Li due to its limited availability with respect to the increasing demand; and safety concerns related to the reactivity of lithium with oxygen. Sodium-ion batteries (NIBs) offer a promising alternative to LIBs. Sodium is abundant and more stable than lithium, making NIBs an attractive option [3]. Additionally, NIBs are safer than LIBs due to sodium's lower reactivity compared to lithium, which mitigates the risks associated with flammable batteries. Actually, NIBs are rechargeable batteries analogous to LIBs, in which both cathodes use the intercalation reaction mechanism. Sodium-based cathodes have high operating potentials and high capacities, potentially comparable to lithium-based ones [4–7]. Consequently, the combination of strong performance and lower costs, along with the ability to leverage existing technology and production methods developed for LIBs without incurring additional expenses, has fueled growing interest in these systems.

Sodium cathode material based on layered transition metal (M) oxide (LTMO), Na_xMO_2, is one of the more promising classes of materials to consider [8,9]. Indeed, Na_xMO_2 materials exhibit reversible phase transitions that enhance the stability of batteries during charge–discharge cycles. LTMO is classified as On and Pn depending on its octahedral and prismatic crystal phase, respectively, where n counts the number of transition metal layers in the unit cell. It has been found that the O3–P3 and P2–O2 phase transitions do not break the M–O bond during the electrochemical cycle, saving the cathode from damages. However, the P2–O2 phase transition requires a minimal transformation of the crystal, preserving the structural integrity of the P2-type phase and ensuring robust cycling stability. Moreover, the spacing between the metal layers is larger in the P2- than in the O3-type phase, allowing the enhancement of the Na+-ion's mobility. On the other hand, the O3-type phase can suffer complex distortion during the charge–discharge process.

Different strategies have been proposed to increase the density of energy and to enhance stability, preventing Jahn–Teller distortions [10,11] in NIB LTMO cathode materials, including structural modulation, surface modification, and elemental doping [12–16].

Machine learning (ML) has become a widely utilized technique, enabling researchers to accelerate their work using trained algorithms. In materials science, ML is applied to predict material properties, significantly reducing the time and computational resources needed compared to traditional numerical physical models, which are often resource-intensive and rely on modern supercomputers [17]. In the present work, we propose a computational study to enhance the stability of Na_xMnO_2 cathode materials by the substitutional doping of Mn elements with two different methods: (1) high-throughput first principles calculations based on density functional theory (DFT); and (2) machine learning techniques based on the training of a convolutional neural network (CNN) from a dataset generated by DFT calculations. The stability is established by the DFT calculations or CNN predictions to determine the formation energy (more stable crystals have lower formation energy). We compare the two methods to assess the predictive capability of the CNN upon training.

Our previous work [18] provided details on the ML methodology and computational architecture employed to predict the formation energy using the Materials Project dataset. The design, training process, and computational tools utilized to achieve accurate predictions on the dataset have been illustrated. However, the focus of the present work diverges from this approach. Instead of exploring the initial training and methodology, this study aims to evaluate the performance of the pre-trained model on an entirely different dataset that was not included in the original training phase. Specifically, the assessment centers on NIB systems, chosen for their distinctive properties, to determine the extent to which the pre-trained model is capable of generalizing to new and unseen data. By doing so, this work seeks to validate the robustness and transferability of the pre-trained model, offering insights into its broader applicability and limitations when applied to diverse datasets.

2. Materials and Methods

2.1. Base Crystal $NaMnO_2$ and Doping

The P2-type crystal cathode materials of the present study were generated from the base P2-$NaMnO_2$ crystal by the substitution of randomly selected Mn atoms with the first-row transition metals Ni and Ti. The procedure of the substitution is shown in Section 2.3 (Computational Workflow). The base crystal where the Mn atoms were substituted was created as the supercell $3 \times 2 \times 1$ of the crystal mp-971647 of the Materials Project database [19] so that 48 atoms were present therein. Preliminary studies indicated that Ti substitution enables the better optimization of the stability and intercalation potential compared to other elements such as V, Cr, Cu, Fe, and Zn [18].

2.2. First-Principles Method

To characterize the structural and electronic properties of the materials under study, a first-principles approach was adopted using the Plane-Wave Self-Consistent Field (PWSCF) code, which is based on density functional theory and is part of the Quantum ESPRESSO suite [20,21]. The PWSCF code conducts various types of self-consistent calculations of electronic structural properties within the framework of DFT [22,23], utilizing a plane-wave (PW) basis set and pseudopotentials (PP).

The electron–ion interaction was described using Rappe–Rabe–Kaxiras–Joannopoulos ultrasoft pseudopotentials with nonlinear core correction, while the exchange–correlation energies were modeled using the generalized gradient approximation (GGA) with the Perdew–Burke–Ernzerhof (PBE) functional [24]. A total of 9, 15, 6, 12, and 10 electrons were explicitly considered in the wave functions for Na, Mn, O, Ti, and Ni, respectively. The electronic wave functions were expanded using a plane-wave basis set with a kinetic energy cut-off of 80 Ry, while the charge density cutoff was set to 640 Ry (eight times greater). A $4 \times 4 \times 4$ Monkhorst-Pack k-point mesh [25] was employed. Additionally, Marzari–Vanderbilt smearing [26] with a width of 0.01 Ry was applied. In our calculations, we considered the spin polarization. Although the generalized gradient approximation (GGA) approach provides accurate predictions for many materials' properties, it is not consistently reliable for all cases. Specifically, GGA functionals struggle with accurately describing electronic correlations in the partially occupied 3*d*, 4*d*, and 5*d* shells of transition metals. This limitation often leads to the underestimation of the energy band gap in transition metal oxides. To address this issue, the Hubbard U self-interaction correction (DFT + U method) was adopted to consider the d–orbitals correlation [27]. We assumed the values $U_{Mn} = 4.7$ eV, $U_{Ni} = 6.95$ eV, and $U_{Ti} = 3.22$ eV for the Hubbard parameters.

Periodic boundary conditions (PBC), which allow for the simulation of an infinitely extended system, were applied in all calculations in combination with the supercell approximation. To determine reliable values for the kinetic energy cutoff, k-point sampling, and smearing width, preliminary test calculations were performed on the pseudopotentials. Total energy, forces, and stress were verified to match the tolerances of 0.001 Ry/atom, 0.0001 Ry/Bohr, and 0.1 kbar, respectively. The conjugate gradient (CG) minimization method was employed for energy minimization, with a convergence threshold for self-consistency set at 10^{-6} Ry. The Broyden–Fletcher–Goldfarb–Shanno (BFGS) quasi-Newton algorithm was used for geometry optimization, which was considered achieved when the following two conditions were met: (i) the energy change between two consecutive self-consistent field (SCF) steps was less than 10^{-4} Ry, and (ii) the maximum force acting on an atom was less than 10^{-3} Ry/Bohr. We also used the BFGS quasi-Newton algorithm for variable cell relaxation, adding the additional condition that all components of the stress tensor should be smaller than 0.05 GPa.

To evaluate the impact of the doping in the starting structure (P2-layered oxide $NaMnO_2$), several structures were built by replacing the Mn atoms (from 1 up to 12) with Ti and Ni atoms. These systems were inspected by calculating the formation energy.

The formation energy E_f is calculated by using the following formula:

$$E_f = \frac{E_{sys}^{tot} - n_{Na}E_{Na}^{bulk} - n_{Mn}E_{Mn}^{bulk} - n_{Ti}E_{Ti}^{bulk} - n_{Ni}E_{Ni}^{bulk} - \frac{1}{2}n_O E_{O_2}}{N_{tot}} \quad (1)$$

where E_{sys}^{tot} is the total energy of the system, N_{tot} is the total number of atoms contained in the supercell, E_X^{bulk} is the total energy per atom of the species *X* (Na, Mn, Ti, and Ni) in the most stable bulk phase, E_{O_2} is the total energy of the oxygen molecule, and n_X is the number of atoms of the species *X* (Na, Mn, Ti, Ni, and O).

2.3. Computational Workflow

The search for optimal cathodic materials in terms of formation energy was addressed here through the systematic screening of numerous materials. To validate the ML model, the values obtained through artificial intelligence (AI) for the formation energy were compared with their counterparts obtained through the DFT calculations, which were conducted for each of the candidates. These DFT calculations, as well as the methodologies employed to generate material variants, were based on well-established procedures, and this circumstance made it possible to automate and integrate them within a computational workflow chain (workchain).

A workchain can be defined as the total or partial automation of a work process, divided into simpler tasks processed sequentially or in parallel according to a logical model. The workchain employed here for the screening of potential cathodic materials was developed within the AiiDA (Automated Interactive Infrastructure and Database) framework [28–30], an open-source workflow management system devised for computational science. The AiiDA infrastructure is implemented in the Python language and has interfaces that allow it to be flexible. The management of DFT calculations within the workflow described here was performed using the plugin for the Quantum ESPRESSO code [31].

The workflow implementation was based on the AiiDA workchain object, which manages the execution of multiple long-duration calculations through the definition of sub-units called steps, ensuring that progress is not lost in case of sudden interruptions.

The tasks for each step are schematically depicted in Figure 1. They consist of the following:

1. Random generator. This involves creating a variation from the starting structure in a random way. Starting from the initial structure S_0 ($NaMnO_2$), alternative structures are randomly generated by applying one or more rules. The rules adopted can be summarized as follows.
    - Doping via the replacement of manganese atoms with transition metals. This rule takes, as the input, the initial structure S_0, a list of N_{tm} transition metals, the minimum number of manganese atoms to be maintained within the structure $min_{(\#Mn)}$, and multiplier *steps* that define the number of atoms to be replaced at a time. The output is S_1, a structure varied by replacing some of the manganese atoms with transition metals belonging to the list of the N_{tm} transition metals. The site positions of the manganese atoms to be replaced are randomly chosen. The number of the replaced manganese atoms is also randomly chosen such that the number of manganese atoms in S_1 is greater or equal to $min_{(\#Mn)}$.
    - Variation in the number of intercalated ions to define structures with fewer or no sodium atoms. This rule takes as the input the initial structure S_0. The output is S_1, a structure varied by removing a randomly chosen number of sodium atoms.
2. Symmetry equivalence check. The symmetry equivalence check task compares the structure generated by the random generator task with all the structures for which the calculation has already been performed to determine whether the structures are symmetrically equivalent. The algorithm is based on the open-source solution reported in the ref. [32] that was implemented in the Python tool Atomic Simulation Environment (ASE) [33].
3. Computing procedure. The computing procedure task submits the DFT + U calculation jobs to the ENEA HPC computing infrastructure [34] to extract the quantities of interest for this study.
4. Data store. The data store task stores the results of the DFT + U calculation in the AiiDA database and in an external Mongo DB instance. To accomplish this, the task consists of two parts:
    - Adapter. This organizes the data extracted from the parser into the chosen format (json in our case).

- Uploader. This loads the data with the new format into the database.

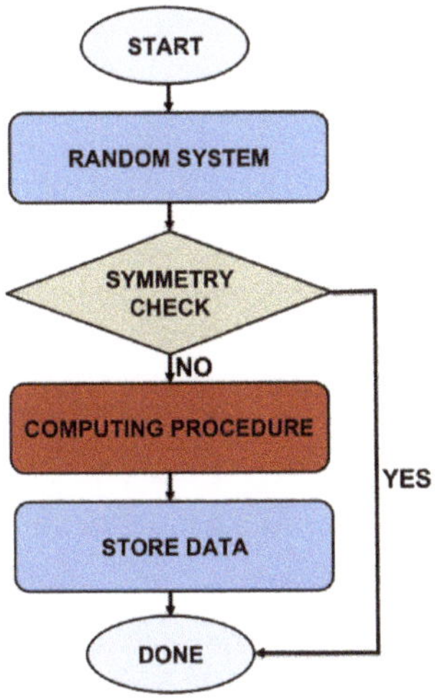

Figure 1. A flowchart of the workchain tasks as implemented in the AiiDA framework.

2.4. Machine Learning Method

In this study, we explored a wide combinatorial space of material candidates for cathode batteries by generating numerous alternatives through the random substitution of crystal elements. To evaluate the properties of this extensive dataset, consisting of systems generated by doping the initial $NaMnO_2$ structure, we adopted a data-driven approach. The geometric crystal graph neural network (GeoCGNN) [35] was used to predict the formation energies of the crystals. The GeoCGNN enhances the earlier crystal graph convolutional neural network (CGCNN) [36] version by integrating detailed spatial geometric information into the model. Both the CGCNN and GeoCGNN architectures leverage message-passing neural networks (MPNNs) [37] and gated convolution.

The input of the GeoCGNN model is a graph generated on top of the crystal structure provided as a crystallographic information file (CIF) where the nodes represent the atoms in the crystal and the edges represent the interactions between them. The edges are defined for atoms separated by less than 8 Å, with a maximum of 12 neighbors considered for each atom. The nodes are encoded as one-hot feature vectors based on atomic properties, while the edge feature vectors are the distance vectors between interacting atoms. Additionally, the model incorporates descriptors that encode lattice vectors and the cell volume. The output is typically a scalar property of the material. In this case, the target property is the formation energy.

The network was trained on a large dataset of materials obtained from the Materials Project (MP) database [19], excluding only structures containing noble gasses. The data were stored on Red Hat Ceph Storage [38], a solution tailored for data analytics and artificial intelligence. We used 80% of the dataset, comprising 126,162 materials, for training, with the remaining 20% split equally for validation and testing. The training process utilized the Stochastic Gradient Descent (SGD) [39] method with a learning rate of 0.001, a batch size of 64, and mean squared error (MSE) as the loss function for the regression task. The training ran for a maximum of 500 epochs, with the best-performing hyperparameters selected based on validation error.

During the inference phase, the trained model was used to estimate the formation energies of the candidate material for cathode batteries. A total of 8071 crystal structures were randomly generated based on the rules outlined in Section 2.3 (Computational Workflow). Since these new structures are not present in the MP database, DFT calculations were performed to compare the obtained formation energy values with those predicted by the model. To efficiently manage the large number of materials, both the random generation of crystal structures and the DFT calculations were automated using the workflow detailed in the previous section (Section 2.3).

3. Results and Discussion

3.1. Base Crystal Structure

As we have reported above, the $NaMnO_2$ base crystal was modeled using the supercell $3 \times 2 \times 1$ of the crystal mp-971647 of the Materials Project database [19]. The resulting supercell was monoclinic with a = 9.30 Å; b = 6.20 Å; c = 10.90 Å; α = 90°; β = 90°; γ = 120°; and N_{tot} = 48 atoms (12 Na atoms, 12 Mn atoms, and 24 O atoms), as depicted in Figure 2. In this P2-layered oxide, the Na and Mn layers alternated at a distance of 2.73 Å (the distance between layers of the same type was doubled at 5.45 Å). Each Mn layer had two O layers (above and below) at a distance of 0.95 Å. The Na layer formed a hexagonal structure with a side a_{Na} = 3.10 Å in which each Na atom had six Na atoms at distance $d_{Na\text{-}Na} = a_{Na}$, and the stacking sequence was ABA. Equally, the Mn layer formed the same hexagonal structure but with an overlapping stacking sequence ($d_{Mn\text{-}Mn} = a_{Mn}$ = 3.10 Å). Each Na(Mn) atom had six Mn(Na) at a distance of $d_{Na\text{-}Mn}$ = 3.26 Å and had six O atoms at a distance of $d_{Na\text{-}O}$ = 2.52 Å ($d_{Mn\text{-}O}$ = 2.03 Å). Upon the substitution of randomly selected Mn atoms, each supercell geometry was fully relaxed, allowing for the optimization of the cell parameters (variable cell relax). As an example, here we report the cell parameters of the base crystal with the Na-ions. Upon the cell relaxation, the structure distorted, increasing the volume by about 2.6% and losing its hexagonal shape. The cell parameters were a = 9.77 Å; b = 5.86 Å; c = 10.92 Å; α = 90°; β = 90°; and γ = 116.7°. The Na and Mn layers' distance remained the same, but the O layers increased their distance from the Mn layers up to 1.12 Å. The Na(Mg) layers lost the hexagonal symmetry and then each Na(Mn) atom had two Na(Mn) atoms at $d_{Na\text{-}Na}$ ($d_{Mn\text{-}Mn}$) = 2.93 Å and four Na(Mn) atoms at $d_{Na\text{-}Na}$ ($d_{Mn\text{-}Mn}$) = 3.26 Å. Likewise, two pairs of Na-Mn atoms approached each other, reducing their distance at $d_{Na\text{-}Mn}$ = 3.23 Å, while four pairs of Na-Mn atoms moved apart each other, increasing their distance at $d_{Na\text{-}Mn}$ = 3.32 Å. For the oxygen, the situation was the opposite, as the distance was reduced to $d_{Mn\text{-}O}$ = 1.97 Å for four pairs of Mn-O atoms and was increased to $d_{Mn\text{-}O}$ = 2.49 Å for two pairs of Mn-O atoms. The distances between the Na and O atoms were reduced to $d_{Na\text{-}O}$ = 2.41 Å for four pairs and to $d_{Na\text{-}O}$ = 2.47 Å for two pairs.

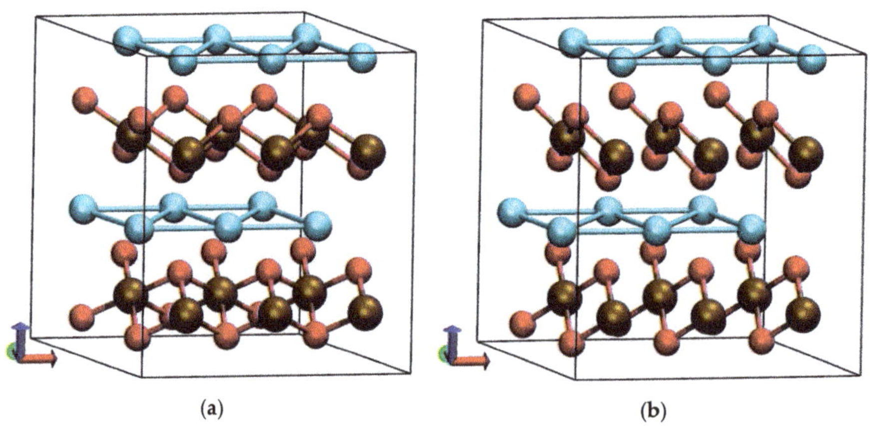

Figure 2. Starting non-relaxed (**a**) and relaxed (**b**) configurations of P2-layered oxide $MnNaO_2$ system. In light blue, brown, and red are Na, Mn, and O atoms, respectively.

3.2. Formation Energy

The formation energy was calculated using Equation (1) for the base crystal $NaMnO_2$; E_f was −2.01 eV for the non-relaxed system and −2.12 eV for the relaxed one. To make a comparison between the predicted and actual formation energies, it was necessary to create a test dataset useful for this case. Creating a good test dataset is crucial for evaluating the performance and the generalizability of a deep learning (DL) model.

Firstly, the test dataset should represent the entire distribution of data that the DL model will encounter in common usage [40,41]. If the test dataset does not represent the diversity of the real usage, the evaluation metrics will be misleading. AiiDA workflow rules have made it possible to represent this diversity by creating a balanced dataset.

Secondly, the test dataset should be large enough to provide reliable performance estimates [40,42]. This was difficult to achieve because of the long times required for the simulations. Despite these difficulties, the dataset contained more than 500 calculations.

Lastly, it should be ensured that the test dataset does not include information from the training dataset [40,41]. In this case, none of our calculated systems were stored in Materials Project database.

The test dataset was created with all these principles in mind by using the AiiDA workflow. In Figure 3, we compare and validate predictions from our ML method with results from the DFT calculations. On the left, the predictions of E_f for the 8071 non-relaxed crystal structures randomly generated are depicted in gray circles, showing a negative range of values between −3 eV and −1 eV. Moreover, some systems are highlighted by circles of different colors and labeled with their specific composition. This evidences that titanium doping increased the stability of the system (more negative E_f), while nickel doping reduced it (less negative E_f). Indeed, $Na_{12}Ti_{12}O_{24}$ (violet circle) was the system with the lowest E_f (−2.78 eV); conversely, $Na_{12}Ni_{12}O_{24}$ was the system (dark green circle) with the highest value of E_f (−1.20 eV). The base crystal $Na_{12}Mn_{12}O_{24}$ (black circle), placed in the middle of the range (−1.88 eV), was in an intermediate state of stability. The substitution of one manganese atom with a titanium atom ($Na_{12}Mn_{11}TiO_{24}$; red circle; E_f = −1.97 eV) decreased the formation energy by 0.09 eV, thereby increasing the system's stability. Conversely, the substitution of a manganese atom with a nickel atom ($Na_{12}Mn_{11}NiO_{24}$; brown circle; E_f = −1.84 eV) increased the formation energy by 0.04 eV, making the structure less stable. In Figure 3, predictions are also reported for the system $Na_{12}Mn_4Ni_4Ti_4O_{24}$ (orange circles) with an equal composition of the three transition metals (Mn, Ti, and Ni) and for the systems in which two transition metals have the same concentration, i.e., $Na_{12}Mn_3Ni_6Ti_3O_{24}$ (magenta circles), $Na_{12}Mn_6Ni_3Ti_3O_{24}$ (green circles), and $Na_{12}Mn_3Ni_3Ti_6O_{24}$ (light blue circles). Since, for these compositions, there were several different configurations, in addition to the range of the variability in the formation energy, the mean value $<E_f>$ and the standard deviation σ_{Ef} were calculated to evaluate statistical uncertainty. For the $Na_{12}Mn_4Ni_4Ti_4O_{24}$ system, the range was from −2.10 to −2.00 eV; $<E_f>$ = −2.06 eV; and σ_{Ef} = 0.01 eV. For the $Na_{12}Mn_3Ni_6Ti_3O_{24}$ system, the range was from −1.87 to −1.80 eV; $<E_f>$ = −1.85 eV; and σ_{Ef} = 0.01 eV. For the $Na_{12}Mn_6Ni_3Ti_3O_{24}$ system, the range was from −2.05 to −1.98 eV; $<E_f>$ = −2.02 eV; and σ_{Ef} = 0.02 eV. For the $Na_{12}Mn_3Ni_3Ti_6O_{24}$ system, the range was from −2.30 to −2.24 eV; $<E_f>$ = −2.27 eV; and σ_{Ef} = 0.01 eV. In Figure 3, statistical uncertainty is indicated by black solid lines centered in the mean values and limited by the standard deviations. On the right side of Figure 3, the calculated values of the formation energy of the non-relaxed systems using the first-principles method are shown by colored cross symbols. At least one of the systems highlighted on the left in the figure was analyzed by DFT calculations by evaluating its formation energy. Notably, we see that the same order of stability as the predicted formation energy was found. The $Na_{12}Ti_{12}O_{24}$ system (violet cross) was the most stable, with E_f = −2.52 eV; the $Na_{12}Ni_{12}O_{24}$ (dark green cross) system was the least stable, with E_f = −1.15 eV. The base crystal $Na_{12}Mn_{12}O_{24}$ (black cross) had a formation energy value equal to −2.01 eV; this value was between the ones corresponding to the $Na_{12}Mn_{11}TiO_{24}$ system (red cross; E_f = −2.03 eV) and to the $Na_{12}Mn_{11}NiO_{24}$ system (brown cross; E_f = −1.93 eV), respectively. The calculated values of the formation energy of the $Na_{12}Mn_4Ni_4Ti_4O_{24}$ system (orange crosses) ranged from −1.98 to −1.94 eV. Finally, we list also the calculated E_f values for the systems in which two transition metals had the same concentration, i.e., $Na_{12}Mn_3Ni_6Ti_3O_{24}$ (magenta crosses; range from −1.80 to −1.77 eV), $Na_{12}Mn_6Ni_3Ti_3O_{24}$ (green crosses; range from −2.07 to −1.96 eV), and $Na_{12}Mn_4Ni_4Ti_4O_{24}$ (light blue crosses; range from −2.15 to −2.12 eV).

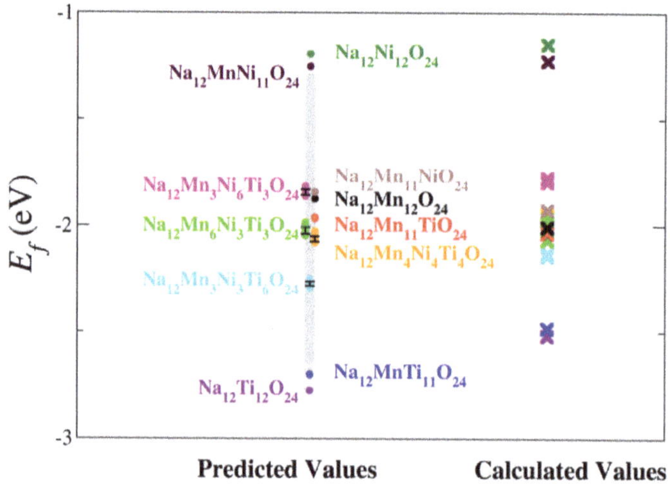

Figure 3. A comparison of the predicted (circles) and calculated (crosses) formation energy for the non-relaxed systems. For the $Na_{12}Mn_3Ni_6Ti_3O_{24}$, $Na_{12}Mn_6Ni_3Ti_3O_{24}$, $Na_{12}Mn_4Ni_4Ti_4O_{24}$, and $Na_{12}Mn_3Ni_3Ti_6O_{24}$ systems, the statistical uncertainty of the predicted formation energies is indicated by black solid lines centered in the mean values and limited by the standard deviations.

To further assess the predictive performance of the machine learning model, the predicted formation energies were compared against the calculated ones obtained through the DFT method managed by the AiiDA workflow. In this second stage, the DFT-calculated E_f was relative to that of the relaxed systems. Figure 4 shows the scatter plots of the predicted versus calculated formation energies for two sets of systems: the non-relaxed systems (red circles) are on panel (a) and the relaxed systems (blue circles) are on panel (b). Each data point represents a unique configuration of the material composition. As illustrated, most of the points lie close to the y = x line (black dashed line), indicating a strong agreement between the predictions and the calculated values. The tight clustering around this line demonstrates the model's accuracy across a wide range of formation energies.

Furthermore, the accuracy of the predictions was quantified using the mean absolute error (MAE), the coefficient of determination (R^2), and Pearson's r correlation, as reported in Table 1. From the comparison of the scatter plots, two key observations emerge: (1) the GeoCGNN could provide reasonably accurate predictions for the formation energy using non-relaxed systems directly from the input CIF; (2) the accuracy improved for the relaxed systems, as indicated by the closer alignment of points with the diagonal in the plot of Figure 4b. The first observation is significant because we trained the DL model to make predictions while saving time, bypassing the need for computationally expensive first-principles calculations. The relaxed structure was only obtained after the calculation was complete (with both the formation energy and the relaxed structure as outputs). As far as we know, no DL models currently exist that could predict the geometry of the relaxed system. Therefore, the standard approach is to predict the formation energy of the non-relaxed crystal structure. Our results show that the GeoCGNN achieved satisfying accuracy in this task. The second observation is equally important, as the best performance of the predictive algorithms for relaxed systems agrees with the fact that Quantum Espresso calculates the formation energy of relaxed structures. This is confirmation that the training of the GeoCGNN was successful.

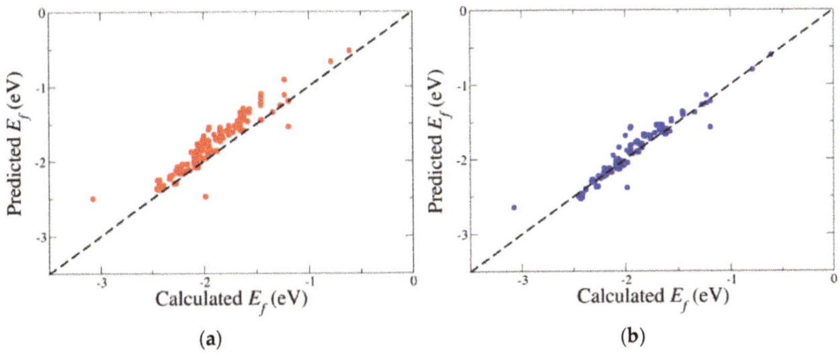

Figure 4. Scatter plots of the predicted vs. calculated formation energies, *Ef*, using non-relaxed (panel (**a**), red circles) and relaxed (panel (**b**), blue circles) systems. The black dashed line represents the ideal y = x relationship.

Table 1. Mean absolute error (MAE), coefficient of determination (R^2), and Pearson's correlation coefficient (r) were calculated by comparing predicted and calculated formation energies for non-relaxed and relaxed systems.

Systems	MAE (eV)	R^2	Pearson's r
Non-relaxed	0.166	0.665	0.960
Relaxed	0.052	0.944	0.975

4. Conclusions

In conclusion, to evaluate the stability of substitutionally doped P2-layered oxide $NaMnO_2$ cathode materials, we adopted two different computational methods: (1) computationally demanding, high-throughput, first-principles DFT calculations; and (2) an ML technique based on the training of a GeoCGNN. In both cases, we explored a wide combinatorial space of material candidates for cathode batteries by generating numerous alternative compositions through the random substitution of Mn atoms. The DFT calculations were automated and integrated into a computational workflow chain developed within the AiiDA framework. The first-principles total energy of the materials was used to evaluate the formation energy. On the other hand, we adopted a data-driven approach where the GeoCGNN was used to predict the formation energy as the target property of the generated crystals upon training on a dataset extracted from the Materials Project database [19]. The input of the GeoCGNN model was a graph supplemented with other information on the crystal lattice. The results revealed that titanium doping improved the system's stability, while nickel doping reduced it.

To evaluate the predictive performance of the DL model, the predicted formation energies were compared to the calculated formation energies obtained from DFT calculations. Our results show that the GeoCGNN achieved satisfying accuracy in predicting the formation energy of non-relaxed crystal structures, with the accuracy improving when relaxed systems were used as the input. This demonstrates that the GeoCGNN was successfully trained to predict results consistent with the DFT calculations, which computed the formation energy of the relaxed structures. However, since the relaxed structures were only available after performing the DFT total energy calculation, the most effective approach without executing DFT calculations was to use the non-relaxed systems as the input for the GeoCGNN to generate predictions.

To further enhance the ML model and expand its applicability, the model's versatility can be improved by incorporating properties such as the band gap, the bulk modulus, the shear modulus, and Poisson's ratio into the training dataset. These properties often correlate with critical performance metrics, such as mechanical stability and electronic characteristics, thereby improving the understanding of material behavior. Additionally, by reusing a pre-trained ML model, such as one developed for predicting electronic properties, it may be possible to fine-tune the model for other related properties, like redox potential, using transfer learning for cross-property prediction techniques. This approach could significantly reduce computational resource requirements while leveraging existing knowledge.

The Hubbard correction improves the description of the d–orbitals correlation in transition metal electronic shells and the energy gap estimation. However, the r2SCAN meta-GGA exchange–correlation functional [43] performs better than GGA in predicting the thermodynamic properties of various 3*d*, 4*d*, and 5*d* transition metals [44]. Therefore, the adoption of the r2SCAN functional in workflows is a promising strategy for future research.

As final observation, it is worth noting that running a typical DFT calculation for a single crystal structure requires a time ranging from 1 to 24 h, while training the model from the dataset requires a few hours, and a single prediction is made in fractions of a second. This demonstrates that the GeoCGNN reduces the time required to predict material properties by several orders of magnitude compared to traditional DFT calculations, while maintaining satisfactory accuracy. Additionally, the reduction in energy consumption is significant and should not be overlooked.

Author Contributions: Conceptualization, F.B.; methodology, F.B., C.R., and S.G.; software, C.R. and S.F.; validation, S.G., F.B., S.M., and C.R.; investigation, C.R., S.M., F.B., S.G., and M.C.; resources, M.C.; data curation, C.R. and S.F.; writing—original draft preparation, C.R., S.M., S.G., and F.B.; writing—review and editing, F.B. and S.G.; supervision, F.B. and M.C.; project administration, M.C.; funding acquisition, M.C. All authors have read and agreed to the published version of the manuscript.

Funding: The Project "Italian Energy Materials Acceleration Platform" (IEMAP)—CUP I62C21000380001 in the Program Mission Innovation is acknowledged for financial support. This work was supported by the European Union through the Next Generation EU funds through the Italian MUR National Recovery and Resilience Plan, Mission 4 Component 2—Investment 1.4—National Center for HPC, Big Data and Quantum Computing—CUP I33C22001270007.

Data Availability Statement: The dataset generated in this study is available at https://iemap.enea.it/ (accessed on 2 December 2024). The code used for this study is available at https://gitlab.brindisi.enea.it/claudio.ronchetti/ai4mat (accessed on 2 December 2024).

Acknowledgments: The computing resources and the related technical support used for this work were provided by the CRESCO/ENEAGRID High-Performance Computing infrastructure and its staff [34]. The CRESCO/ENEAGRID High-Performance Computing infrastructure is funded by ENEA, the Italian National Agency for New Technologies, Energy and Sustainable Economic Development and by Italian and European research projects; see http://www.cresco.enea.it/english (accessed on 30 April 2024) for information. The authors acknowledge the extensive use of the ENEA FARO facility and the support of its management team [45].

Conflicts of Interest: Author Claudio Ronchetti was employed by the company Telespazio S.p.A. The remaining authors declare that the research was conducted in the absence of any commercial or financial relationships that could be construed as a potential conflict of interest.

References

1. Bhuiyan, M.R.A. Overcome the future environmental challenges through sustainable and renewable energy resources. *Micro Nano Lett.* **2022**, *17*, 402–416. [CrossRef]
2. Chakraborty, A.; Kunnikuruvan, S.; Kumar, S.; Markovsky, B.; Aurbach, D.; Dixit, M.; Major, D.T. Layered Cathode Materials for Lithium-Ion Batteries: Review of Computational Studies on $LiNi_{1-x-y}Co_xMn_yO_2$ and $LiNi_{1-x-y}Co_xAl_yO_2$. *Chem. Mater.* **2020**, *32*, 915–952. [CrossRef]
3. Kim, S.-W.; Seo, D.-H.; Ma, X.; Ceder, G.; Kang, K. Electrode Materials for Rechargeable Sodium-Ion Batteries: Potential Alternatives to Current Lithium-Ion Batteries. *Adv. Energy Mater.* **2012**, *2*, 710–721. [CrossRef]

4. Clement, R.J.; Bruce, P.G.; Grey, C.P. Review—Manganese-Based P2-Type Transition Metal Oxides as Sodium-Ion Battery Cathode Materials. *J. Electrochem. Soc.* **2015**, *162*, A2589. [CrossRef]
5. Liu, Q.; Hu, Z.; Chen, M.; Zou, C.; Jin, H.; Wang, S.; Chou, S.-L.; Dou, S.-X. Recent Progress of Layered Transition Metal Oxide Cathodes for Sodium-Ion Batteries. *Small* **2019**, *15*, 1805381. [CrossRef]
6. Zhang, J.; Wang, W.; Wang, W.; Wang, S.; Li, B. Comprehensive Review of P2-Type $Na_{2/3}N\ i_{1/3}Mn_{2/3}O_2$, a Potential Cathode for Practical Application of Na-Ion Batteries. *ACS Appl. Mater. Interfaces* **2019**, *11*, 22051–22066. [CrossRef]
7. Rostami, H.; Valio, J.; Suominen, P.; Tynjälä, P.; Lassi, U. Advancements in Cathode Technology, Recycling Strategies, and Market Dynamics: A Comprehensive Review of Sodium Ion Batteries. *Chem. Eng. J.* **2024**, *495*, 153471. [CrossRef]
8. Lee, D.H.; Xu, J.; Meng, Y.S. An advanced cathode for Na-ion batteries with high rate and excellent structural stability. *Phys. Chem. Chem. Phys.* **2013**, *15*, 3304–3312. [CrossRef]
9. Prosini, P.P.; Carewska, M.; Cento, C.; Tarquini, G.; Maroni, F.; Birrozzi, A.; Nobili, F. Tin-Decorated Reduced Graphene Oxide and $NaLi_{0.2}Ni_{0.25}Mn_{0.75}O_\delta$ as Electrode Materials for Sodium-Ion Batteries. *Materials* **2019**, *12*, 1074. [CrossRef]
10. Sun, Y.-K.; Jeon, Y.-S.; Leeb, H.J. Overcoming Teller-Teller Distortion for Spinel Mn Phase. *Electrochem. Solid-State Lett.* **1999**, *3*, 7–9. [CrossRef]
11. Thackeray, M.M. Manganese oxides for lithium batteries. *Prog. Solid State Chem.* **1997**, *25*, 1–71. [CrossRef]
12. Gupta, P.; Pushpakanth, S.; Ali Haider, M.; Basu, S. Understanding the Design of Cathode Materials for Na-Ion Batteries. *ACS Omega* **2022**, *7*, 5605–5614. [CrossRef] [PubMed]
13. Nguyen, T.P.; Kim, I.T. Recent Advances in Sodium-Ion Batteries: Cathode Materials. *Materials* **2023**, *16*, 6869. [CrossRef] [PubMed]
14. Wang, H.; Gao, X.; Zhang, S.; Mei, Y.; Ni, L.; Gao, J.; Liu, H.; Hong, N.; Zhang, B.; Zhu, F.; et al. High-Entropy Na-Deficient Layered Oxides for Sodium-Ion Batteries. *ACS Nano* **2023**, *17*, 12530–12543. [CrossRef] [PubMed]
15. Yoshida, H.; Yabuuchi, N.; Kubota, K.; Ikeuchi, I.; Garsuch, A.; Schulz-Dobrick, M.; Komaba, S. P2-Type $Na_{2/3}N\ i_{1/3}Mn_{2/3-x}T\ i_xO_2$ as a New Positive Electrode for Higher Energy Na-Ion Batteries. *Chem. Commun.* **2014**, *50*, 3677–3680. [CrossRef]
16. Cao, Z.; Li, L.; Zhou, C.; Ma, X.; Wang, H. Contribution of titanium substitution on improving the electrochemical properties of P2-$Na_{0.67}Ni_{0.33}Mn_{0.67}O_2$ cathode material for sodium-ion storage. *Funct. Mater. Lett.* **2020**, *13*, 2051010.
17. Lombardo, T.; Duquesnoy, M.; El-Bouysidy, H.; Årén, F.; Gallo-Bueno, A.; Bjørn Jørgensen, P.; Bhowmik, A.; Demortière, A.; Ayerbe, E.; Alcaide, F.; et al. Artificial Intelligence Applied to Battery Research: Hype or Reality? *Chem. Rev.* **2022**, *122*, 10899–10969. [CrossRef]
18. Ronchetti, C.; Puccini, M.; Ferlito, S.; Giusepponi, S.; Palombi, F.; Buonocore, F.; Celino, M. Machine learning techniques for data analysis in materials science. In Proceedings of the 2022 AEIT International Annual Conference (AEIT), Rome, Italy, 3–5 October 2022; pp. 1–6. [CrossRef]
19. Jain, A.; Ping Ong, S.; Hautier, G.; Wei, C.; Richards, W.D.; Dacek, S.; Cholia, S.; Gunter, D.; Skinner, D.; Ceder, G.; et al. Commentary: The Materials Project: A materials genome approach to accelerating materials innovation. *APL Mater.* **2013**, *1*, 011002. [CrossRef]
20. Giannozzi, P.; Baroni, S.; Bonini, N.; Calandra, M.; Car, R.; Cavazzoni, C.; Ceresoli, D.; Chiarotti, G.L.; Cococcioni, M.; Dabo, I.; et al. QUANTUM ESPRESSO: A modular and open-source software project for quantum simulations of materials. *J. Phys.* **2009**, *21*, 395502. [CrossRef]
21. Giannozzi, P.; Andreussi, O.; Brumme, T.; Bunau, O.; Nardelli, M.B.; Calandra, M.; Car, R.; Cavazzoni, C.; Ceresoli, D.; Cococcioni, M.; et al. Advanced capabilities for materials modelling with Quantum ESPRESSO. *J. Phys. Condens. Matter* **2017**, *29*, 465901. [CrossRef]
22. Hohenberg, P.; Kohn, W. Inhomogeneous electron gas. *Phys. Rev.* **1964**, *136*, B864. [CrossRef]
23. Kohn, W.; Sham, L.J. Self-consistent equations including exchange and correlation effects. *Phys. Rev.* **1965**, *140*, A1133. [CrossRef]
24. Perdew, J.P.; Burke, K.; Ernzerhof, M. Generalized gradient approximation made simple. *Phys. Rev. Lett.* **1996**, *77*, 3865–3868. [CrossRef] [PubMed]
25. Monkhorst, H.J.; Pack, J.D. Special points for Brillouin-zone integrations. *Phys. Rev. B* **1976**, *13*, 5188. [CrossRef]
26. Marzari, N.; Vanderbilt, D.; De Vita, A.; Payne, M.C. Thermal contraction and disordering of the Al(110) surface. *Phys. Rev. Lett.* **1999**, *82*, 3296. [CrossRef]
27. Himmetoglu, B.; Floris, A.; Gironcoli, S.; Cococcioni, M. Hubbard-corrected DFT energy functionals: The LDA+U description of correlated systems. *Int. J. Quantum Chem.* **2014**, *114*, 14–49. [CrossRef]
28. Pizzi, G.; Cepellotti, A.; Sabatini, R.; Marzari, N.; Kozinsky, B. AiiDA: Automated interactive infra-structure and database for computational science. *Comput. Mater. Sci.* **2016**, *111*, 218–230. [CrossRef]
29. Huber, S.P.; Zoupanos, S.; Uhrin, M.; Talirz, L.; Kahle, L.; Häuselmann, R.; Gresch, D.; Müller, T.; Yakutovich, A.V.; Andersen, C.W.; et al. AiiDA 1.0, a scalable computational infrastructure for automated reproducible workflows and data provenance. *Sci. Data* **2020**, *7*, 300. [CrossRef]
30. Uhrin, M.; Huber, S.P.; Yu, J.; Marzari, N.; Pizzi, G. Workflows in AiiDA: Engineering a high-throughput, event-based engine for robust and modular computational workflows. *Comput. Mater. Sci.* **2021**, *187*, 110086. [CrossRef]
31. Available online: https://github.com/aiidateam/aiida-quantumespresso (accessed on 30 April 2023).
32. Lonie, D.; Zurek, E. Identifying duplicate crystal structures: XTALCOMP, an open-source solution. *Comput. Phys. Commun.* **2012**, *183*, 690–697. [CrossRef]

33. Larsen, A.H.; Mortensen, J.J.; Blomqvist, J.; Castelli, I.E.; Christensen, R.; Dułak, M.; Friis, J.; Groves, M.N.; Hammer, B.; Hargus, C.; et al. The atomic simulation environment—A Python library for working with atoms. *J. Phys. Condens. Matter* **2017**, *29*, 273002. [CrossRef]
34. Iannone, F.; Ambrosino, F.; Bracco, G.; De Rosa, M.; Funel, A.; Guarnieri, G.; Migliori, S.; Palombi, F.; Ponti, G.; Santomauro, G.; et al. CRESCO ENEA HPC clusters: A working example of a multi-fabric GPFS Spectrum Scale layout. In Proceedings of the 2019 International Conference on High Performance Computing & Simulation (HPCS), Dublin, Ireland, 15–19 July 2019; pp. 1051–1052.
35. Cheng, J.; Zhang, C.; Dong, L. A geometric-information-enhanced crystal graph network for predicting properties of materials. *Commun. Mater.* **2021**, *2*, 92. [CrossRef]
36. Xie, T.; Grossman, J.C. Crystal graph convolutional neural networks for an accurate and interpretable prediction of material properties. *Phys. Rev. Lett.* **2018**, *120*, 145301. [CrossRef] [PubMed]
37. Gilmer, J.; Schoenholz, S.S.; Riley, P.F.; Vinyals, O.; Dahl, G.E. Neural message passing for quantum chemistry. In Proceedings of the 34th International Conference on Machine Learning, Sydney, Australia, 6–11 August 2017; PMLR, Volume 70, pp. 1263–1272.
38. Weil, S.A.; Brandt, S.A.; Miller, E.L.; Long, D.D.; Maltzahn, C. Ceph: A scalable, high-performance distributed file system. In Proceedings of the 7th Conference on Operating Systems Design and Implementation (OSDI '06), Seattle, WA, USA, 6–8 November 2006; pp. 307–320.
39. Amari, S.I. Backpropagation and stochastic gradient descent method. *Neurocomputing* **1993**, *5*, 185–196. [CrossRef]
40. Goodfellow, I.; Bengio, Y.; Courville, A. *Deep Learning*; MIT Press: Cambridge, MA, USA, 2016.
41. Géron, A. *Hands-On Machine Learning with Scikit-Learn, Keras, and TensorFlow*, 2nd ed.; O'Reilly Media: Sebastopol, CA, USA, 2019.
42. Bishop, C.M. *Pattern Recognition and Machine Learning*; Springer: Berlin/Heidelberg, Germany, 2006.
43. Furness, J.W.; Kaplan, A.D.; Ning, J.; Perdew, J.P.; Sun, J. Accurate and Numerically Efficient r2SCAN Meta-Generalized Gradient Approximation. *J. Phys. Chem. Lett.* **2020**, *11*, 8208–8215. [CrossRef]
44. Kothakonda, M.; Kaplan, A.D.; Isaacs, E.B.; Bartel, C.J.; Furness, J.W.; Ning, J.; Wolverton, C.; Perdew, J.P.; Sun, J. Testing the r2SCAN Density Functional for the Thermodynamic Stability of Solids with and without a van der Waals Correction. *ACS Mater. Au* **2023**, *3*, 102–111. [CrossRef]
45. Mariano, A.; D'amato, G.; Ambrosino, F.; Aprea, G.; Buonocore, F.; Celino, M.; Colavincenzo, A.; Fina, M.; Funel, A.; Giusepponi, S.; et al. Fast Access to Remote Objects 2.0 a renewed gateway to ENEAGRID distributed computing resources. *Future Gener. Comput. Syst.* **2019**, *94*, 920–928. [CrossRef]

Disclaimer/Publisher's Note: The statements, opinions and data contained in all publications are solely those of the individual author(s) and contributor(s) and not of MDPI and/or the editor(s). MDPI and/or the editor(s) disclaim responsibility for any injury to people or property resulting from any ideas, methods, instructions or products referred to in the content.

batteries

MDPI

Article

Concentrated, Gradient Electrolyte Design for Superior Low-Temperature Li-Metal Batteries

Jason S. Packard, Ethan A. Adams and Vilas G. Pol *

Davidson School of Chemical Engineering, Purdue University, West Lafayette, IN 47907, USA
* Correspondence: vpol@purdue.edu

Abstract: Improving the low-temperature performance of lithium-ion batteries is critical for their widespread adoption in cold environments. In this study, we designed a novel LHCE featuring a solvent polarity gradient, designed to maximize both room- and low-temperature ion mobility. Extremely polar fluoroethylene carbonate (FEC) and low-freezing-point, −135 °C, non-polar nonaflurobutyl methyl ether (NONA) were supplemented by two intermediate solvents with incremental step-downs in polarity. The intermediate solvents consist of methyl (2,2,2-triflooethyl) carbonate (FEMC) and either diethylene carbonate (DEC), ethyl methyl carbonate (EMC), or dibutyl carbonate (DBC). The four solvents were combined with *1 M* lithium bis(fluorosulfonyl)amide (LiFSI) salt and were able to accommodate 37.5% diluent volume, resulting in ultra-low electrolyte freezing points below −120 °C. This contrasts with our previously investigated three-solvent LHCE, which only allowed for a 14% diluent volume and a −85 °C freezing point. Localized high salt concentrations were shown by less than 3% of FSI- anions being free in solution. The gradient LHCEs also showed room-temperature ionic conductivities above 10–3 S/cm and maintained high ion mobility below −40 °C. Lithium metal coin cells with LiFePO4 (LFP) cathodes featuring the gradient LHCEs, a reference three-solvent LHCE, and commercial (*1 M* LiPF6 in 1:1 EC:DEC) electrolyte were constructed. All gradient LHCEs outperformed both the three-solvent and commercial electrolytes at all temperatures, with the DEC-based gradient LHCE showing the best performance of 159.7 mAh/g at 25 °C and 109.2 mAh/g at −50 °C, corresponding to a 68% capacity retention. These findings highlight the potential of LHCE systems to improve battery performance in low-temperature environments and propose a new gradient design strategy for electrolytes to yield advantages of both polar and weakly polar solvents.

Keywords: lithium-metal battery; concentrated electrolyte; ionic conductivity; solid electrolyte interface

1. Introduction

Since their commercialization in 1991, lithium-ion batteries (LIBs) have profoundly impacted society. Their high-energy density and long life cycles have proliferated the portable electronics, electric vehicle, and battery grid storage industries, and now play an imperative role in the transition to sustainable energy [1]. Despite their incremental improvement, typical LIBs with ethylene carbonate (EC) electrolytes suffer from poor low-temperature performance and lose over 50% of their room-temperature capacity at −20 °C [2–5]. To improve low-temperature performance, often LIB packs are maintained within a narrow, optimal temperature range using a thermal management system [6] or provide polyethylene glycol cooling solvent circulation around the battery pack. These methods increase cost, reduce energy density, and are insufficient for more extreme environments. The development of intrinsic material solutions is necessary to enable the mass adoption of sustainable energy technologies in harsh climates.

Limited low-temperature performance of LIBs arises primarily from sluggish cell kinetics including, reduced electrolyte ionic conductivity [7,8], high resistances for charge transfer [9,10] and solid electrolyte interface (SEI) diffusion processes [8,11]. Strategies to increase ionic conductivity and decrease charge transfer and Li+ diffusion resistances through electrolyte modifications have been extensively studied. High salt concentration electrolytes (HCE) have been shown to create advantageous anion-derived SEIs, which lower interfacial resistances [12–15]. However, increasing the salt concentration increases viscosity and decreases ionic conductivity, limiting ultra-low-temperature performance. Locally high-concentrated electrolytes (LHCEs) are a promising method to maintain the superior anion-derived SEIs of HCEs while lowering viscosity and increasing ionic conductivity by adding an electrochemically inert, low-polarity and freezing-point co-solvent [16–21].

LHCE studies have often used coordinating solvents with moderate polarity, such as acetonitrile (AN), 1,2-dimethoxyethane (DME), or tetrahydrofuran (THF) [18,20]. Although advantageous for low-temperature performance, through improved desolvation and anion participation in the solvation shell, these lower polarity solvents limit ionic conductivity and performance at ambient temperatures. This problem could potentially be addressed by maximizing the polarity of coordinating solvent, while minimizing diluent polarity, to maintain low-temperature performance. This could allow for room-temperature ion mobility comparable to standard EC-based LIB electrolytes, while the diluent severely lowers the solution freezing point, viscosity and allows for anion participation in the solvation shell. Our recent work investigated this strategy using very polar fluoroethylene carbonate (FEC), methyl 2,2,2-trifluoroethyl carbonate (FEMC), and nearly non-polar nonafluorobutyl methyl ether (NONA) in a 2:4:1 ratio with *1 M* Lithium bis(fluorosulfonyl)imide (LiFSI) salt [22]. This LHCE showed a freezing point of −85 °C and 61% retention of room-temperature capacity at −50 °C for Li-metal cells. However, creating an electrolyte solution with such a stark difference in polarity of its solvents gives rise to miscibility issues, analogous to mixing oil and water. The LHCE was limited to only 14% diluent by volume; a further reduction in freezing point and increase in low-temperature performance could have been achieved through greater diluent incorporation. In this work, we developed a gradient LHCE, featuring three different coordinating solvents, each with an incremental drop in polarity. This step-down design of solvent polarity allows for maximum miscibility and 37.5% incorporation of the diluent, despite using very polar FEC and non-polar NONA.

2. Materials and Methods

2.1. Materials

Fluoroethylene carbonate (FEC) was purchased from Acros Organics, Antwerp, Belgium. Methyl-2,2,2-trifluoroethyl carbonate (FEMC) was obtained from Synquest Labs, Alachua, FL, USA. Polyvinylidene fluoride (PVDF) polymer was purchased from Alfa Aesar, Ward Hill, MA, USA. The COM electrolyte, which included 1 *M* LiPF6 in 1:1 $v/v\%$ EC:DEC, nonaflurobutyl methyl ether (NONA), N-Methyl-2-pyrrolidone (NMP), lithium bis(fluorosulfonyl)imide (LiFSI), and Lithium bis(trifluoromethanesulfonyl)imide (LiTFSI), was purchased from Sigma Aldrich, Waukesha, WI, USA. Lithium iron phosphate (LFP) was purchased from Sigma Aldrich. Chemicals were used as purchased without further purification.

2.2. Electrolyte Synthesis

The experimental electrolytes were synthesized in an argon-filled glovebox, NEXUS II Vacuum Atmospheres Co. (Hawthorne, CA, USA), with oxygen concentrations below 1.0 ppm and water concentrations below 0.5 ppm. Solvents were first measured and stirred in previously mentioned ratios at room temperature in vials. Then, salts were added in *1 M* concentrations and stirred until a homogenous mixture was formed. Electrolytes were stored in the glove box and out of light at room temperature.

2.3. Electrode and Coin Cell Fabrication

LFP powder was first mixed with PVDF and Super P in a 8:1:1 mass ratio. NMP was added to a desired viscosity and the solution was mixed in a planetary Thinky mixer for

20 min. The resulting slurry was coated onto aluminum foil using a doctor blade on an MTI laminate coater. The coated foil was then dried at 60 °C under vacuum overnight and punched into 12 mm electrodes with an average cathode mass loading of 3.5 mg/cm^2. The utilized mass loading is lower than typical commercial LIBs, >10 mg/cm^2, which allows for reduced transport resistances in our cathode. This enables greater isolation of the electrolyte's effect on kinetics and low-temperature performance. Lithium, a Celgard 2500 separator wetted with the electrolyte, and the LFP were inserted into CR-2032 coin cells, along with a 1.0 mm spacer, 0.5 mm spacer, and spring to apply constant stack pressure.

2.4. Electrochemical Characterization

Cycling studies were performed between 2.5 and 4.2 V vs. Li/Li+ with an Arbin BT-2000 Galvanostat. Cells were first formed with five full cycles at C/10. Electrochemical impedance spectroscopy (EIS) was conducted on lithium metal half cells that completed four full charge and discharge cycles, with a scanning range of 1 MHz to 0.01 Hz. Stainless steel symmetric cells were used to measure ionic conductivities through impedance spectroscopy from 25 °C to −40 °C. Lower temperature measurements, such as −50 °C, were limited by ice and impedance buildup in the circuit, limiting measurement accuracy. Linear sweep voltammetry (LSV) utilized a stainless steel | | Li electrode setup while cyclic voltammetry used an LFP | | Li half cell. For both CV and LSV, a 0.5 mV/s scan rate was utilized.

3. Results and Discussion

The developed gradient electrolytes consist of LiFSI salt, FEC, FEMC, NONA, and an experimental solvent, which is either diethylene carbonate (DEC), ethyl methyl carbonate (EMC), or dibutyl carbonate (DBC). The use of fluorinated carbonate solvents, FEC, and FEMC enables high ionic conductivity and an advantageous SEI formation [23]. The incorporation of NONA, which has an extremely low freezing point of −135 °C, significantly expands the electrolyte's liquid temperature range. LiFSI salt further enhanced SEI kinetics [22]. The electrolytes are abbreviated as F-FXFN, where X represents D for DEC, E for EMC, and B for DBC. As shown in Figure 1, The carbonate solvents have incrementally decreased dielectric constants of 89.6 for FEC, 9.56 for FEMC, and approximately 3.0 for DEC, EMC, and DBC. These gradient LHCEs were prepared using a *1 M* LiFSI concentration with a 1:2:2:3 (FEC:FEMC:X:NONA) solvent volume ratio, and exhibited ultra-low freezing points below −120 °C. For comparison, we also prepared an electrolyte without the third coordinating solvent, abbreviated as F-FFN. This electrolyte was mixed in a 2:4:1 (FEC:FEMC:NONA) volume ratio. Both the gradient and F-FFN electrolytes' incorporation of the diluent, 37.5% and 14%, respectively, were near their respective miscibility limit.

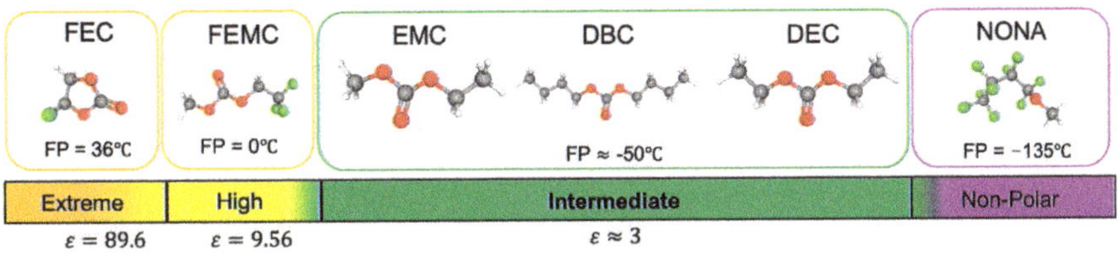

Figure 1. Graphical abstract of the step-down, gradient solvent design for localized high-concentrated electrolytes (FPs = freezing points).

To characterize these electrolytes, we thoroughly explored the solvation structure, electrochemical stability, and cell kinetics. Finally, we evaluated their effect on Li-metal cell performance, where the gradient LHCEs demonstrated superior low- and ultra-low-temperature performance compared to both commercial (COM) and F-FFN electrolytes.

3.1. Electrolyte Characterization

Raman spectroscopy was employed to characterize the behavior of anions within the electrolyte solvation shell. The spectra were deconvoluted to quantify the relative abundance of free anions ($FSIs^-$)—those not coordinated in solution—contact-ion pairs (CIPs), which consist of shells with a single coordinating anion, and ion aggregates (AGGs), which contain two or more coordinating anions. The peaks corresponding to $FSIs^-$, CIPs, and AGGs are observed at 720, 732, and 746 cm^{-1}, respectively [15]. The deconvoluted spectra and tabulated results are shown in Figure 2 and Table 1, respectively, for the experimental electrolytes.

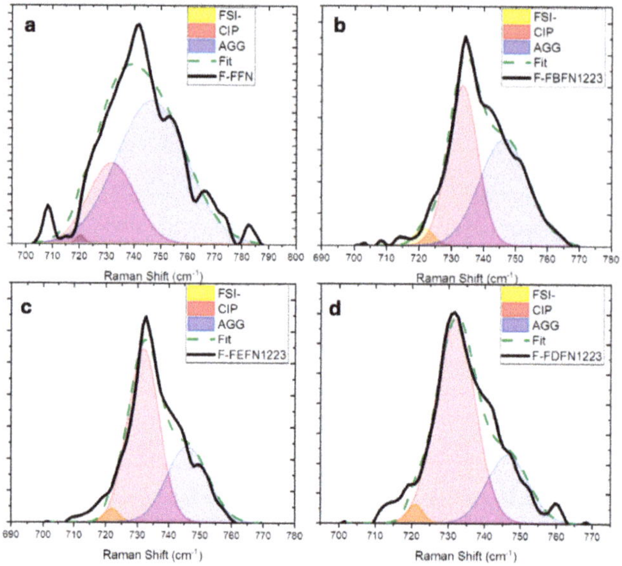

Figure 2. Solvation shell characterization with Raman spectroscopy for (**a**) F-FFN, (**b**) F-FBFN, (**c**) F-FEFN and (**d**) F-FDFN electrolytes.

Table 1. Relative abundance of $FSIs^-$, CIPs, and AGGs for gradient and three-solvent electrolytes.

Electrolyte	FSI^- (%)	CIP (%)	AGG (%)
F-FFN	2	27	71
F-FDFN	2.4	71.8	25.6
F-FEFN	2	62.1	35.8
F-FBFN	3	46.2	50.7

The behavior of free FSI^- anions remain relatively consistent across both the gradient and F-FFN electrolytes, with 2–3% participation. This consistency suggests that anion participation in the solvation shell is preserved by adding the fourth solvent. However, the participation levels of CIPs and AGGs vary significantly. When comparing F-FFN to the gradient electrolytes, F-FFN exhibits a higher proportion of AGGs, with an integrated peak area of 71%. Among the gradient electrolytes, F-FDFN, F-FEFN, and F-FBFN show AGG peak areas of 25.6%, 35.8%, and 50.7%, respectively. The lower proportion of ion aggregates in the gradient electrolytes seems surprising given the reduced highly polar FEC content but may be explained by F-FFN's sterically simpler solvation structure [24,25]. The fourth solvent, EMC, DBC, or DEC, added to make F-FFN into the gradient electrolytes, are much bulkier molecules than FEC. In turn, this limits the amount of anion coordination possible, explaining the increased number of CIPs and reduced AGGs. When examining the differences in AGGs within the different gradient electrolytes, steric effects no longer explain the trend. EMC and DEC have very similar and well-studied dielectric constants of 2.4 and 2.8, respectively [26,27]. DBC's dielectric constant is unavailable in the literature but is likely lower, <2.8, due to the larger alkyl chains. With this, the dielectric trend is inverse to the observed trend in AGG content. F-FDFN's lower AGG content is explained by DEC's stronger ability to solvate lithium ions and prevent aggregates. F-FEFN's AGG content is slightly higher than F-FDFN due to the slight decrease in dielectric constant from DEC to EMC. F-FBFN shows the greatest AGG content due to DBC's weaker solvating ability.

To characterize the electrolyte's electrochemical stability, cyclic voltammetry (CV) on Li| |LFP half cells was performed from 2.5 V to 4 V at a scan rate of 0.5 mV/s (Figure A1a). After five cycles, the F-FDFN gradient LHCE shows a negligible, less than 10%, increase in the oxidation peak at 3.6 V and no change in the current density of the reduction peak at 3.25 V, indicating its electrochemical stability. Linear sweep voltammetry (LSV) was performed on Li| |stainless steel cells, at a scan rate of 0.5 mV/s from OCV to 5 V (Figure A1b). All gradient LHCEs were stable up to 4.2 V. Above 4.2 V, the electrolytes show significant oxidation and current densities greater than the three-solvent, F-FFN, electrolyte, and much greater than the COM electrolyte. These results imply adequate stability for the gradient electrolytes within voltage limits for a Li| |LFP cell. The gradient LHCE stability is further characterized later with long-term, low-temperature cycling studies.

3.2. Enhanced Kinetics

To investigate the effect of temperature on internal resistances and energy barriers, electrochemical impedance spectroscopy (EIS) was performed on LFP| |Li half cells. Impedance measurements were taken at various temperatures ranging from −40 °C to 20 °C, with the frequency range spanning from 1 MHz to 0.01 Hz. Results were presented with Nyquist plots, where cell processes were observed across characteristic frequency ranges and modeled using the Randle's equivalent circuit (Figure A2). Through fitting the model, resistances associated with the desolvation–charge–transfer process (R_{CT}) and diffusion of lithium ions through the cathode solid-electrolyte interface (R_{SEI}) were calculated. After determining the resistances at different temperatures, the Arrhenius plots depicting R_{SEI} and R_{CT} across the temperature range were constructed (Figure 3a,b, respectively). Activation energies were then calculated by using linear regression and are shown in Table 2.

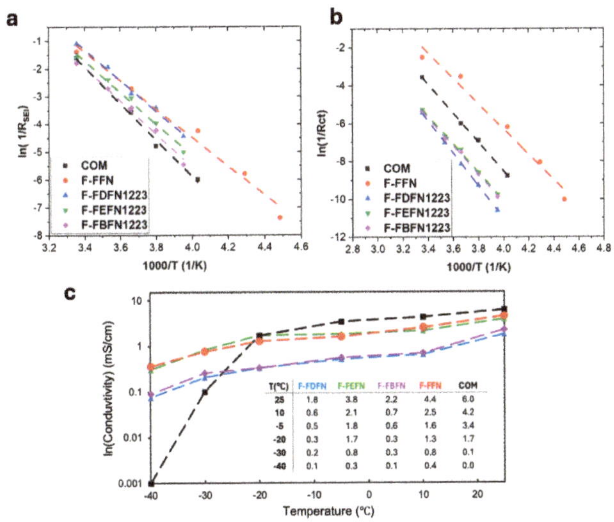

Figure 3. Kinetic characterizations of electrolytes with (**a**) Arrhenius plot for Li resistance through CEI, (**b**) Arrhenius plot for charge transfer process, and (**c**) ionic conductivity from 20 to −40 °C.

Table 2. Activation energies for $E_{a,SEI}$ and $E_{a,CT}$ for gradient, three-solvent and COM electrolytes.

Electrolyte	$E_{a,SEI}$ (kJ/mol)	$E_{a,CT}$ (kJ/mol)
F-FFN	42.9	55.7
F-FDFN	46.2	59.6
F-FEFN	47.4	62.1
F-FBFN	50.3	62.4
COM	54.8	64.3

The three-solvent electrolyte, F-FFN, shows the lowest $E_{a,CT}$ at 55.7 KJ/mol. The gradient electrolytes all show greater $E_{a,CT}$ values, with F-FDFN, F-FEFN, and F-FBFN displaying values of 59.6, 62.1, and 62.3 kJ/mol, respectively. This result is expected considering F-FFN's greater contribution of AGG in its solvation shell, which lowers the desolvation energy. Similarly, $E_{a,SEI}$ follows the same trend: F-FFN has the lowest value, 42.8 kJ/mol, while F-FDFN, F-FEFN, and F-FBFN show values of 46.02, 47.4, and 47.4, respectively. This result once again reflects the F-FFN's increased AGG contribution and the more anion-derived SEI. Interestingly, F-FDFN shows lower $E_{a,CT}$ and $E_{a,SEI}$ despite having the smallest AGG peak. This observation contradicts the Raman data, which indicated that F-FDFN had the lowest AGG contribution. This discrepancy suggests that competing effects may influence the interfacial kinetics; F-FDFN's DEC content may offer advantages in SEI formation and desolvation.

In addition to characterizing Li SEI diffusion and charge transfer processes, the ionic conductivity of all electrolytes was measured (Figure 3c). EIS was performed on symmetric cells with stainless steel electrodes. The COM electrolyte demonstrated the highest ionic conductivity at −20 °C and above, but its conductivity rapidly decreased at lower temperatures as it approached its freezing point. Interestingly, the F-FFN and F-FEFN electrolytes exhibited similar ionic conductivities, both of which were higher than F-FDFN's and F-FBFN's. Despite this small discrepancy, both the F-FFN and the gradient electrolytes followed a similar trend with temperature, maintaining ion mobility at −40 °C and below.

3.3. Electrochemical Performance

Galvanostatic charge–discharge studies were performed on LiFePO4 (LFP) | | Li coin cells to evaluate the impact the electrolytes had on cell performance. Capacity retention through incremental temperature reductions, long-term low-temperature cycling, and ultra-low-temperature charge–discharge profiles are shown in Figure 4.

Figure 4a presents the capacity retention at −50, −40, −25, 0, and 25 °C for the gradient, F-FFN, and COM electrolytes, all cycled at C/50. At 25 °C and 0 °C, all experimental electrolytes exhibit similar capacities of approximately 160 and 140 mAh/g, respectively. At 25 °C, the gradient and F-FFN electrolytes outperform the commercial electrolyte. This arises from the improved interfacial kinetics, $E_{a,SEI}$ and $E_{a,CT}$, of the LHCEs while maintaining similar room-temperature ionic conductivities to COM electrolytes. Among the subzero temperatures, F-FDFN displayed the highest capacity, achieving 109.2 mAh/g at −50 °C, which corresponds to 68% retention of its room-temperature capacity. Although F-FEFN and F-FBFN demonstrated lower subzero capacity retention compared to F-FDFN, they still outperformed F-FFN. F-FEFN and F-FBFN showed 97.2 and 92.2 mAh/g at −50 °C, which corresponds to 60.1% and 58.5% retention, respectively. F-FFN showed a capacity of 88.2 mAh/g at −50 °C, retaining 56% of its room-temperature capacity. In contrast, the COM electrolyte, while achieving high capacities at 25 °C, experienced a significant decline at lower temperatures. Its capacity decreased from 111.6 mAh/g at 0 °C to 42.1 mAh/g at −25 °C.

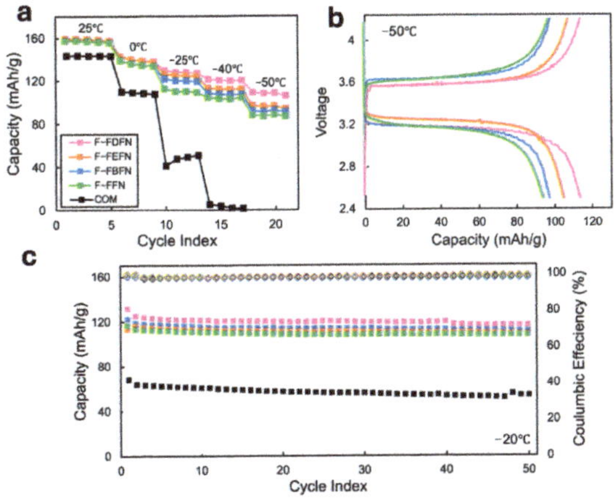

Figure 4. Performance of LFP | | Li coin cells with F-FFN (green), F-FDFN (pink), F-FEFN (orange), F-FBFN (blue) and COM (black) electrolytes under various tests: (**a**) incremental temperature drop from 25 °C to −40 °C at C/50 rate; (**b**) charge–discharge curves at ultra-low-temperature, −50 °C and C/50 rate; and (**c**) discharge capacity and coulombic efficiency over many cycles at −20 °C and C/20.

Charge–discharge curves at ultra-low-temperature, −50 °C, and a C/50 rate are shown in Figure 4b. F-FDFN demonstrates the highest capacity, followed by F-FEFN, F-FBFN, and F-FFN. F-FDFN shows a capacity of 114.1 mAh/g with an overpotential of about 0.125 volts. With a similar overpotential is F-FEFN, but with slightly lower capacity of 105.5 mAh/g. The F-FBFN and F-FFN electrolytes exhibit similar polarizations of approximately 0.15 V, and capacities of 97.8 mAh/g and 94.3 mAh/g, respectively. The decreased overpotential correlates well with the $E_{a,SEI}$ values where F-FEFN and F-FDFN showed lower energy requirements than F-FBFN. This result could demonstrate an increased contribution of diffusion resistance to poor ultra-low-temperature performance. The COM electrolyte was unable to cycle at this temperature due to electrolyte freezing.

The long-term cyclability of the electrolytes was tested at −20 °C with a C/20 rate; the results are shown in Figure 4c. F-FDFN exhibited the highest capacity retention after 50 cycles, maintaining 116.8 mAh/g. All electrolytes, including COM, demonstrated stable cycling behavior and high coulombic efficiencies. F-FFN, F-FDFN, F-FEFN, F-FBFN, and the COM electrolyte recorded average coulombic efficiencies of 99.2%, 98.1%, 99.3%, 98.2%, and 98.4%, respectively. F-FEFN showed the best capacity retention, retaining 95.9% of its first cycle capacity on cycle 50. F-FFN was slightly behind at 92.6%. F-FDFN and F-FBFN showed capacity retentions of 88.4% and 90.9%, respectively, with COM far behind at 78.7%. This indicates that the gradient electrolytes form stable passivation layers and are not continuously consumed at the electrode surfaces. In contrast, the COM electrolytes' greater loss in capacity may be due to dendrite formation and subsequent electrolyte consumption, arising from its worse interfacial kinetics. Similarly, the F-FFN's greatest capacity retention may be explained; its interfacial kinetics are superior to that of the gradient electrolytes, suppressing dendritic growth. At −20 °C, the COM electrolyte showed less than half the capacity of the gradient electrolytes, with F-FDFN and the COM electrolyte achieving 116.8 mAh/g and 53.8 mAh/g, respectively, at cycle 50. The F-FDFN electrolyte once again showed superior capacity compared to F-FEFN and F-FBFN which showed 109.2 and 111.7 mAh/g, respectively. For all electrolytes, there is a minor change in capacity during the first three cycles, arising from electrolyte consumption and SEI formation. On cycle 40, F-FDFN has a slight drop from 120.0 mAh/g to 118.0 mAh/g. COM also has a gain in capacity, 51.9 mAh/g to 55.6 mAh/g on cycle 47. Both changes can be explained by minor dendritic growth and additional SEI growth on the lithium anode surface. Electrolyte consumption could lead to both a temporary increase in capacity, in the case of COM, or a decrease in lithium reserve and subsequent capacity, in the case of F-FDFN.

All the gradient electrolytes demonstrated superior performance at low and ultra-low temperatures compared to the COM and F-FFN electrolyte. This result is despite having higher $E_{a,CT}$ and $E_{a,SEI}$ values and similar ionic conductivities. This result suggests that while improving interfacial kinetics is important, they do not solely determine low-temperature performance. The enhanced low-temperature performance of the gradient electrolytes can be attributed to their substantially lowered freezing point, achieved by incorporating 37.5% diluent by volume. Within the gradient electrolytes, the fourth solvent made a substantial difference in solvation shell, interfacial kinetics and subsequent performance.

4. Conclusions

Four solvent, gradient LHCEs were created, allowing for enhanced miscibility between extremely polar FEC and non-polar NONA. This improved miscibility allowed for 37.5% diluent by volume, which lowered the electrolyte freezing point below −120 °C. Within the gradient LHCE design, three different solvents featuring intermediate dielectric constants and variable steric hindrance were investigated. The gradient LHCEs maintained a localized high salt concentration, shown by Raman spectroscopy with less than 3% free anion in the solution. In addition, an anion-derived SEI was shown by the reduced $E_{a,CT}$ and $E_{a,SEI}$ compared to the COM electrolyte. Due to the incorporation of FEC, room-temperature ionic conductivities similar to that of the EC-based COM electrolyte were observed, while the substantial incorporation of NONA (FP = −135 °C) allowed for ion mobility below −40 °C. Within Li | | LFP coin cells, all three gradient electrolytes outperformed both the three-solvent F-FFN electrolyte as well as a COM electrolyte. The F-FDFN gradient LHCE showed the best performance, retaining 68% of its 25 °C capacity at −50 °C. This electrolyte design shows a promising avenue for expanding the low-temperature operating range while maintaining advantageous room-temperature performance.

Author Contributions: Conceptualization, E.A.A. and J.S.P.; methodology, E.A.A.; formal analysis, J.S.P. and E.A.A.; investigation, J.S.P. and E.A.A.; resources, V.G.P.; data curation, J.S.P. and E.A.A.; writing—original draft preparation, J.S.P.; writing—review and editing, J.S.P., E.A.A. and V.G.P.; visualization, J.S.P. and E.A.A.; supervision, V.G.P.; project administration, V.G.P.; funding acquisition, V.G.P. All authors have read and agreed to the published version of the manuscript.

Funding: This research received no external funding.

Data Availability Statement: Data are available from the authors upon request.

Conflicts of Interest: The authors declare no conflicts of interest.

Appendix A

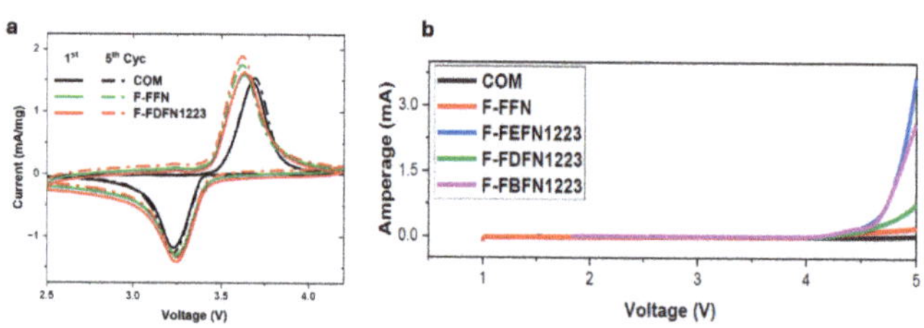

Figure A1. Electrochemical stability evaluated with (**a**) cyclic voltammetry, (**b**) linear sweep voltammetry on Li | | LFP half cells. The gradient electrolytes all show similar electrochemical stability as the COM and F-FFN electrolytes, up to about 4 V.

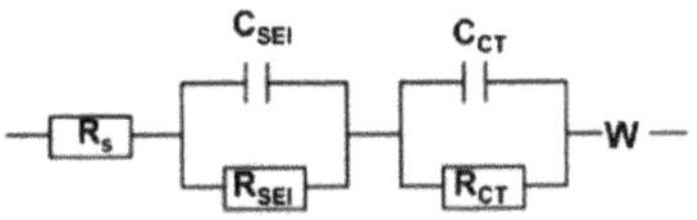

Figure A2. Equivalent Randles circuit used to evaluate charge transfer and SEI diffusion resistances.

References

1. Chu, S.; Cui, Y.; Liu, N. The Path towards Sustainable Energy. *Nat. Mater.* **2017**, *16*, 16–22. [CrossRef] [PubMed]
2. Hou, J.; Yang, M.; Wang, D.; Zhang, J. Fundamentals and Challenges of Lithium Ion Batteries at Temperatures between −40 and 60 °C. *Adv. Energy Mater.* **2020**, *10*, 1904152. [CrossRef]
3. Zhu, G.; Wen, K.; Lv, W.; Zhou, X.; Liang, Y.; Yang, F.; Chen, Z.; Zou, M.; Li, J.; Zhang, Y.; et al. Materials Insights into Low-Temperature Performances of Lithium-Ion Batteries. *J. Power Sources* **2015**, *300*, 29–40. [CrossRef]
4. Feng, Y.; Zhou, L.; Ma, H.; Wu, Z.; Zhao, Q.; Li, H.; Zhang, K.; Chen, J. Challenges and Advances in Wide-Temperature Rechargeable Lithium Batteries. *Energy Environ. Sci.* **2022**, *15*, 1711–1759. [CrossRef]
5. Zhang, N.; Deng, T.; Zhang, S.; Wang, C.; Chen, L.; Wang, C.; Fan, X. Critical Review on Low-Temperature Li-Ion/Metal Batteries. *Adv. Mater.* **2022**, *34*, 2107899. [CrossRef] [PubMed]
6. An, Z.; Jia, L.; Ding, Y.; Dang, C.; Li, X. A Review on Lithium-Ion Power Battery Thermal Management Technologies and Thermal Safety. *J. Therm. Sci.* **2017**, *26*, 391–412. [CrossRef]
7. Smart, M.C.; Ratnakumar, B.V.; Surampudi, S. Electrolytes for Low-Temperature Lithium Batteries Based on Ternary Mixtures of Aliphatic Carbonates. *J. Electrochem. Soc.* **1999**, *146*, 486. [CrossRef]
8. Li, J.; Yuan, C.F.; Guo, Z.H.; Zhang, Z.A.; Lai, Y.Q.; Liu, J. Limiting Factors for Low-Temperature Performance of Electrolytes in $LiFePO_4$/Li and Graphite/Li Half Cells. *Electrochim. Acta* **2012**, *59*, 69–74. [CrossRef]
9. Liao, X.-Z.; Ma, Z.-F.; Gong, Q.; He, Y.-S.; Pei, L.; Zeng, L.-J. Low-Temperature Performance of $LiFePO_4$/C Cathode in a Quaternary Carbonate-Based Electrolyte. *Electrochem. Commun.* **2008**, *10*, 691–694. [CrossRef]
10. Zhang, S.S.; Xu, K.; Jow, T.R. A New Approach toward Improved Low Temperature Performance of Li-Ion Battery. *Electrochem. Commun.* **2002**, *4*, 928–932. [CrossRef]
11. Zhang, S.S.; Xu, K.; Jow, T.R. The Low Temperature Performance of Li-Ion Batteries. *J. Power Sources* **2003**, *115*, 137–140. [CrossRef]
12. Yamada, Y.; Furukawa, K.; Sodeyama, K.; Kikuchi, K.; Yaegashi, M.; Tateyama, Y.; Yamada, A. Unusual Stability of Acetonitrile-Based Superconcentrated Electrolytes for Fast-Charging Lithium-Ion Batteries. *J. Am. Chem. Soc.* **2014**, *136*, 5039–5046. [CrossRef]
13. Suo, L.; Hu, Y.-S.; Li, H.; Armand, M.; Chen, L. A New Class of Solvent-in-Salt Electrolyte for High-Energy Rechargeable Metallic Lithium Batteries. *Nat. Commun.* **2013**, *4*, 1481. [CrossRef] [PubMed]
14. Wang, J.; Yamada, Y.; Sodeyama, K.; Chiang, C.H.; Tateyama, Y.; Yamada, A. Superconcentrated Electrolytes for a High-Voltage Lithium-Ion Battery. *Nat. Commun.* **2016**, *7*, 12032. [CrossRef]
15. Yamada, Y.; Yaegashi, M.; Abe, T.; Yamada, A. A Superconcentrated Ether Electrolyte for Fast-Charging Li-Ion Batteries. *Chem. Commun.* **2013**, *49*, 11194–11196. [CrossRef]
16. Dong, X.; Lin, Y.; Li, P.; Ma, Y.; Huang, J.; Bin, D.; Wang, Y.; Qi, Y.; Xia, Y. High-Energy Rechargeable Metallic Lithium Battery at −70 °C Enabled by a Cosolvent Electrolyte. *Angew. Chem. Int. Ed.* **2019**, *58*, 5623–5627. [CrossRef] [PubMed]
17. Chen, J.; Zhang, H.; Fang, M.; Ke, C.; Liu, S.; Wang, J. Design of Localized High-Concentration Electrolytes via Donor Number. *ACS Energy Lett.* **2023**, *8*, 1723–1734. [CrossRef]
18. Yamada, Y.; Wang, J.; Ko, S.; Watanabe, E.; Yamada, A. Advances and Issues in Developing Salt-Concentrated Battery Electrolytes. *Nat. Energy* **2019**, *4*, 269–280. [CrossRef]
19. Kim, S.; Seo, B.; Ramasamy, H.V.; Shang, Z.; Wang, H.; Savoie, B.M.; Pol, V.G. Ion–Solvent Interplay in Concentrated Electrolytes Enables Subzero Temperature Li-Ion Battery Operations. *ACS Appl. Mater. Interfaces* **2022**, *14*, 41934–41944. [CrossRef] [PubMed]
20. Kim, S.; Pol, V.G. Tailored Solvation and Interface Structures by Tetrahydrofuran-Derived Electrolyte Facilitates Ultralow Temperature Lithium Metal Battery Operations. *ChemSusChem* **2023**, *16*, e202202143. [CrossRef]
21. Jiang, L.-L.; Yan, C.; Yao, Y.-X.; Cai, W.; Huang, J.-Q.; Zhang, Q. Inhibiting Solvent Co-Intercalation in a Graphite Anode by a Localized High-Concentration Electrolyte in Fast-Charging Batteries. *Angew. Chem. Int. Ed.* **2021**, *60*, 3402–3406. [CrossRef] [PubMed]
22. Adams, E.; Parekh, M.; Gribble, D.; Adams, T.; Pol, V.G. Novel Ternary Fluorinated Electrolyte's Enhanced Interfacial Kinetics Enables Ultra-Low Temperature Performance of Lithium-Ion Batteries. *Sustain. Energy Fuels* **2023**, *7*, 3134–3141. [CrossRef]
23. Cho, Y.-G.; Li, M.; Holoubek, J.; Li, W.; Yin, Y.; Meng, Y.S.; Chen, Z. Enabling the Low-Temperature Cycling of NMC| |Graphite Pouch Cells with an Ester-Based Electrolyte. *ACS Energy Lett.* **2021**, *6*, 2016–2023. [CrossRef]
24. Li, X.; Pan, Y.; Liu, Y.; Jie, Y.; Chen, S.; Wang, S.; He, Z.; Ren, X.; Cheng, T.; Cao, R.; et al. Understanding Steric Hindrance Effect of Solvent Molecule in Localized High-Concentration Electrolyte for Lithium Metal Batteries. *Carb Neutrality* **2023**, *2*, 34. [CrossRef]
25. Chen, Y.; Yu, Z.; Rudnicki, P.; Gong, H.; Huang, Z.; Kim, S.C.; Lai, J.-C.; Kong, X.; Qin, J.; Cui, Y.; et al. Steric Effect Tuned Ion Solvation Enabling Stable Cycling of High-Voltage Lithium Metal Battery. *J. Am. Chem. Soc.* **2021**, *143*, 18703–18713. [CrossRef]

26. Ue, M.; Mori, S. Mobility and Ionic Association of Lithium Salts in a Propylene Carbonate-Ethyl Methyl Carbonate Mixed Solvent. *J. Electrochem. Soc.* **1995**, *142*, 2577. [CrossRef]
27. McEwen, A.B.; McDevitt, S.F.; Koch, V.R. Nonaqueous Electrolytes for Electrochemical Capacitors: Imidazolium Cations and Inorganic Fluorides with Organic Carbonates. *J. Electrochem. Soc.* **1997**, *144*, L84. [CrossRef]

Disclaimer/Publisher's Note: The statements, opinions and data contained in all publications are solely those of the individual author(s) and contributor(s) and not of MDPI and/or the editor(s). MDPI and/or the editor(s) disclaim responsibility for any injury to people or property resulting from any ideas, methods, instructions or products referred to in the content.

batteries

MDPI

Article

Lithium Tracer Diffusion in Li_xCoO_2 and $Li_xNi_{1/3}Mn_{1/3}Co_{1/3}O_2$ (x = 1, 0.9, 0.65)-Sintered Bulk Cathode Materials for Lithium-Ion Batteries

Erwin Hüger [1,2,*], Daniel Uxa [2] and Harald Schmidt [1,2]

1 Clausthal Center for Materials Technology, Clausthal University of Technology, 38678 Clausthal-Zellerfeld, Germany; harald.schmidt@tu-clausthal.de

2 Solid State Kinetics Group, Institute of Metallurgy, Clausthal University of Technology, 38678 Clausthal-Zellerfeld, Germany; daniel.uxa@tu-clausthal.de

* Correspondence: erwin.hueger@tu-clausthal.de

Abstract: The knowledge of Li diffusivities in electrode materials of Li-ion batteries (LIBs) is essential for a fundamental understanding of charging/discharging times, maximum capacities, stress formation and possible side reactions. The literature indicates that Li diffusion in the cathode material $Li(Ni,Mn,Co)O_2$ strongly increases during electrochemical delithiation. Such an increased Li diffusivity will be advantageous for performance if it is present already in the initial state after synthesis. In order to understand the influence of a varying initial Li content on Li diffusion, we performed Li tracer diffusion experiments on Li_xCoO_2 (LCO) and $Li_xNi_{1/3}Mn_{1/3}Co_{1/3}O_2$ (NMC, x = 1, 0.9, 0.65) cathode materials. The measurements were performed on polycrystalline sintered bulk materials, free of additives and binders, in order to study the intrinsic properties. The variation of Li content was achieved using reactive solid-state synthesis using pressed Li_2CO_3, NiO, Co_3O_4 and/or MnO_2 powders and high temperature sintering at 800 °C. XRD analyses showed that the resultant bulk samples exhibit the layered LCO or NMC phases with a low amount of cation intermixing. Moreover, the presence of additional NiO and Co_3O_4 phases was detected in NMC with a pronounced nominal Li deficiency of x = 0.65. As a tracer source, a ^6Li tracer layer with the same chemical composition was deposited using ion beam sputtering. Secondary ion mass spectrometry in depth profile mode was used for isotopic analysis. The diffusivities followed the Arrhenius law with an activation enthalpy of about 0.8 eV and were nearly identical within error for all samples investigated in the temperature range up to 500 °C. For a diffusion mechanism based on structural Li vacancies, the results indicated that varying the Li content does not result in a change in the vacancy concentration. Consequently, the design and use of a cathode initially made of a Li-deficient material will not improve the kinetics of battery performance. The possible reasons for this unexpected result are discussed.

Keywords: Li-ion battery cathode material; $LiNi_{1/3}Mn_{1/3}Co_{1/3}O_2$; $LiCoO_2$; Li diffusion; Li off-stoichiometry

1. Introduction

The diffusion of lithium (Li) in electrode materials [1–11] of rechargeable Li-ion batteries (LIBs) is a determining factor for the incorporation and removal of Li into/from electrodes during electrochemical cycling. Diffusion directly influences charging/discharging times, power densities, maximum capacities, stress development and possible side reactions. In this context, the Li diffusion in electrode materials is of importance for the optimization of battery performance. Within the class of cathodes, $LiNi_yMn_zCo_{1-y-z}O_2$ ($y + z < 1$) active materials are important for applications and fundamental studies (see references [12–15]). Noteworthy compositions include those with y = z = 0 ($LiCoO_2$, referred to as LCO), y = 1, z = 0 ($LiNiO_2$, referred to as LNO) and y = z = 1/3 ($LiNi_{1/3}Mn_{1/3}Co_{1/3}O_2$, referred to as NMC) [16].

A review of the literature indicates a substantial body of theoretical and computational research focusing on diffusion in cathode materials [17–24]. In contrast, experimental investigations on Li tracer diffusion, including the determination of diffusivities, activation energies, pre-exponential factors and the mechanism of diffusion are relatively limited for these materials. The knowledge on chemical diffusion in cathode materials is predominantly derived from electrochemical measurements such as galvanostatic and potentiostatic intermittent titration techniques (GITT and PITT) and electrochemical impedance spectroscopy (EIS) [5,25–31]. It should be noted that these non-Li selective methods are typically only applicable at, or in the proximity of, room temperature. Consequently, they offer limited insights into activation energies, pre-exponential factors, the underlying defect structure and the diffusion mechanism. Electrochemical measurements of diffusivities are known to be limited by a number of factors, including interfacial processes, parasitic electronic currents, ohmic resistances and side reactions. These factors lead to a substantial variation in diffusivities [21] across different studies and methods, as well as during lithiation/delithiation [5,21].

In contrast, tracer methods provide a direct measurement of the Li tracer diffusivity, eliminating the need for model-dependent correction factors. For a basic introduction into diffusion science, we refer to references [32–35]. Recently, we conducted preliminary research on Li tracer diffusion in sintered bulk samples of NMC with stable isotopes ($^6Li/^7Li$) and isotope depth profiling using secondary ion mass spectrometry (SIMS) [2]. Li tracer diffusivities between 110 and 350 °C are temperature dependent and show an activation enthalpy of 0.85 eV. This was interpreted as the migration energy of a single Li vacancy and it was concluded that Li diffusion takes place via structural vacancies, where the concentration is fixed by a low Li deficiency. At room temperature, the Li tracer diffusivity [5] is considerably lower than the chemical diffusivities reported in the literature from electrochemical measurements, which can be attributed to a significant thermodynamic factor correlating both types of diffusivities [5,35]. In a separate publication, it was observed that the NMC and LCO compounds exhibited the same Li diffusivities within error bars [10]. This observation shows that a change in the transition metal (M) within the $LiMO_2$ structure does not result in a significant influence on Li tracer diffusion.

Moreover, a comparative study of Li tracer diffusion in LCO single crystals and polycrystalline material was performed [3]. It has been evidenced that the Li diffusivities in the ab-plane of single crystals are the same as those observed in polycrystalline LCO. This finding indicates only a small impact of grain boundaries on the predominant Li-ion migration pathways within the material. Along the c-axis, diffusivities are significantly lower. This experimentally evidenced the often-postulated slow Li diffusion along the c-axis. We suggest that Li diffusion along the c-axis of the 2D ion conductor is controlled by a slow transfer of Li ions between the Li-O layers across the Co-O layers [3].

Recently [36], we reported Li tracer diffusivities in sputter-deposited 1-micrometer-thin amorphous and crystallized NMC films. The diffusivities in crystalline thin films and in sintered bulk materials were found to be nearly identical. In contrast, the Li diffusivities in the amorphous state were found to be lower due to a higher activation enthalpy. The higher activation enthalpy was attributed to hindered Li diffusion in the disordered amorphous state, emphasizing the aspect of disorder for Li diffusion.

In the present study, we investigated Li diffusion in polycrystalline bulk Li_xCoO_2 and $Li_xNi_{1/3}Mn_{1/3}Co_{1/3}O_2$ (x = 1, 0.9, 0.65) cathode materials. In order to illustrate the choice of these materials, we would like to mention that LCO and NMC are among the most used materials for positive electrodes of commercial LIBs. Beneath their suitability from a chemical point of view, the active material is typically used in the form of micrometer- or nanometer-sized crystalline particles surrounded by binder and conductive agents to form micrometer-thick polycrystalline layers. These are polycrystalline but consist of single particles bonded together. In order to determine intrinsic Li transport, it is necessary to measure the Li diffusivity in sintered LCO and NMC bulk pellets without additives. High temperature sintering at 800 °C for 24 h was performed to ensure the formation of a crystallographic phase with a layered structure, necessary for the smooth intercalation and deintercalation of Li ions. This will be discussed in detail in the third section.

The aim of the study was to understand the influence of varying the Li content on Li diffusion, which is important for the overall performance of an LIB in two ways.

First, in a recent study on bulk sintered NMC based on SIMS tracer diffusion experiments, we demonstrated that electrochemical delithiation of about 10% charge (Li deficient state) increases the Li tracer diffusivity by orders of magnitude [5]. Moreover, Figure 1 shows Li chemical diffusivities as a function of the working electrode potential, E_{we}, on the lower x-axis, and the corresponding Li deficiency, expressed as $\delta = 1 - x$, in $Li_xNi_{1/3}Mn_{1/3}Co_{1/3}O_2$ on the upper x-axis. The Li deficiency was realized using electrochemical delithiation. The symbols in Figure 1 indicated as "Hüger" are Li chemical diffusivities obtained in our laboratory using PITT and EIS on the same type of sintered NMC pellets (x = 1) as used in the current work for SIMS diffusion experiments [5] (see Section 2). The remaining diffusivities on NMC in Figure 1 correspond to values reported by other research groups. As indicated in Figure 1, the data are from Kang et al. [30] (binder- and additive-free films; GITT), Charbonneau et al. [29] (films with NMC particles embedded in binders and conductive additives; EIS) and Wu et al. [26] (films with NMC particles embedded in binders and conductive additives; GITT). These chemical diffusivities also show an increase of several orders of magnitude when significant Li is extracted from the cathode material (around 3.6 V and $\delta = 1 - x \approx 0.1$). Moreover, reference [5] demonstrates that the Li chemical diffusivities are in good agreement with the Li tracer diffusivities obtained from SIMS experiments when the thermodynamic factor is taken into account. Electrochemical investigations [37,38] on Li-deficient $Li_xNi_{1/3}Mn_{1/3}Co_{1/3}O_2$ (x < 1) indicate improved properties as a cathode material.

Such an enhanced Li diffusivity can be advantageous for LIB operation if the NMC electrode shows it already in the initial state right after synthesis. In order to achieve this, the electrode has to be produced with a Li deficiency using appropriate methods like a solid-state reaction. The difference between Li-deficient electrode samples made using a solid-state reaction and electrochemical delithiation has to be elucidated. A possible difference may lie in a high-temperature treatment and the resulting more homogeneous structure of the sample prepared using the solid-state synthesis. The electrochemically obtained delithiations were performed at room temperature, which might result in more Li trapping defects and a higher inhomogeneity due to the presence of two-phase delithiation processes (see e.g., Figure 5 in reference [39] and Figure S4 in reference [5]). Therefore, the solid-state synthesis can be expected to minimize these effects, and this is discussed in detail in the Section 3.

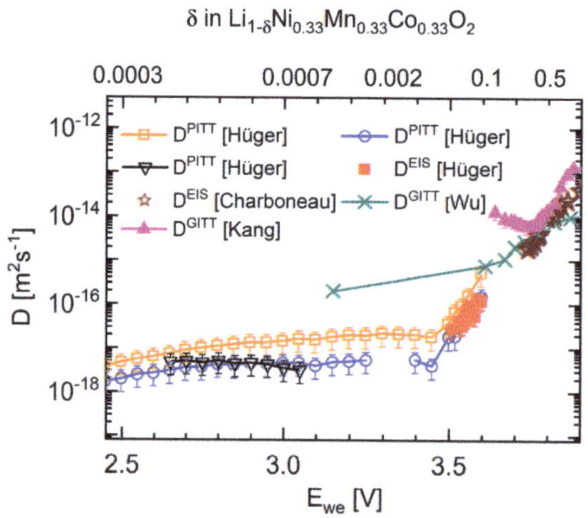

Figure 1. Literature data on Li chemical diffusivities as a function of potential (E_{we}) and Li deficiency ($\delta = 1 - x$) in $Li_xNi_{0.33}Mn_{0.33}Co_{0.33}O_2$ as obtained from EIS, PITT and GITT measurements. The data of Hüger correspond to an NMC (x = 1.0) sample that was also used for tracer diffusion studies [5]. The other literature data are from Charboneau et al. [29], Kang et al. [30] and Wu et al. [26]. Reference is made to the main text for further description. Reproduced with permission of the authors under their license CC-BY 4.0.

Further, a recent review has identified the critical role of Li off-stoichiometry for the performance of LNO batteries [16]. This may also be relevant for the case of isostructural NMC and LCO. The presence of a slight initial Li deficiency in LNO may result in a reduction in Li diffusivities due to the Ni atoms present in the Li_2O interlayers, which can impede Li diffusion [16]. It is recommended that such electrodes are avoided for optimized LIB cycling. Further discussion on this subject is given in the third section. Accordingly, the latter concept suggests that Li diffusion in initially Li-deficient NMC cathodes may be hindered [16], rather than enhanced as proposed above [5].

It might also be argued that a reduced initial Li content in the cathode material is not beneficial because it reduces the Li content of a full cell, especially for x = 0.65. A high deficiency can also make the material less stable. This can lead to a lower energy density, shorter cycle life, reduced stability and reduced safety.

The present study exactly evaluates this contradiction in the literature by investigating the Li diffusion properties in polycrystalline bulk Li_xCoO_2 and $Li_xNi_{1/3}Mn_{1/3}Co_{1/3}O_2$ cathode materials with nominally no Li deficiency ($x \approx 1$), medium Li deficiency ($x \approx 0.9$) and even high Li deficiency ($x \approx 0.65$) to examine the effect of Li deficiency on Li tracer diffusion. The current study focuses mainly on the differences between x = 0.9 and x = 1.0. For a sake of completeness and to address fundamental aspects, we included a strongly off-stoichiometric sample with x = 0.65. A priori, this sample was not considered as a perspective for LIB operation, but at such a high Li off-stoichiometry, "inactive domains" [40] may appear in NMC during delithiation, which do not re-lithiate and reduce the LIB capacity [16]. Although the present study was not performed during LIB cycling, the investigation of Li diffusion in LCO and NMC cathode materials with Li off-stoichiometry may provide information on the influence of diffusion on the kinetic limitation during re-lithiation.

This paper is structured as follows: After the introductory section, the second section describes the measurement techniques and the preparation of the LCO and NMC materials. The third section starts with a part on structural and chemical sample characterization

(Section 3.1). The analysis is closely linked to literature which should help to understand Li diffusion. The measurement, analysis and discussion of the Li diffusion data are presented in Section 3.2. The final section summarizes the results.

2. Experimental Details

The samples were prepared via solid-state synthesis, utilizing a two-step reaction route. Powders of Li_2CO_3 were obtained from Sigma Aldrich (Taufkirchen, Germany), while powders of NiO, MnO_2 and Co_3O_4 were obtained from Carl Roth (Karlsruhe, Germany). The powder materials were introduced into an agate ball milling tank (WiseMix Ball Mill, Wisd Laboratory Instruments, Wertheim, Germany) with respect to the required Li to metal ratio and with the addition of an appropriate amount of ethanol as a dispersing agent. The mixture was subjected to a ball-milling process at a rate of 150 rpm (rotations per minute) for a duration of 1 h. Afterwards, the dried powder mixture was subjected to a thermal treatment at 800 °C (16 h) at a rate of 3 K/min in pure oxygen. The powder mixture thus obtained was subjected to high-energy ball milling (SPEX 8000M shaker mill, SPEX, CertiPrep, SampleSpec, Metuchen, NJ, USA) for a total of 4 min at 1080 cycles per minute with zirconia ceramic balls, and a pellet was subsequently pressed at 100 MPa and sintered at 800 °C in air for a long period of 24 h. This process resulted in polycrystalline, dense samples with relative densities of approximately 97% [2]. The thickness of the LCO and NMC pellets was approximately 1 mm and the diameter was about 8 mm. For two types of samples (LCO, x = 1.0 and NMC, x = 0.9), a modified procedure was applied. LCO and NMC powders were bought from Alfa Aesar (Kandel, Germany) and Sigma-Aldrich (Taufkirchen, Germany), respectively. Afterwards, the high energy milling and pressing steps were carried out accordingly.

The relative Li concentration was determined by analyzing the samples using inductively coupled plasma-optical emission spectroscopy (ICP-OES) with a Plasma Quant 9100 machine (Analytik Jena, Jena, Germany). By varying the initial Li_2CO_3 fraction, we obtained Li_xCoO_2 and $Li_xNi_{1/3}Mn_{1/3}Co_{1/3}O_2$ samples with nominal Li amounts of approximately 1.0, 0.9 and 0.65.

The sputter targets for tracer layer deposition were produced in an analogous manner using isotope-enriched $^6Li_2CO_3$ (95% enriched) as the starting material instead. Due to the larger diameter of the sputter target (about 20 mm), a higher pressure of 330 MPa was applied. The deposition of the tracer was achieved with the help of ion-beam coating, performed with a commercially available apparatus (IBC 681, Gatan, Irvine, CA, USA). The thicknesses of the tracer layers were between 70 and 110 nm. The deposition was carried out at an energy of 5 keV and a current of 200 μA with argon sputter gas at 5×10^{-3} mbar. Given that the ^{6}Li tracer layer and the bulk sample have a similar chemical composition (resulting from the same production route), it was anticipated that pure isotope interdiffusion would occur during the diffusion experiments. The sputtered samples were annealed at elevated temperatures in air using a commercial rapid annealing setup (AO 600, MBE Komponenten, Weil der Stadt, Germany). The structural investigation of the bulk samples was carried out using X-ray Diffractometry (XRD) on a Bruker D8 Discover diffractometer (Bruker AXS Advanced X-ray Solutions GmbH, Karlsruhe, Germany) in $\theta/2\theta$ mode with CuK_α radiation at 40 keV and 40 mA.

Depth profiles of the two Li isotopes and of the Ni, Mn and Co species were recorded using SIMS. For this investigation, a Cameca IMS 3f/4f machine (CAMECA SAS, Gennevilliers Cedex, France) with O^- primary ions was applied. SIMS was used in depth profile mode. A beam of O^- primary ions was directed onto the sample and scanned over the surface in an area of approximately 250 μm × 250 μm to sputter away material, creating a zone with a crater-like depression termed "sputter crater". This crater was measured with a mechanical stylus profilometer (Tencor Alphastep 500, Milpitas, CA, USA) for depth calibration. The relative ^{6}Li isotope fraction, c, was calculated from the ^{6}Li and ^{7}Li signal intensities, I, as measured using SIMS.

$$c = \frac{I(^6Li)}{I(^6Li) + I(^7Li)} \tag{1}$$

3. Results and Discussion

3.1. Structural Characterization and Relative Li Concentration

Before presenting the diffusion results, which is the main task of the current study, we present investigations on the structure and chemical composition of the LCO and NMC samples and discuss the expected effects on Li diffusion.

For the determination of the relative Li concentration, the samples were analyzed using ICP-OES and the results are shown in Table 1. By varying the initial Li_2CO_3 fraction, we obtained Li_xCoO_2 and $Li_xNi_{1/3}Mn_{1/3}Co_{1/3}O_2$ samples with nominal Li amounts of approximately x = 1.0, 0.9 and 0.65, where x corresponds to the Li to metal ratio L/M.

Table 1. Relative metal element concentrations as analyzed using ICP-OES of the samples under investigation. L/M denotes the Li to metal ratio.

Sample	Li (at.%)	Ni (at.%)	Mn (at.%)	Co (at.%)	L/M
LCO (x = 1)	24.64 ± 0.13	--	--	24.94 ± 0.21	0.99 ± 0.01
LCO (x = 0.9)	22.72 ± 0.02	--	--	25.08 ± 0.17	0.91 ± 0.01
NMC (x = 1)	24.09 ± 0.10	7.72 ± 0.03	7.82 ± 0.01	8.32 ± 0.04	1.01 ± 0.01
NMC (x = 0.9)	22.50 ± 0.10	7.42 ± 0.02	9.88 ± 0.06	8.05 ± 0.04	0.89 ± 0.01
NMC (x = 0.65)	18.20 ± 0.10	9.66 ± 0.13	9.29 ± 0.14	8.92 ± 0.16	0.65 ± 0.01

X-ray diffraction patterns of the sintered sample are shown in Figure 2a. Most of the Bragg peaks correspond to the rhombohedral (or hexagonal) α-$NaFeO_2$-type layered structure belonging to the R-3m space group (see indexing). The results show that LCO and NMC samples with x = 1 und 0.9 show no obvious impurity phases. This indicates an upper limit of possible impurity phases of 3–5% corresponding to the detection limit. For the NMC sample with x = 0.65, additionally, amounts of NiO and Co_3O_4 phases in the ten percent range can be detected due to the strongly reduced Li content. The exact amount of impurity phases was not of importance for this diffusion study, as long as no percolation path was formed, which is the case.

Beneath the detected hexagonal high temperature structure, low-temperature spinel (cubic, Fd3m), and rock-salt (cubic, random occupation of the cation sites) phases have been reported for LCO and NMC [11,16,39,41–64]. Note that the difference in the XRD patterns between the hexagonal phase and the spinel phase is very small [42]. The main indication is a slightly different 2θ position of the strongest reflection, (003) for the layered and (111) for the spinel phase, a splitting of the spinel reflection (222) into the layered (006) and (012) reflections, and the splitting of the spinel (440) reflection into the layered (018) and (110) reflections [39,44].

For Cu K_α radiation and LCO, the layered (003) reflection was located at 2θ = 18.94° and the spinel (111) reflection was located at 2θ = 19.17° [44]. The measured XRD patterns of the current study with x = 1 and 0.9 in Li_xCoO_2 show only a single strong reflection located at 2θ = 18.9° and no reflection at 2θ = 19.2° (see Figure 2b). The reflection splitting of (006), (012), (018) and (110) was also well-resolved in the measurement (see Figure 2c,d). This indicates that the LCO pellets consist of the layered hexagonal phase only, which can be expected due the sintering for 24 h at the high temperature of 800 °C. Cho et al. [38] have shown that the phase transition of NMC to its layered phase only occurs after the lithium carbonate (Li_2CO_3) has fully reacted at a temperature above 660 °C. Figure 2 shows no evidence of Li_2CO_3, which is also expected for LCO only for $x \geq 1$, as reported in the literature [42]. An excess of Li is commonly aimed to minimize off-stoichiometry and to compensate for the loss of Li during the calcination process [16]. However, the presence of the cubic phases in very small amounts below the detection limits of XRD cannot be excluded, but they are not expected to form a percolation path and will not influence Li diffusion.

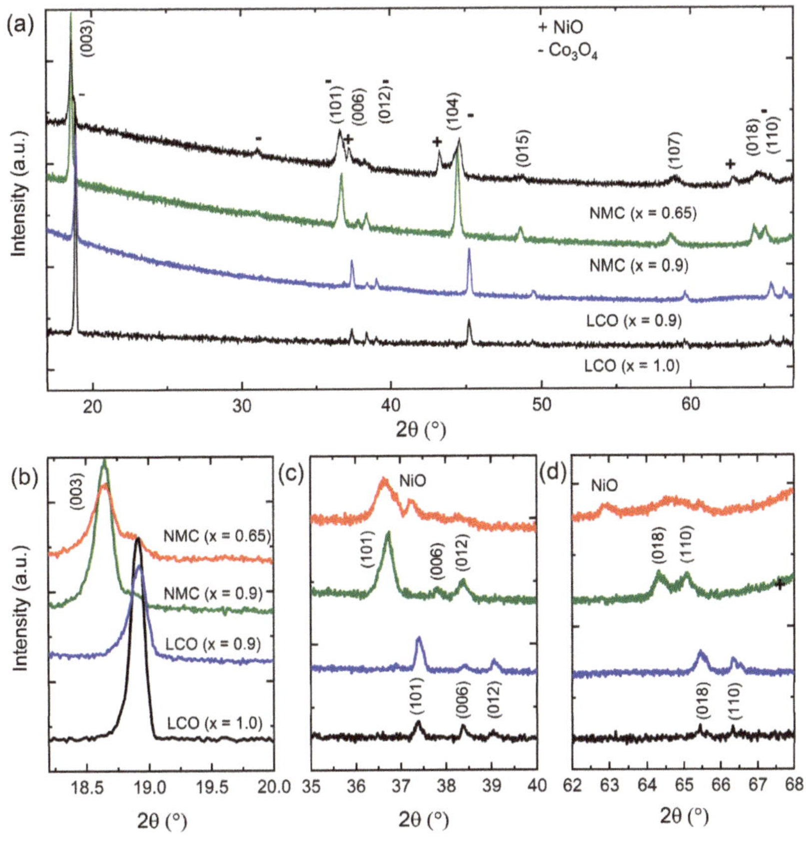

Figure 2. X-ray diffraction patterns of bulk NMC and LCO samples under investigation. Bragg peaks corresponding to the main NMC or LCO phase are indexed (hexagonal scheme) and impurity phases are marked: (**a**) overview and (**b**–**d**) enlarged views.

For further discussion, note that the hexagonal phase has a structure of Li layers alternating with transition metal (e.g., Co) layers between oxygen layers. LCO and NMC both obey this layered structure type, but show a subtle difference. The Ni and Mn content in NMC more often leads to the mixing of Li and Ni cations on the different sub-lattices as anti-site defects also termed cation mixing, which occurs both during synthesis and cycling. This means that the transition metal is located in the Li layer, replacing a Li ion there, and that Li is located in the transition metal oxide. This type of defect may have a strong influence on Li diffusion [16,57]. A high cation mixing rate in the cathode materials is expected to give a lower Li diffusivity and a higher activation enthalpy [49]. Li/transition metal anti-site defect disorder also directly correlates to the electrochemical properties [59]. The most homogeneous samples have the highest capacity, lowest voltage polarization, lowest capacity fade and highest initial coulombic efficiency [59]. The Li/transition metal exchange was visualized using TEM [39,54,55,57]. For LCO, this effect is expected to only be small [45,47,49,51,52,56,57].

XRD studies developed two simple criteria that may indicate cation mixing in the layered phase. If the integrated intensity ratio I(003)/I(104) < 1.2 and the positions of the (018)/(110) pairs of reflections begin to merge, then transition metal (e.g., Ni) and Li ion mixing is very likely occurring [39,53]. Here, I(003)/I(104) ≈ 6.6 was high for LCO and x = 1, indicating nearly no Co/Li ion mixing. The integrated intensity of I(003)/I(104) ≈ 2 for x = 0.9 was still above the threshold of 1.2, indicating a low mixing rate, but slightly more Li/Co antisite defects. Calculations [52,56] predict an anti-site defect concentration of only 0.002% in LCO at 1000 K. The introduction of Mn and Ni for NMC should cause an increase in the anti-site defect concentration [52], which is highest for LNO at up to almost 13%. A recent review on why anti-site defects occur can be found in [57]. Attempts have been made to reduce the amount of anti-site defects using surface engineering (deposition of surface layers) and/or volume engineering by doping NMC with other transition metals to stabilize the transition metal–oxygen skeleton [11,16,57,60,62,63].

In the case of NMC (x = 0.9) (Figure 2), the (003) hexagonal reflection is located at 2θ = 18.7°, which corresponds to the value reported in the literature [50]. There is also a small reflection at 2θ = 18.9°, which could be from LCO, but is more likely from a spinel phase. Co_3O_4 can also not be excluded, but is unlikely, since no other higher intensity peaks of that phase occur here. However, the position of the layered (006), (012), (018) and (110) reflections is also well-resolved in the measurement. The ratio of the integrated intensity of $I_{(003)}/I_{(104)} \approx 1.3$ for NMC with x = 0.9 was close to the threshold of 1.2. These results show that the NMC sample had a layered hexagonal phase with, as expected, a higher, but still low, concentration of cation mixing.

For a high Li deficiency (x = 0.65), beneath the already mentioned impurity phases, the XRD pattern showed a ratio of $I_{(003)}/I_{(104)} \approx 1.6$, which was again close to threshold of 1.2. The (006), (012), (018) and (110) reflections of the layered phase can also be assumed to exist separately, but were not well-resolved. Therefore, we assume that this sample also shows the layered hexagonal phase together with the NiO- and Co_3O_4-impurity phases.

The Li/transition metal mixing is expected to be strongly reduced through a long heating time (more than 5 h) at temperatures as high as 800 °C [48,53], as was performed in this study. However, diffraction experiments on NMC [45] found that, even after annealing at 900 °C, a small proportion (≈2%) of the Ni may be located at the Li site, displacing an equal amount of Li to the transition metal site, in agreement with our findings.

A strategy to reduce the number of anti-site defects is the preparation of Li-rich samples. However, experimental work on optimizing the NMC electrode material by adjusting the Li/transition metal ratio always resulted in the presence of a low amount of anti-site defects, where ≈2% seems to be a threshold [47]. A Li deficiency is expected to first produce Li/transition metal anti-site defects [16,41,43]), followed by the appearance of the rock salt phase for a higher Li deficiency (see, e.g., Figure 1 in reference [54]). As recently discussed for LNO [16], due to the ability of Ni to induce cation mixing, this material tends to be under-stoichiometric in Li and contains excess Ni^{2+} located in the Li layers, which strongly affects its physicochemical properties [16]. The synthesis of LNO samples yields compounds that always contain 0.99–0.98 in Li_xNiO_2 even under very well-controlled conditions [16].

With regard to other point defects, it is worth mentioning that oxygen vacancies are thought to promote a proper atomic rearrangement and a more defect-free layered phase using diffusion during annealing at high temperatures. LCO films deposited using laser ablation at low oxygen pressures produce oxygen vacancies that enhance cation mobility and favor the hexagonal phase [44]. A high-temperature treatment is necessary to allow the transition metal (e.g., Ni) to diffuse from the Li/Ni anti-site position to the regular position. The role of oxygen vacancies also plays a role in the reverse case, i.e., in the conversion of the layered phase into other phases, e.g., into the spinel phase [65], during electrochemical Li extraction. There, the transition metal ions must diffuse from the transition metal layer into the Li layer [65]. This has to occur by squeezing the transition metal through a densely packed oxygen layer. This process would be accelerated due to the presence of oxygen vacancies. Furthermore, oxygen vacancies in cathode materials can also play a beneficial role for LIB cycling of Li-rich NMC cathode materials [16,66–68]. It has been found that oxygen depletion of the surface layer is one way to improve the cyclability of this family of cathode materials [16,66–68]. Theoretical calculations and experimental characterizations show that oxygen vacancies provide a favorable environment for ionic diffusion in the bulk and significantly suppress oxygen gas release from the surface [68]. These facts indicate that not only the knowledge of Li diffusion is important for NMC materials, but also the diffusion of oxygen and transition metals.

In summary, the long annealing time at 800 °C of the LCO and NMC samples resulted in a hexagonal layered phase with a low number of Li/transition metal anti-site defects. NMC with x = 0.65 showed additional NiO and Co_3O_4 impurity phases, also as expected from the results published in the literature [42]. Impurity phases (including the spinel phase) are not expected to form a percolation path and will not affect Li diffusion. XRD suggested that the number of anti-site defects and, consequently, the Li diffusion should decrease from LCO with x = 1 to LCO with x = 0.9, to NMC with x = 1, to NMC with x = 0.9 and to NMC with x = 0.65. This trend is in contrast to the electrochemical delithiation experiments in Figure 1, which show the opposite behavior. The Li tracer diffusivity measurements of the following section should give more insights in this regard.

3.2. Li Tracer Diffusion Measurements

Figure 3 provides an illustration of typical SIMS depth profiles (relative ^{6}Li fraction vs. depth) observed after tracer deposition and diffusion annealing at elevated temperatures for a specified time for each sample type. For further examples of typical depth profiles, we refer to references [2,10].

After tracer deposition, we obtained a well-localized ^{6}Li distribution, restricted to the first 100 nm of the depth profile superimposed on the natural ^{6}Li background of the sample. This corresponds to the as-deposited tracer layer. After annealing, we noted a clear penetration of the ^{6}Li tracer deep into the sample. This redistribution of ^{6}Li was caused by isotope diffusion and can be used to determine the Li tracer diffusivity by comparing the experimental data to appropriate solutions of Fick's second law.

Without diffusion annealing, the depth profile may be fitted with the thick-film solution of Fick's second law [69] as follows:

$$c(x,t) = c_\infty + \frac{(c_0 - c_\infty)}{2}\left[\mathrm{erf}\left(\frac{h+x}{R}\right) + \mathrm{erf}\left(\frac{h-x}{R}\right)\right]. \quad (2)$$

In Equation (2), c_∞ is defined as the residual ^{6}Li isotope fraction in the NMC or LCO bulk sample, while c_0 signifies the ^{6}Li fraction present in the ^{6}Li-enriched layer. The thickness of the ^{6}Li-enriched layer is denoted as h, and R = R_0 is the broadening of the Li distribution at the interface. The diffusion process of the ^{6}Li tracer into the specimen, initiated during the annealing procedure, results in the broadening of the ^{6}Li distribution, denoted by R. To determine the Li tracer diffusivities, the profiles of diffusion-annealed specimens were also fitted with Equation (2). In this context, R serves as a fit parameter, while the value of h is fixed to its initial value. The Li diffusivity (D) is thus determined from the difference in the respective broadening of the ^{6}Li distribution of the annealed profile (R) and of the as-deposited profile (R_0). This is expressed as follows: $D = (R^2 - R_0{}^2)/4t$, where t is the annealing time. The results of the fitting process are displayed as continuous red lines in Figure 3.

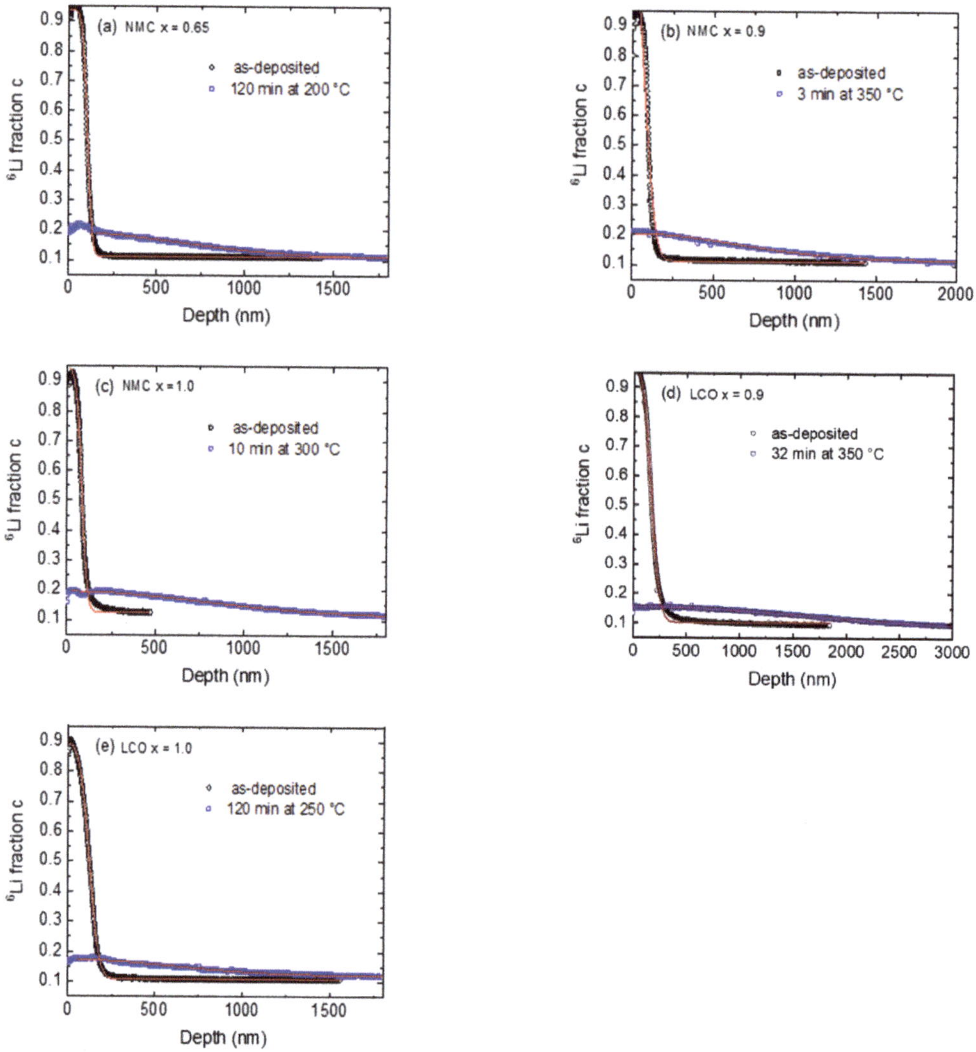

Figure 3. ^{6}Li fraction, c, as a function of depth for various types of samples directly after tracer deposition and after annealing of (**a**) NMC with x = 0.65, (**b**) NMC with x = 0.9, (**c**) NMC with x = 1.0, (**d**) LCO with x = 0.9, and (**e**) LCO with x = 1.0. The fit of curves according to Equation (2) is plotted with red continuous lines.

In Figure 4, all Li tracer diffusivities determined are depicted as a function of the reciprocal temperature. Evidently, the Li diffusivities of all samples are in agreement with each other within error limits, despite statistical deviations. This result clearly demonstrates that the Li diffusivities are independent of the Li fraction between x = 1 and 0.65

for the present LCO and NMC samples, which is in contradiction to the expectations of Sections 1 and 3.1.

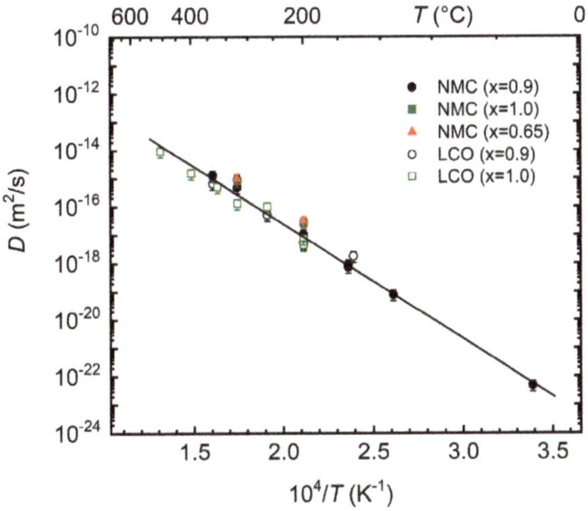

Figure 4. Li tracer diffusivities of Li_xCoO_2 (LCO) and $Li_xNi_{1/3}Mn_{1/3}Co_{1/3}O_2$ (NMC) as a function of reciprocal temperature for different values of x, as measured using SIMS.

The diffusivities follow the Arrhenius law:

$$D = D_0 \cdot \exp(-\Delta H/kT) \tag{3}$$

where ΔH is the activation enthalpy of diffusion and D_0 is the pre-exponential factor. A least-squares fit of Equation (3) to the diffusivities gives an activation enthalpy of $\Delta H = (0.80 \pm 0.03)$ eV and a pre-exponential factor of $D_0 = 3.1^{+2.9}_{-1.5} \times 10^{-9}$ m^2/s, where preliminary data were published in references [2,3,5,10].

Following the arguments of ref. [2], it was hypothesized that all samples under investigation were at least slightly Li sub-stoichiometric, including the samples considered to be nominally $x \approx 1$. This sub-stoichiometry can be attributed to the presence of structural vacancies in the Li sub-lattice. Consequently, the calculated activation enthalpy of diffusion is equivalent to the migration enthalpy of diffusion of the Li ions, $\Delta H = \Delta H_m \approx 0.8$ eV. In comparison to the results of first principles calculations employing the pseudopotential method within the local density approximation of density functional theory, a monovacancy mechanism can be postulated based on a direct next-neighbor hop of a Li-ion into a Li vacancy. Migration enthalpies of 0.8 eV [70] and 0.7 eV [71] were calculated, which is in good agreement with the measured activation enthalpy.

For structural Li vacancies, the pre-exponential factor is given by

$$D_0 = f_V a^2 \nu_0 \exp\left(\frac{\Delta S_m}{k_B}\right) \tag{4}$$

neglecting the correlation factor. In this context, it is important to note that a = 0.286 nm names the shortest Li-Li atomic distance, while ν_0 is approximately equal to 1×10^{13} s^{-1} and is a characteristic vibration frequency. Furthermore, f_V denotes the fraction of structural vacancies, while ΔS_m is approximately equal to 0–1 kB and denotes the migration component of the entropy of diffusion. The application of Equation (4) to the experimentally determined pre-exponential factor enables the assessment of a mole fraction of vacancies within the range of $f_V = 10^{-2} - 10^{-3}$.

This low f_V value obtained for all samples investigated means that the mole fraction of vacancies was not correlated to the fraction of Li ions present in the samples. For example, for x = 0.65, a fraction of vacancies of $f_V = 1 - x = 0.35$ would be expected, if it is assumed that each one lacking Li results in a vacancy. Consequently, it can be assumed that during sample synthesis with the solid-state reaction route, as was performed in this study, the same number of structural vacancies is always formed within the material. The vacancy concentration corresponds to the border of the stability range of the sub-stoichiometric NMC phase. Removing more Li will lead to the formation of impurity phases like NiO and Co_3O_4, as detected for the sample with a nominal composition of x = 0.65.

These results are in contrast to the aspects outlined in the Introduction section, that is, that the removal of Li from the cathode material (Li deficient material) should lead to a strong increase in Li diffusivities/ion conductivities and, consequently, vacancies. This indicates that the vacancy-rich structural state with enhanced diffusion formed during electrochemical delithiation cannot be achieved using a sample synthesis with solid-state reactions.

A second aspect is that the possible presence of cation mixing is expected to block Li diffusion pathways. According to the discussion in Section 3.1, cation mixing and consequently diffusion should decrease with a growing Li deficiency and for NMC compared to LCO. This was also not observed in the results of the SIMS experiments. Consequently, a possible increase in the amount of cation mixing with off-stoichiometry present in the samples is not sufficient to influence diffusion to a measurable extent. However, a blocking of diffusion can already be present for $x \approx 1$ (nominal), where a small amount of transition metal might be present in the Li interlayer. As mentioned for LNO [16], the synthesis of real electrode samples will result in compounds with $x \approx 0.99$, and at least 1% of residual Ni is located in the Li layer even under very well-controlled synthesis conditions. A reduction in x will not influence Li tracer diffusion any more.

If Li diffusion in LCO and NMC would be predominantly determined based on the degree of cation mixing, then it can be argued that the higher Li diffusivity found at the onset of electrochemical delithiation (Figure 1) is due to a reduction in cation mixing in the Li layers. This is curious, because during delithiation Li is extracted from the Li layers, which would cause the transition metal (i.e., Ni) to jump into the Li layers, increasing rather than decreasing the concentration of anti-site defects.

As mentioned, the decrease in charge capacity during re-lithiation can be described in terms of the formation of 'inactive domains' [16]. Based on the results of the present study, it can be assumed that NMC material is composed of two kinds of domains: active domains that show fast Li diffusion in unblocked Li interlayers (and a reversible Li charge capacity) and inactive domains that show slow diffusion in blocked interlayers. A variation in x may change the number of these domains. Since the diffusivity does not depend on the Li off-stoichiometry, the electrode with $x \approx 0.65$ should have percolation paths composed of active domains. This insight, obtained exclusively from Li tracer diffusion experiments, is in agreement with the XRD measurements discussed in Section 3.1 (Figure 2). The XRD analysis indicated the presence of the layered hexagonal crystallographic phase with a low concentration of cation mixing in all electrodes studied, even in NMC with x = 0.65. The lower Li content only led to the detection of Li-free Co_3O_4 and NiO phases. The diffusion experiments showed that these impurity phases do not block the total Li diffusion. Although the impurity phases may have low Li diffusivities, the Li diffusion is not completely blocked because the well-ordered layered phase may still form percolation regions for Li diffusion, even at a Li deficiency corresponding to x = 0.65. It is briefly mentioned that slow diffusion in the inactive regions has an impact on the re-lithiation of the NMC [16]. Slow diffusion requires long time intervals for Li to diffuse across the inactive domains, which is typically achieved using constant voltage steps during cycling.

The advantage of the solid-state reaction method with heating at high temperatures may lie in the production of more extended ordered Li layers with a lower degree of cation mixing. The higher the Li deficiency, the higher the proportion of impurity phases. The layered phase creates a percolation path for Li diffusion and thus determines the measured Li tracer diffusivity similar for all the Li-deficient samples.

In this context, it should be mentioned that another, obviously more direct explanation for the results of the SIMS experiments is that thermal annealing of the diffusion samples at temperatures of only 150 °C and above destroy a vacancy-rich state, which will be investigated in the future by annealing electrochemically delithiated samples.

Another aspect is that it is known that even a small removal of Li from LCO and NMC enhances the electronic conductivity and the corresponding activation energy is reduced [30]. At room temperature, the electronic conductivity in NMC increases by more than two orders of magnitude when 10% of Li is electrochemically removed [25]. An explanation for this might be that the sluggish kinetics at the stoichiometric composition (x = 1.0) of NMC in the electrochemical cell are charge transfer limited [30]. Based on our results, this was not the case for the present samples, whose Li deficiency was produced using solid-state synthesis and high-temperature annealing.

Nevertheless, the cycling of LCO and NMC not to full Li content may remain a suitable strategy for better LIB cycling from the point of view of Li kinetics.

4. Conclusions

According to the literature, Li diffusivities in the cathode material $Li(Ni,Mn,Co)O_2$ are initially low but increase by orders of magnitude by reducing the Li content during electrochemical LIB operation (delithiation). Such an increase in Li diffusivity will be advantageous for LIB operation if the electrode already shows this enhanced diffusivity in the initial state right after synthesis. Consequently, we produced Li_xCoO_2 and $Li_xNi_{1/3}Mn_{01/3}Co_{1/3}O_2$ cathode materials with different Li fractions of x = 1, 0.9 and 0.65 using a solid-state reaction, and carried out Li tracer diffusion experiments in order to understand the influence of varying the Li content on Li diffusion. The measurements were performed on polycrystalline sintered bulk materials, free of additives and binders in order to probe intrinsic properties.

The variation of the Li content x was achieved through a reactive solid-state synthesis using defined amounts of Li_2CO_3, NiO, Co_3O_4 and/or MnO_2 powders. LCO and NMC pellets were sintered at 800 °C for 24 h. The resulting samples exhibited the layered hexagonal phase with only low cation intermixing. Unreacted phases of NiO and Co_3O_4 were detected in NMC, with the most pronounced Li deficiency of x = 0.65.

The diffusivities followed the Arrhenius law, with an activation enthalpy of about 0.8 eV, and were nearly identical within error limits for all samples investigated in the temperature range up to 500 °C. Nominally Li-deficient cathode materials prepared using the solid-state route did not show a modification in Li diffusion compared to the near-stoichiometric case. If a diffusion mechanism based on structural Li vacancies was operating, the results show that Li deficiency does not result in a modification to the vacancy concentration. This indicates that the vacancy-rich structural state obtained during electrochemical delithiation cannot be achieved using a synthesis with thermally induced solid-state reactions. This is in strong contrast to what happens during electrochemical delithiation, where diffusion is accelerated through a Li-deficient state formed through Li removal from the electrode [5]. Consequently, the design and use of a cathode initially made of a Li-deficient material will not improve the kinetics of battery performance.

Author Contributions: Conceptualization, H.S. and E.H.; methodology, H.S. and D.U.; validation, H.S., D.U. and E.H.; formal analysis, H.S., D.U. and E.H.; investigation, D.U. and E.H.; resources, H.S.; data curation, H.S. and E.H.; writing—original draft preparation, H.S. and E.H.; writing—review and editing, H.S. and E.H.; visualization, H.S. and E.H.; supervision, H.S.; project administration, H.S.; funding acquisition, H.S. All authors have read and agreed to the published version of the manuscript.

Funding: This research was funded by the Deutsche Forschungsgemeinschaft (DFG, German Research Foundation) under contract SCHM 1569/33-2 (413672097). The financial support is gratefully acknowledged.

Data Availability Statement: Dataset available on request from the authors.

Acknowledgments: We are indebted to G. Zander for ICP-OES analysis and R. Deichmann for producing the sputter targets.

Conflicts of Interest: The authors declare no conflicts of interest.

References

1. Strauß, F.; Hüger, E.; Julin, J.; Munnik, F.; Schmidt, H. Lithium Diffusion in Ion-Beam Sputter-Deposited Lithium–Silicon Layers. *J. Phys. Chem. C* **2020**, *124*, 8616–8623. [CrossRef]
2. Uxa, D.; Holmes, H.J.; Meyer, K.; Dörrer, L.; Schmidt, H. Lithium tracer diffusion in $LiNi_{0.33}Mn_{0.33}Co_{0.33}O_2$ cathode material for lithium-ion batteries. *Phys. Chem. Chem. Phys.* **2021**, *23*, 5992–5998. [CrossRef]
3. Uxa, D.; Hüger, E.; Meyer, K.; Dörrer, L.; Schmidt, H. Lithium-Ion Diffusion in Near-Stoichiometric Polycrystalline and Monocrystalline $LiCoO_2$. *Chem. Mater.* **2023**, *35*, 3307–3315. [CrossRef]
4. Uxa, D.; Schmidt, H. Lithium tracer diffusion in near stoichiometric $LiNi_{0.5}Mn_{1.5}O_4$ cathode material for lithium-ion batteries. *Z. Für Phys. Chem.* **2022**, *236*, 979–989. [CrossRef]
5. Hüger, E.; Uxa, D.; Schmidt, H. Electrochemical (PITT, EIS) and Analytical (SIMS) Determination of Li Diffusivities at the Onset of Charging $LiNi_{0.33}Mn_{0.33}Co_{0.33}O_2$ Electrodes. *J. Phys. Chem. C* **2024**, *128*, 7408–7423. [CrossRef]
6. Hasegawa, G.; Kuwata, N.; Tanaka, Y.; Miyazaki, T.; Ishigaki, N.; Takada, K.; Kawamura, J. Tracer diffusion coefficients of Li+ ions in c-axis oriented LixCoO2 thin films measured by secondary ion mass spectrometry. *Phys. Chem. Chem. Phys.* **2021**, *23*, 2438–2448. [CrossRef]
7. Kuwata, N.; Hasegawa, G.; Maeda, D.; Ishigaki, N.; Miyazaki, T.; Kawamura, J. Tracer Diffusion Coefficients of Li Ions in $Li_xMn_2O_4$ Thin Films Observed by Isotope Exchange Secondary Ion Mass Spectrometry. *J. Phys. Chem. C* **2020**, *124*, 22981–22992. [CrossRef]
8. Schwab, C.; Höweling, A.; Windmüller, A.; Gonzalez-Julian, J.; Möller, S.; Binder, J.R.; Uhlenbruck, S.; Guillon, O.; Martin, M. Bulk and grain boundary Li-diffusion in dense LiMn2O4 pellets by means of isotope exchange and ToF-SIMS analysis. *Phys. Chem. Chem. Phys.* **2019**, *21*, 26066–26076. [CrossRef]
9. Takai, S.; Yoshioka, K.; Iikura, H.; Matsubayashi, M.; Yao, T.; Esaka, T. Tracer diffusion coefficients of lithium ion in $LiMn_2O_4$ measured by neutron radiography. *Solid State Ion.* **2014**, *256*, 93–96. [CrossRef]
10. Uxa, D.; Holmes, H.J.; Meyer, K.; Dörrer, L.; Schmidt, H. Lithium Tracer Diffusion in Sub-Stoichiometric Layered Lithium-Metal-Oxide Compounds. *DDF* **2021**, *413*, 125–135. [CrossRef]
11. Hüger, E.; Riedel, L.; Zhu, J.; Stahn, J.; Heitjans, P.; Schmidt, H. Lithium niobate for fast cycling in Li-ion batteries: Review and new experimental data. *Batteries* **2023**, *9*, 244. [CrossRef]
12. Li, Z.; Chernova, N.; Roppolo, M.; Upreti, S.; Petersburg, C.; Alamgir, F.; Whittingham, M. Comparative Study of the Capacity and Rate Capability of LiNi y Mn y Co1−2y O2 (y = 0.5, 0.45, 0.4, 0.33). *J. Electrochem. Soc.* **2011**, *158*, A516. [CrossRef]
13. Kim, M.-H.; Shin, H.-S.; Shin, D.; Sun, Y.-K. Synthesis and electrochemical properties of Li[Ni0.8Co0.1Mn0.1]O2 and Li[Ni0.8Co0.2]O2 via co-precipitation. *J. Power Sources* **2006**, *159*, 1328–1333. [CrossRef]
14. Wu, F.; Wang, M.; Su, Y.; Bao, L.; Chen, S. Surface of $LiCo_{1/3}Ni_{1/3}Mn_{1/3}O_2$ modified by CeO_2-coating. *Electrochim. Acta* **2009**, *54*, 6803–6807. [CrossRef]
15. McNulty, R.C.; Hampson, E.; Cutler, L.N.; Grey, C.P.; Dose, W.M.; Johnson, L.R. Understanding the limits of Li-NMC811 half-cells. *J. Mater. Chem. A* **2023**, *11*, 18302–18312. [CrossRef]
16. Bianchini, M.; Roca-Ayats, M.; Hartmann, P.; Brezesinski, T.; Janek, J. There and Back Again-The Journey of $LiNiO_2$ as a Cathode Active Material. *Angew. Chem. (Int. Ed. Engl.)* **2019**, *58*, 10434–10458. [CrossRef]
17. Ceder, G.; Doyle, M.; Arora, P.; Fuentes, Y. Computational Modeling and Simulation for Rechargeable Batteries. *MRS Bull.* **2002**, *27*, 619–623. [CrossRef]

18. Bhatt, M.D.; O'Dwyer, C. Recent progress in theoretical and computational investigations of Li-ion battery materials and electrolytes. *Phys. Chem. Chem. Phys.* **2015**, *17*, 4799–4844. [CrossRef]
19. Fallahzadeh, R.; Farhadian, N. Molecular dynamics simulation of lithium ion diffusion in $LiCoO_2$ cathode material. *Solid State Ion.* **2015**, *280*, 10–17. [CrossRef]
20. Haruyama, J.; Sodeyama, K.; Tateyama, Y. Cation Mixing Properties toward Co Diffusion at the $LiCoO_2$ Cathode/Sulfide Electrolyte Interface in a Solid-State Battery. *ACS Appl. Mater. Interfaces* **2017**, *9*, 286–292. [CrossRef]
21. Hüger, E.; Schmidt, H. Li Chemical and Tracer Diffusivities in LiCoO2 Sintered Pellets. *Batteries* **2024**, *10*, 446. [CrossRef]
22. Islam, M.S.; Fisher, C.A.J. Lithium and sodium battery cathode materials: Computational insights into voltage, diffusion and nanostructural properties. *Chem. Soc. Rev.* **2014**, *43*, 185–204. [CrossRef]
23. Urban, A.; Seo, D.-H.; Ceder, G. Computational understanding of Li-ion batteries. *NPJ Comput. Mater.* **2016**, *2*, 1126. [CrossRef]
24. van der Ven, A.; Bhattacharya, J.; Belak, A.A. Understanding Li diffusion in Li-intercalation compounds. *Acc. Chem. Res.* **2013**, *46*, 1216–1225. [CrossRef]
25. Amin, R.; Chiang, Y.-M. Characterization of Electronic and Ionic Transport in Li 1-x Ni 0.33 Mn 0.33 Co 0.33 O 2 (NMC 333) and Li 1-x Ni 0.50 Mn 0.20 Co 0.30 O 2 (NMC 523) as a Function of Li Content. *J. Electrochem. Soc.* **2016**, *163*, A1512–A1517. [CrossRef]
26. Wu, S.-L.; Zhang, W.; Song, X.; Shukla, A.K.; Liu, G.; Battaglia, V.; Srinivasan, V. High Rate Capability of Li(Ni1/3Mn1/3Co1/3)O_2 Electrode for Li-Ion Batteries. *J. Electrochem. Soc.* **2012**, *159*, A438. [CrossRef]
27. Li, X.; Liu, J.; Banis, M.N.; Lushington, A.; Li, R.; Cai, M.; Sun, X. Atomic layer deposition of solid-state electrolyte coated cathode materials with superior high-voltage cycling behavior for lithium ion battery application. *Energy Environ. Sci.* **2014**, *7*, 768–778. [CrossRef]
28. Zheng, W.; Shui, M.; Shu, J.; Gao, S.; Xu, D.; Chen, L.; Feng, L.; Ren, Y. GITT studies on oxide cathode $LiNi_{1/3}Co_{1/3}Mn_{1/3}O_2$ synthesized by citric acid assisted high-energy ball milling. *Bull. Mater. Sci.* **2013**, *36*, 495–498. [CrossRef]
29. Charbonneau, V.; Lasia, A.; Brisard, G. Impedance studies of Li+ diffusion in nickel manganese cobalt oxide (NMC) during charge/discharge cycles. *J. Electroanal. Chem.* **2020**, *875*, 113944. [CrossRef]
30. Kang, S.D.; Kuo, J.J.; Kapate, N.; Hong, J.; Park, J.; Chueh, W.C. Galvanostatic Intermittent Titration Technique Reinvented: Part II. Experiments. *J. Electrochem Soc.* **2021**, *168*, 120503. [CrossRef]
31. Tsai, P.-C.; Wen, B.; Wolfman, M.; Choe, M.-J.; Pan, M.S.; Su, L.; Thornton, K.; Cabana, J.; Chiang, Y.-M. Single-particle measurements of electrochemical kinetics in NMC and NCA cathodes for Li-ion batteries. *Energy Environ. Sci.* **2018**, *11*, 860–871. [CrossRef]
32. Mehrer, H. *Diffusion in Solids. Fundamentals, Methods, Materials, Diffusion-Controlled Processes*, 1st ed.; Springer Series in Solid-State Sciences; Springer: Berlin/Heidelberg, Germany; New York, NY, USA, 2007; Volume 155, ISBN 978-3-540-71486-6.
33. Maier, J. *Physical Chemistry of Ionic Materials: Ions and Electrons in Solids*, 1st ed.; John Wiley & Sons: Chichester, UK, 2004; Print ISBN 9780471999911, Online ISBN 9780470020227. [CrossRef]
34. Maier, J. On the correlation of macroscopic and microscopic rate constants in solid state chemistry. *Solid State Ion.* **1998**, *112*, 197–228. [CrossRef]
35. Maier, J. Mass Transport in the Presence of Internal Defect Reactions-Concept of Conservative Ensembles: I, Chemical Diffusion in Pure Compounds. *J. Am. Ceram. Soc.* **1993**, *76*, 1212–1217. [CrossRef]
36. Hüger, E.; Schmidt, H. Lithium tracer diffusion in ion-beam sputtered nano-crystalline and amorphous $LiNi_{0.33}Mn_{0.33}Co_{0.33}O_2$ films. *Solid State Ion.* **2024**, *417*, 116702. [CrossRef]
37. Cho, S.; Uddin, M.; Alaboina, P.; Han, S.; Nandasiri, M.; Choi, Y.; Hu, E.; Nam, K.; Schwarz, A.; Nune, S.; et al. Exploring Lithium Deficiency in Layered Oxide Cathode for Li-Ion Battery. *Adv. Sustain. Syst.* **2017**, *1*, 1700026. [CrossRef]
38. Cho, S.; Chung, C.; Podowitz-Thomas, S.; Jones, J. Understanding the lithium deficient LixNiyMnzCo1-y-zO2 (x < 1) cathode materials structure. *Mater. Chem. Phys.* **2019**, *228*, 32–36. [CrossRef]
39. Lu, X.; Sun, Y.; Jian, Z.; He, X.; Gu, L.; Hu, Y.; Li, H.; Wang, Z.; Chen, W.; Duan, X.; et al. New Insight into the Atomic Structure of Electrochemically Delithiated O3-$Li_{(1-x)}CoO_2$ ($0 \leq x \leq 0.5$) Nanoparticles. *Nano Lett.* **2012**, *12*, 6192–6197. [CrossRef]
40. Arai, H. Reversibility of $LiNiO_2$ cathode. *Solid State Ion.* **1997**, *95*, 275–282. [CrossRef]
41. Dahn, J.R.; von Sacken, U.; Michal, C.A. Structure and electrochemistry of $Li_1 \pm yNiO_2$ and a new Li_2NiO_2 phase with the $Ni(OH)_2$ structure. *Solid State Ion.* **1990**, *44*, 87–97. [CrossRef]
42. Rossen, E.; Reimers, J.; Dahn, J. Synthesis and electrochemistry of spinel LT-$LiCoO_2$. *Solid State Ion.* **1993**, *62*, 53–60. [CrossRef]
43. Antaya, M.; Dahn, J.; Preston, J.; Rossen, E.; Reimers, J. Preparation and Characterization of $LiCoO_2$ Thin Films by Laser Ablation Deposition. *J. Electrochem. Soc.* **1993**, *140*, 575–578. [CrossRef]
44. Antaya, M.; Cearns, K.; Preston, J.; Reimers, J.; Dahn, J. In situ growth of layered, spinel, and rock-salt $LiCoO_2$ by laser ablation deposition. *J. Appl. Phys.* **1994**, *76*, 2799–2806. [CrossRef]

45. Whitfield, P.S.; Davidson, I.J.; Cranswick, L.M.D.; Swainson, I.P.; Stephens, P.W. Investigation of possible superstructure and cation disorder in the lithium battery cathode material $LiMn_{1/3}Ni_{1/3}Co_{1/3}O_2$ using neutron and anomalous dispersion powder diffraction. *Solid State Ion.* **2005**, *176*, 463–471. [CrossRef]
46. Shaju, K.M.; Bruce, P.G. Macroporous $Li(Ni_{1/3}\ Co_{1/3}Mn_{1/3})O_2$: A high-rate positive electrode for rechargeable lithium batteries. *J. Power Sources* **2007**, *174*, 1201–1205. [CrossRef]
47. Zhang, X.; Jiang, W.J.; Mauger, A.; Qilu; Gendron, F.; Julien, C. Minimization of the cation mixing in $Li1{+}x(NMC)1{-}xO_2$ as cathode material. *J. Power Sources* **2010**, *195*, 1292–1301. [CrossRef]
48. Wang, D.; Kou, R.; Ren, Y.; Sun, C.; Zhao, H.; Zhang, M.; Li, Y.; Huq, A.; Ko, J.; Pan, F.; et al. Synthetic Control of Kinetic Reaction Pathway and Cationic Ordering in High-Ni Layered Oxide Cathodes. *Adv. Mater.* **2017**, *29*, 1606715. [CrossRef]
49. Sun, G.; Yin, X.; Yang, W.; Song, A.; Jia, C.; Yang, W.; Du, Q.; Ma, Z.; Shao, G. The effect of cation mixing controlled by thermal treatment duration on the electrochemical stability of lithium transition-metal oxides. *Phys. Chem. Chem. Phys.* **2017**, *19*, 29886. [CrossRef]
50. Reynaud, M.; Casas-Cabanas, M. Order and disorder in NMC layered materials: A FAULTS simulation analysis. *Powder Diffr.* **2017**, *32*, S213–S220. [CrossRef]
51. Xiao, Y.; Liu, T.; Liu, J.; He, L.; Chen, J.; Zhang, J.; Luo, P.; Lu, H.; Wang, R.; Zhu, W.; et al. Insight into the origin of lithium/nickel ions exchange in layered Li(NixMnyCoz)O2 cathode materials. *Nano Energy* **2018**, *49*, 77–85. [CrossRef]
52. Kim, Y. Theoretical investigation of the cation antisite defect in layer-structured cathode materials for Li-ion batteries. *Phys. Chem. Chem. Phys.* **2019**, *21*, 24139. [CrossRef] [PubMed]
53. Zheng, J.; Ye, Y.; Liu, T.; Xiao, Y.; Wang, C.; Wang, F.; Pan, F. Ni/Li Disordering in Layered Transition Metal Oxide: Electrochemical Impact, Origin, and Control. *Acc. Chem. Res.* **2019**, *52*, 2201–2209. [CrossRef] [PubMed]
54. Park, H.; Mesnier, A.; Lee, S.; Jarvis, K.; Manthiram, A.; Warner, J.H. Intrinsic Li Distribution in Layered Transition-Metal Oxides Using Low-Dose Scanning Transmission Electron Microscopy and Spectroscopy. *Chem. Mater.* **2021**, *33*, 4638–4650. [CrossRef]
55. Orlova, E.D.; Savina, A.A.; Abakumov, S.A.; Morozov, A.V.; Abakumov, A.M. Comprehensive Study of Li+/Ni2+ Disorder in Ni-Rich NMCs Cathodes for Li-Ion Batteries. *Symmetry* **2021**, *13*, 1628. [CrossRef]
56. Boev, A.O.; Fedotov, S.S.; Abakumov, A.M.; Stevenson, K.J.; Henkelman, G.; Aksyonov, D.A. The role of antisite defect pairs in surface reconstruction of layered AMO2 oxides: A DFT+U study. *Appl. Surf. Sci.* **2021**, *537*, 147750. [CrossRef]
57. Wei, H.; Tang, L.; Huang, Y.; Wang, Z.; Luo, Y.; He, Z.; Yan, C.; Mao, J.; Dai, K.; Zheng, J. Comprehensive understanding of Li/Ni intermixing in layered transition metal oxides. *Mater. Today* **2021**, *51*, 365–392. [CrossRef]
58. Aksyonov, D.A.; Boev, A.O.; Fedotov, S.S.; Abakumov, A.M. Computational insights into ionic conductivity of transition metal electrode materials for metal-ion batteries—A review. *Solid State Ion.* **2023**, *393*, 116170. [CrossRef]
59. Tahmasebi, M.; Obrovac, M.N. Quantitative Measurement of Compositional Inhomogeneity in NMC Cathodes by X-ray Diffraction. *J. Electrochem. Soc.* **2023**, *170*, 080519. [CrossRef]
60. Fan, F.; Zeng, R.; Zheng, T.; Xu, H.; Wen, X.; Wang, X.; Tian, G.; Wang, S.; Zheng, C.; Wei, X.; et al. Cation-ordered Ni-rich positive electrode material with superior chemical and structural stability enabled by atomic substitution for lithium-ion batteries. *Chem. Eng. J.* **2023**, *477*, 147181. [CrossRef]
61. Ben-Barak, I.; Obrovac, M.N. All-Dry Synthesis of NMC from [Ni,Mn,Co]3O4 Spinel Precursors. *J. Electrochem. Soc.* **2024**, *171*, 040535. [CrossRef]
62. Zeng, C.; Fan, F.; Zheng, R.; Wang, X.; Tian, G.; Liu, S.; Liu, P.; Wang, C.; Wang, S.; Shu, C. Structure and Charge Regulation Strategy Enabling Superior Cycling Stability of Ni-Rich Cathode Materials. *ACS Appl. Mater. Interfaces* **2024**, *16*, 11377–11388. [CrossRef]
63. Zeng, C.; Zheng, R.; Fan, F.; Wang, X.; Tian, G.; Liu, S.; Liu, P.; Wang, C.; Wang, S.; Shu, C. Phase compatible surface engineering to boost the cycling stability of single-crystalline Ni-rich cathode for high energy density lithium-ion batteries. *Energy Storage Mater.* **2024**, *72*, 103788. [CrossRef]
64. Lei, X.; Zhao, J.; Wang, J.; Su, D. Tracking lithiation with transmission electron microscopy. *Sci. China Chem.* **2024**, *67*, 291–311. [CrossRef]
65. Ceder, G.; Van der Ven, A. Phase diagrams of lithium transition metal oxides: Investigations from first principles. *Electrochim. Acta* **1999**, *45*, 131–150. [CrossRef]
66. Sharifi-Asl, S.; Lu, J.; Amine, K.; Shahbazian-Yassar, R. Oxygen Release Degradation in Li-Ion Battery Cathode Materials: Mechanisms and Mitigating Approaches. *Adv. Energy Mater.* **2019**, *9*, 1900551. [CrossRef]
67. Cho, Y.; Lee, S.; Lee, Y.; Hong, T.; Cho, J. Spinel-Layered Core-Shell Cathode Materials for Li-Ion Batteries. *Adv. Energy Mater.* **2011**, *1*, 821. [CrossRef]
68. Qiu, B.; Zhang, M.; Wu, L.; Wang, J.; Xia, Y.; Qian, D.; Liu, H.; Hy, S.; Chen, Y.; An, K.; et al. Gas–solid interfacial modification of oxygen activity in layered oxide cathodes for lithium-ion batteries. *Nat. Commun.* **2016**, *7*, 12108. [CrossRef] [PubMed]
69. Crank, J. *The Mathematics of Diffusion*; Oxford University Press: Oxford, UK, 1979.

70. Van der Ven, A.; Ceder, G. Lithium Diffusion in Layered LixCoO2. *Electrochem. Solid-State Lett.* **2000**, *3*, 301. [CrossRef]
71. Hoang, K.; Johannes, M.D. Defect chemistry in layered transition-metal oxides from screened hybrid density functional calculations. *J. Mater. Chem. A* **2014**, *2*, 5224–5235. [CrossRef]

Disclaimer/Publisher's Note: The statements, opinions and data contained in all publications are solely those of the individual author(s) and contributor(s) and not of MDPI and/or the editor(s). MDPI and/or the editor(s) disclaim responsibility for any injury to people or property resulting from any ideas, methods, instructions or products referred to in the content.

batteries

MDPI

Article

Ternary PEO/PVDF-HFP-Based Polymer Electrolytes for Li-Ion Batteries

Hoang Bao Tran Nguyen *, Ling Ding, Björn Pohle, Toni Schmeida, Hoang Bao An Nguyen and Daria Mikhailova *

Leibniz Institute for Solid State and Materials Research (IFW) Dresden e.V., Helmholtzstraße 20, 01069 Dresden, Germany

* Correspondence: h.b.t.nguyen@ifw-dresden.de (H.B.T.N.); d.mikhailova@ifw-dresden.de (D.M.)

Abstract: The impetus to study and develop polymer electrolytes for metal-ion batteries is due to their enhanced safety compared to flammable organic liquid electrolytes, promising ionic conductivity, and broad electrochemical stability window, making them to viable candidates for battery application. In the current work, we present a simple fabrication procedure and a comprehensive physico–chemical study of various PVDF-HFP-based electrolyte formulations with a sufficient addition of PEO polymer, LiTFSI conducting salt, and EMIMTFSI ionic liquid. The ionic conductivity, activation energy for ionic movement and thickness of the resulting polymer electrolyte show a non-linear dependency on the PVDF-HFP/PEO ratio. The electrolyte composition with a 0.35PEO-0.65PVDF-HFP/1LiTFSI/1EMIMTFSI mass fraction exhibits the highest ionic conductivity among the compositions, revealing 7.7×10^{-5} S cm^{-1} at 30 °C. Electrochemical tests in half full and full Li-ion batteries with a $LiFePO_4$ cathode and $Li_4Ti_5O_{12}$ anode also emphasized this composition as the most promising one, providing an initial capacity in full cells of 120 mAh g^{-1} and a capacity retention of about 75% after 50 charge/discharge cycles at a 0.1 C current rate. In the PEO/PVDF-HFP polymer blend with EMIMTFSI as a plasticizer, the amount of crystalline parts, which are detrimental to a fast ionic diffusion, is significantly reduced.

Keywords: polymer blend; EMIMTFSI plasticizer; casting method; lithium transport mechanism

1. Introduction

Rechargeable batteries are currently in high demand due to their high performance and sustainability, e.g., for use in mobile and portable electronic devices and electric vehicles. Lithium-ion batteries (LIBs) are predominantly used for this purpose. The strong increase in demand for LIBs over the last decades necessitates improvement of electrode materials and electrolyte compositions. The electrolyte should not only be chemically stable over a wide voltage and temperature range, but should also have a high Li^+ conductivity and good compatibility with the electrode materials [1,2]. Besides the classical liquid electrolytes, which are mostly used at present for commercial batteries, the investigation and commercialisation of solid-state electrolytes, e.g., polymer electrolytes, is a new research field. The non-flammability of polymer electrolyte components, heat control, and thus the reduced risk of explosions represent significant advantages in comparison to liquid electrolytes. Furthermore, LIBs with polymer electrolytes can achieve a lower weight and a higher gravimetric energy density, as no additional separator is needed [1,3].

Lithium-polymer batteries are a special form of LIBs, in which the electrolyte is based on a lithium salt and a polymer or polymer blend. Polymer electrolytes used in

LIBs, have to fulfil the following conditions: electrical insulation, high ionic conductivity, high mechanical strength, a wide electrochemical stability window, and good chemical and thermal stability [4–6]. In recent years, numerous polymers, such as polyethylene oxide (PEO) [7–12], polyvinylidene fluoride (PVDF) [13] and its copolymer polyvinylidene fluoride-co hexafluoropropylene (PVDF-HFP) [14–17], polyacrylonitrile (PAN) [18], polymethyl methacrylate (PMMA) [19], polyvinylpyrrolidone (PVP) [20], and polyvinyl chloride (PVC) [21], have been investigated for the development of polymer electrolytes. All of them have their own advantages and disadvantages.

Based on the monomer ethylene oxide, PEO is a semi-crystalline polymer whose crystalline phase corresponds to 80% of the entire structure [22]. PEO is the most widely studied polymer for solid electrolyte applications due to its easy complexation with lithium salts, high dielectric constant, and strong solvating ability. In addition, PEO has a low glass transition temperature, good flexibility, and high mechanical stability [2,23,24]. However, the mostly crystalline nature of PEO is a major disadvantage, resulting in poorer conductivity at room temperature [22]. PVDF-HFP is a copolymer based on a crystalline vinylidene fluoride (-VDF) part and an amorphous hexafluoropropylene (-HFP) structure. The crystalline regions offer high mechanical stability, while the lithium ions in the amorphous phase can diffuse easily [25,26]. Furthermore, with a high dielectric constant (ε = 8.4) due to the strong electron-withdrawing -C-F group, PVDF-HFP enables completed dissociation of various lithium salts and, thus, a high concentration of the charge carrier [15,27]. In addition, PVDF-HFP can easily be dissolved in many solvents and has a comparatively low degree of crystallinity of about 29% at room temperature [23].

A combination of both polymers can facilitate their advantages and mitigate their disadvantages. A lot of studies on the blending of PEO and PVDF-HFP have been published over the last few years. Zhang et al. [28] prepared PVDF-HFP-based polymer electrolytes with PEO and SiO_2. Here, SiO_2 was added as an inorganic filler to increase the ionic conductivity and improve the thermal stability of polymer electrolytes. The ionic conductivity reached 1.12×10^{-3} S cm^{-1}, and the Li-cell containing $LiFePO_4$/gel polymer electrolyte/Li showed a high discharge capacity of 146.3 mAhg^{-1} at 0.1 C-rate (fully charging the cell in 10 h) after 200 cycles. In the work of Manjunatha et al. [29], the polymer blend matrix PEO/PVDF-HFP with 20 wt% LiTFSI salt demonstrated a lower ionic conductivity of 3.14×10^{-6} S cm^{-1}. To improve the ionic conductivity, this electrolyte system was irradiated with oxygen ions. Another method for enhancement of the ionic conductivity is adding a plasticiser. Prabakaran and Manimuthu [6] investigated polymer electrolytes based on a PEO/PVDF-HFP blend matrix with different lithium salts and propylene carbonate (PC) as the plasticiser. The highest ionic conductivity of 8.20×10^{-4} S cm^{-1} at 30 °C was found in the PEO (6.25 wt%)/PVDF-HFP (18.75 wt%)/PC (67 wt%)/LiTFSI (8 wt%) composition. However, the studies of both Manjunatha et al. [29] and Prabakaran [6] did not show any application of the polymer electrolyte in Li-ion batteries. During the search for a suitable conducting salt for polymer electrolyte systems, many known candidates were tested. According to the work of Ue et al. [30], lithium bis(trifluoromethanesulfonyl)-imide (LiTFSI) has the lowest mobility but the highest dissociation degree compared to $LiAsF_6$, $LiPF_6$, $LiClO_4$, $LiBF_4$, and LiTf. Furthermore, Kurzweil and Dietlmeier [31] established the following order of ionic conductivity based on data from the study [30] mentioned above: $LiPF_6$, $LiAsF_6$ > LiTFSI, $LiClO_4$ > $LiBF_4$ > LiTf. The salts with the highest conductivity are inapplicable for polymer electrolytes, since $LiAsF_6$ is very toxic [32] and $LiPF_6$ is extremely water-sensitive, irreversibly decomposing to hydrofluoric acid in presence of water [33–35]. Therefore, we chose LiTFSI as a conducting salt for the polymer electrolyte system. As a plasticiser to reduce the crystallinity of polymers and thereby to increase the ionic conductivity, 1-Ethyl-3-methylimidazolium bis(trifluoromethanesulfonyl)imide (EMIMTFSI) ionic liquid was selected, which contains the same anion $TFSI^-$ to avoid undesirable interactions in the electrolyte. EMIMTFSI exhibits high thermal stability (decomposition temperature above 400 °C) [36], a wide electrochemical stability window from 1 V to 5.3 V vs. Li^+/Li [37], and comparatively high ionic conductivity [38,39].

Ionic mobility occurs mostly in the amorphous phase of the host polymers. The transport mechanism of Li^+ cations in PEO-based polymer electrolytes has been intensively investigated, while the interaction of Li^+ with PVDF-HFP and the corresponding ionic transport are rarely described in the literature. Depending on the concentration of the lithium salt and type of the anion, different PEO_X:Li^+ complexes can be formed [9,40]. Here, a Lewis acidic Li^+ cation can be coordinated by five ether oxygen atoms of PEO as Lewis bases, thus reducing the mobility of Li^+. However, local segmental movement of the polymer chains is still possible, whereby the bonding of Li^+ with the oxygen atoms can be broken and Li^+ migrates to the other side of the same polymer chain (intrachain hopping) or into other polymer chains (interchain hopping) (Figure 1a) [2,41,42].

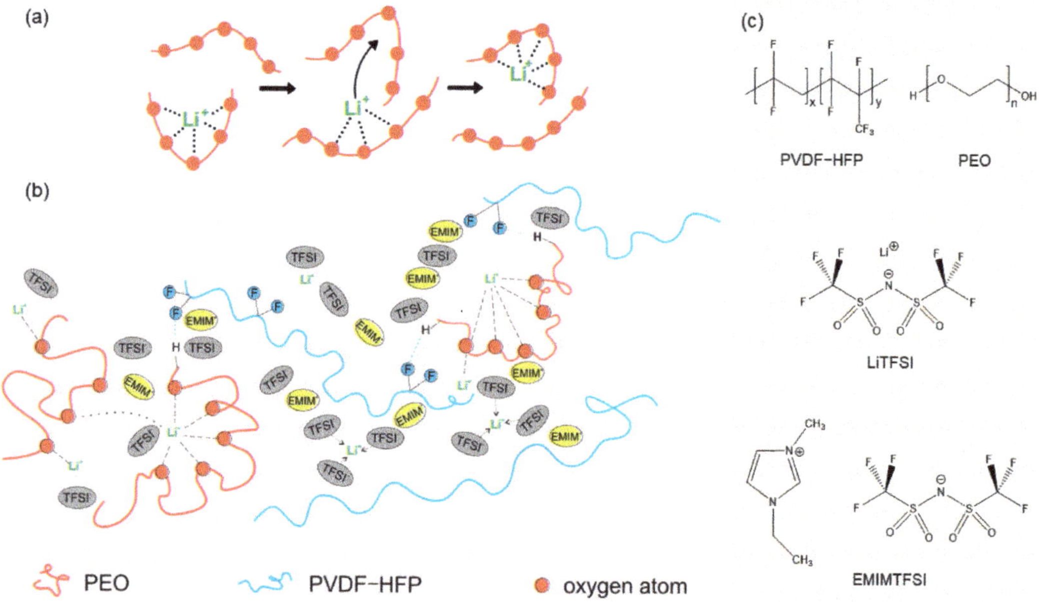

Figure 1. (**a**) Transport mechanism of lithium cations along PEO chains. The figure was adopted from Meyer [41]. (**b**) Schematic representation of lithium-ion transport in the ternary PVDF-HFP, PEO, LiTFSI, and EMIMTFSI polymer electrolyte system. (**c**) Structure of single components PVDF-HFP, PEO, LiTFSI, and EMIMTFSI.

In contrast to PEO, lithium ions are not directly coordinated by the PVDF-HFP polymer chains. On the one hand, due to the chemical inertness of PVDF-HFP, complex formation with lithium cations does not occur. Here, the -C-F group in the alkyl chains does not play a role as a Lewis base [3], and the interaction between F-anions and Li^+ is electrostatic in nature and cannot thoroughly compensate for the charge of Li cations. Therefore, some clustering of cations and anions of the salt is possible, since anions are needed for shielding the charge. Here, Li-ions can be at least partially coordinated by anions, which are also present in the ionic liquid. On the other hand, due to the weaker F–Li interaction, this inactive polymer can support a higher ionic conduction of Li^+ cations. This makes the polymer chains more flexible, which can lead to high segmental movement and subsequent high ionic conductivity [43].

Considering the structures of PEO and PVDF-HFP, an interaction between the two polymer chains can take place. In this case, the hydrogen atoms of PEO can interact with the fluorine atoms of PVDF-HFP, forming intermolecular hydrogen bonds. This results in a more chaotic structure of the polymer blend, which can destroy the crystalline phases and increase the amount of the amorphous phase [26]. Furthermore, ion transport is also affected by the ionic liquid (IL). On the one hand, the electrolyte will be diluted by the addition of IL, resulting in a slight deterioration of the lithium cations' ability to move between the polymer chains. On the other hand, the segmental movement of the polymer chains will be enhanced, improving the overall diffusivity of the lithium ions [44]. Figure 1b shows a possible schematic representation of lithium cation transport in the ternary polymer electrolyte systems under investigation, with all the previously mentioned mechanisms shown together.

In this study, we prepared and characterised a ternary polymer electrolyte (tPE) system containing a PEO/PVDF-HFP polymer blend with LiTFSI and EMIMTFSI. Since the Li^+ coordination and transport mechanisms along both polymers are different, some non-linearity in the properties of the resulting polymer electrolyte is expected. The amounts of lithium salt and ionic liquid in the electrolyte were kept constant, while the mass ratio between PEO and PVDF-HFP was varied as follows: 25 wt%:75 wt%, 30 wt%:70 wt%, 35 wt%:65 wt%, 40 wt%:60 wt%, 45 wt%:55 wt%, and 50 wt%:50 wt%. This corresponds to 1.16 m LiTFSI concentration in the resulting electrolyte formulation. Since two components of the polymer blend have the same molecular weight, the variation in the composition does not affect the total mass of the polymer electrolyte. Various electrolyte compositions were tested in half Li-cells with $LiFePO_4$ (LFP) and $Li_4Ti_5O_{12}$ (LTO) working electrodes and metallic Li as a counter electrode, and in full cells with a LFP cathode and LTO anode. We clearly observed a positive impact of PEO addition to the polymer blend, which leads, however, to nonmonotoneous changes in properties. This polymer blend system represents a good candidate as a non-toxic, environmentally friendly and easy to fabricate material for applications in lithium-ion batteries.

2. Experimental Section

2.1. Preparation of the Ternary Polymer Electrolytes

The ternary polymer electrolytes were prepared using the casting method (Figure 2). The electrolytes consist of the following components: PEO (M_w = 400 000 g/mol (Sigma Aldrich, St. Louis, MO, USA), PVDF-HFP (M_w = 400 000 g/mol, Sigma Aldrich), conducting salt LiTFSI (>99%, Ossila, UK) and ionic liquid EMIMTFSI (>99%, Ionic Liquid Technologies, Heilbronn, Germany). The mass ratio between polymer, Li-salt and IL corresponded to 1:1:1. Depending on the selected composition, mass proportions of PEO and PVDF-HFP were varied. The prepared and investigated compositions of the ternary polymer electrolytes (**y**PEO/**z**PVDF-HFP):LiTFSI:EMITFSI (y, z in wt-%) are listed in Table 1. Both polymers were dissolved separately in acetone (99%, VWR Chemicals, Darmstadt, Germany), with PEO representing a viscous solution at 40 °C, and PVDF-HFP at room temperature. The Li-salt and IL were then added and stirred at 40 °C until a homogeneous viscous mixture was obtained.

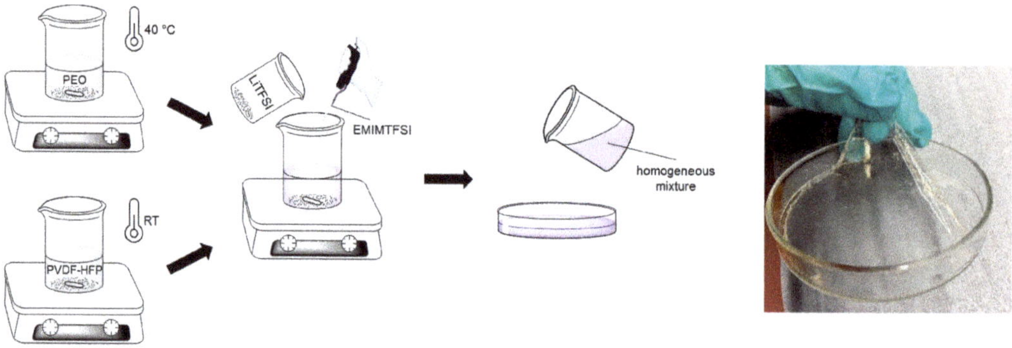

Figure 2. Fabrication scheme of ternary polymer electrolytes using the casting method.

Table 1. Electrolyte compositions and their battery cell designations used in this study. In all compositions, a 1.16 m LiTFSI concentration was realised.

Electrolyte Composition			Name
25% PEO/75% PVDF-HFP	1 LiTFSI	1 EMITFSI	tPE $PEO_{25}PVDF\text{-}HFP_{75}$
30% PEO/70% PVDF-HFP			tPE $PEO_{30}PVDF\text{-}HFP_{70}$
35% PEO/65% PVDF-HFP			tPE $PEO_{35}PVDF\text{-}HFP_{65}$
40% PEO/60% PVDF-HFP			tPE $PEO_{40}PVDF\text{-}HFP_{60}$
45% PEO/55% PVDF-HFP			tPE $PEO_{45}PVDF\text{-}HFP_{55}$
50% PEO/50% PVDF-HFP			tPE $PEO_{50}PVDF\text{-}HFP_{50}$

2.2. Preparation of the Electrodes

For the electrode mixture, 400 mg active material $LiFePO_4$ (LFP, Phostech, Elmhurst, IL, USA) or $Li_4Ti_5O_{12}$ (LTO, >99%, Sigma Aldrich), 50 mg carbon SUPER C65 (TIMCAL, Bodio, Switzerland), and 50 mg binder PVDF (Solvay Speciality Polymers, Alorton, IL, USA) were first mixed and finely ground. The mass ratio corresponded to 80 wt% active material, 10 wt% carbon and 10 wt% PVDF. The finely ground mixture was then suspended for 30 min in a vibration mill from the manufacturer Retsch® (Haan, Germany) at a frequency of 20 Hz with 3 mL of solvent *N*-methyl-2-pyrrolidone (NMP, 99%, Sigma Aldrich). The suspension was then poured onto an aluminium foil of about DIN A4 size for LFP or a copper foil for LTO. The thickness for the electrodes was set to 200 µm, and a squeegee was used to flatten the suspension and to air-dry it for one day. The next day, the moisture was removed from the coated foils overnight at 80 °C in a drying oven. Before the battery cells were assembled, the electrodes were punched out with a diameter of 12 mm and further dried in the oven in a glovebox for one night at 80 °C.

2.3. Characterisation Techniques

The X-ray diffraction (XRD) measurements were performed using a STADI P diffractometer from STOE in a transmission geometry. The X-ray used corresponded to Cu-$K\alpha_1$ radiation with a wavelength of 1.54060 Å. The diffractograms were recorded at 2θ angles, which varied from 5° to 90°. The morphology of the ternary polymer electrolytes was analysed using a scanning electron microscope (SEM) (Leo Gemini 1530, Carl Zeiss AG, Oberkochen, Germany) at an acceleration voltage of 5 kV with an SE2 detector. The optical microscopy images were recorded using a digital microscope VHX-7000 from KEYENCE (KEYENCE DEUTSCHLAND GmbH). The Fourier-transform infrared spectroscopy (FTIR) measurements were performed using a Nicolet 6700 (Thermo Fischer Scientific, Dreieich, Germany) in the wavenumber range of 4000–500 cm^{-1}. Thermogravimetry (TG) measurements and differential thermal analysis (DTA) were carried out using a STA 449 C Jupiter device from NETZSCH (Selb, Germany). The measurements were performed in a nitrogen flow with a heating rate of 10 $Kmin^{-1}$ in the temperature range from room temperature to 600 °C. The thickness of polymer electrolytes was determined using a digital micrometre. For each polymer electrolyte, the thickness was measured at least four different positions.

2.4. Electrochemical Tests

Galvanostatic cycling with potential limitation (GCPL) tests were carried out using battery coin cells of the CR2025 type (Gelon, China). The cell assembly took place under an argon atmosphere in a glove box, with the H_2O and O_2 content less than 0.1 ppm. Ternary polymer electrolytes were used as a separator. Lithium chips (ChemPUR, Karlsruhe, Germany) with a diameter of 12 mm and a thickness of 250 μm were used as the anode material in half cells. All measurements for the half cells of Li/LFP in a voltage range of 2.0–4.1 V (vs. Li^+/Li), and Li/LTO 1.0–3.0 V (vs. Li^+/Li) were performed using a Biologic VMP3 potentiostat in a temperature-controlled climate chamber at 25 °C. The full cells of LTO/LFP were tested on a LAND potentiostat from 0.5–3.0 V. Here, the C-rate for all GCPL measurements was 0.1 C. The current required to fully charge or discharge the battery within one hour is defined as 1 C. The mass balancing of both active materials in the full cells was LFP:LTO = ~1:1.2 (*w*/*w*), and for all GCPL measurements, and the specific capacities were calculated based on the mass of LFP. Cyclic voltammetry (CV) measurements for determination of the stability window for polymer electrolytes were carried out in different potential ranges, depending on materials, with a scan rate of 0.1 mVs^{-1}. For the potential range from 0 V to 5 V, an asymmetrical cell Al/tPE/Li and a symmetrical cell with stainless steel (SS) were assembled. The asymmetrical SS/tPE/Li cell was measured in the potential range from −2 V to 5 V. The electrochemical impedance spectroscopy (EIS) measurements for calculations of ionic conductivity were performed using an AUTOLAB potentiostat from the manufacturer RHD Instruments (Darmstadt, Germany). The measuring temperature was in the range of −20 °C to 80 °C, with a temperature interval of 10 °C. At the end of the experiments, the impedance spectrum was recorded again at 20 °C to control for possible changes in characteristics during measurements. The applied frequencies were varied from 0.1 Hz to 1 MHz and the potential amplitude was 100 mV. To determine the lithium transference number, a symmetrical cell configuration with lithium chips was assembled. Potentiostatic polarisation tests were conducted using chronoamperometry (CA), with an applied voltage of 10 mV for 6 h. EIS was performed before and after the polarisation, utilizing a frequency range of 0.01 Hz to 1 MHz and a potential amplitude of 10 mV.

3. Results and Discussion

3.1. Structure Characterisations

The XRD patterns of pure polymers, their mixture in a 1:1 ratio, fabricated using acetone solution, the conducting salt LiTFSI, mixtures of PEO/LiTFSI/EMIMTFSI (1:1:1), and PVDF-HFP/LiTFSI/EMIMTFSI (1:1:1), and the prepared ternary polymer electrolytes are shown in Figure 3a. Both PEO and PVDF-HFP are semi-crystalline materials in ambient conditions, showing the typical significant crystalline peaks of pure PEO at 19.2° and 23.4°, and of pure PVDF-HFP at 18.4° and 19.9° [24]. The crystalline nature of LiTFSI salt is also confirmed. After mixing of PEO and PVDF-HFP in acetone, followed by evaporation, the crystallinity of PEO is preserved, while reflections of PVDF-HFP drastically decreased, pointing to its pronounced amorphisation. Both the mixtures of PEO/LiTFSI/EMIMTFSI (1:1:1) and PVDF-HFP/LiTFSI/EMIMTFSI (1:1:1), with two broad peaks at 13° and 20°, reflect amorphous behaviour.

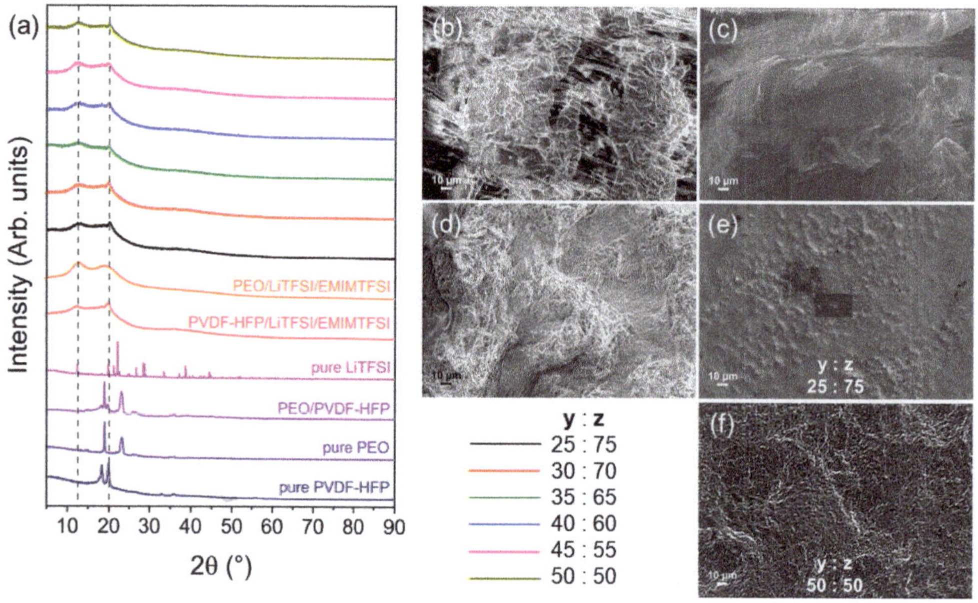

Figure 3. (**a**) XRD patterns of pure components PVDF-HFP, PEO, LTFSI, polymer blend PEO/PVDF-HFP (1:1), mixtures of PEO/LiTFSI/EMIMTFSI (1:1:1) and PVDF-HFP/LiTFSI/EMIMTFSI (1:1:1), and ternary polymer electrolytes tPE $PEO_yPVDF\text{-}HFP_z$, recorded with Cu-Kα_1 radiation; (**b**–**f**) SEM pictures of: (**b**) pure PEO, (**c**) pure PVDF-HFP, (**d**) polymer blend PEO/PVDF-HFP (1:1), (**e**) tPE $PEO_{25}PVDF\text{-}HFP_{75}$, (**f**) tPE $PEO_{50}PVDF\text{-}HFP_{50}$.

After ternary electrolyte preparation, the obtained structure is also almost amorphous for all compositions, with two broad peaks at around 13° and 20°. The small crystalline peak at $2\theta = 19.9°$, probably of PVDF-HFP, appears in the electrolyte systems, but the intensity is significantly reduced compared to the pure polymer. Moreover, the crystalline peaks of Li salt, observed at about 30°–50° in the diffraction pattern, and the crystalline peaks of PEO are not visible in the XRD patterns of the polymer electrolytes. This confirms that the composition of both polymers and the addition of the ionic liquid leads to the suppression of the crystallinity.

The surface of the pure polymers, polymer blend (1:1 mass ratio in the polymer matrix), prepared by casting the mixture with acetone and drying at room temperature, and ternary polymer electrolytes were investigated by SEM at 500 times magnification (Figure 3b–f). PEO shows a mosaic/fibre-like structure (Figure 3b), while a smooth structure is observed for PVDF-HFP (Figure 3c). Thus, many polymer fibres appear in the polymer blend matrix due to the presence of a significant amount of PEO (Figure 3d). Accordingly, tPE $PEO_{25}PVDF\text{-}HFP_{75}$ demonstrates a smoother surface than the surface of tPE $PEO_{50}PVDF\text{-}HFP_{50}$ due to the high amount of the PEO phase at room temperature. In the presence of LiTFSI and EMIMTFSI, the surface morphology of tPE reflects a more homogeneous structure compared to the PEO/PVDF-HFP polymer blend.

The interactions in the polymer electrolyte can be followed up using FT-IR spectroscopy, with the intensities and shifts of spectral features being interpreted. Figure 4 shows the measured vibrational spectra for all investigated polymer electrolyte compositions: ionic liquid EMIMTFSI, the lithium salt LiTFSI, pure polymers, and the polymer mixture in the wavenumber range of 4000–500 cm^{-1}.

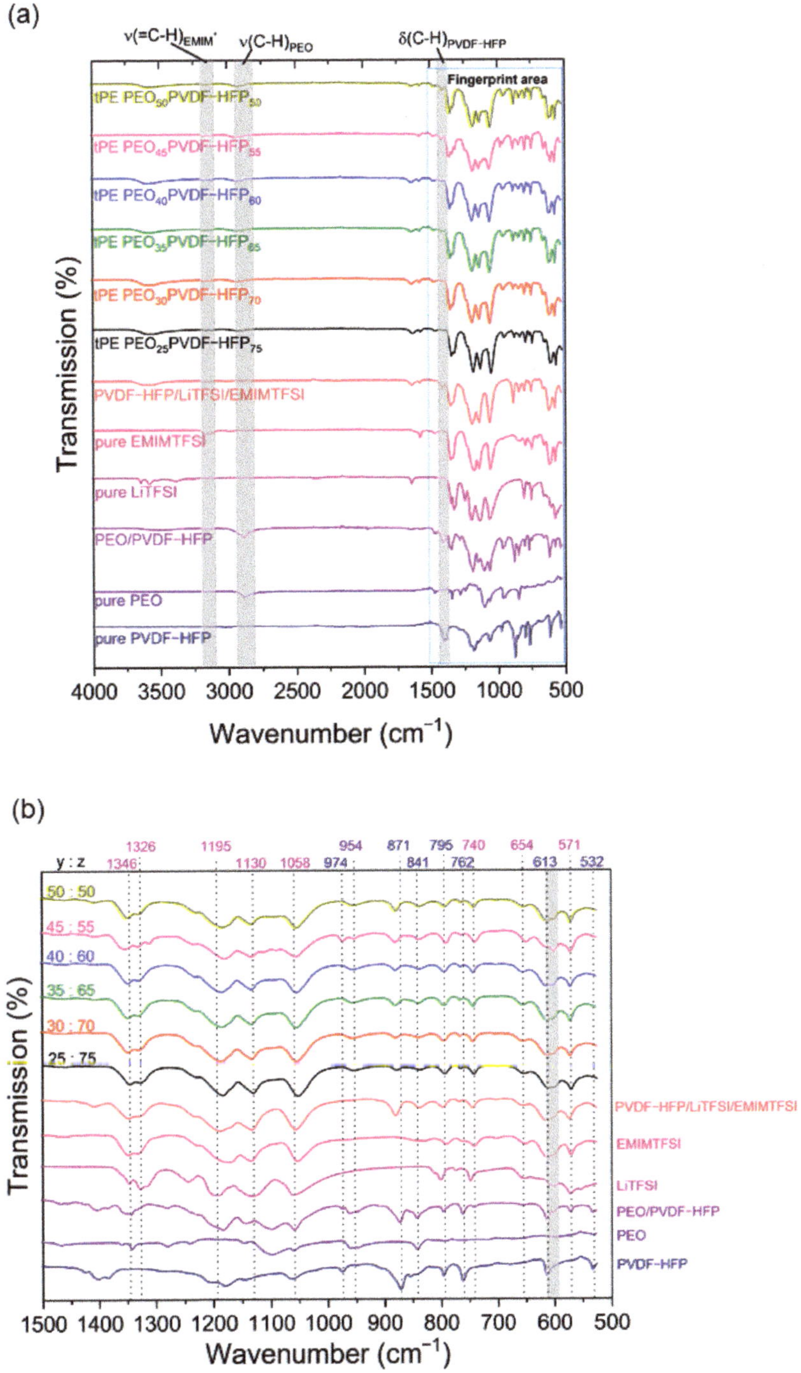

Figure 4. FTIR spectra of tPE $PEO_yPVDF\text{-}HFP_z$, EMITFSI, LiTFSI, PEO/PVDF HFP polymer blend with the mass ratio of 1:1, PVDF-HFP/LiTFSI/EMIMTFSI (1:1:1) and individual PEO and PVDF-HFP from 4000–500 cm^{-1} (**a**), and in the fingerprint region (**b**).

A pronounced band at 2886 cm^{-1} represents the C-H stretching vibration of PEO [45]. This band appears at 2882 cm^{-1} in the polymer blend and is shifted to 2920, 2922, 2925, 2927, 2933, 2927 cm^{-1} in the ternary polymer electrolyte systems with increasing proportion of PEO. Other significant bands at 1359, 1342, and 1100 cm^{-1} define CH_2 wagging, CH_2 bending, and symmetric and asymmetric C-O-C stretching vibration of PEO [29,46]. These bands are not observed in the tPE spectra. The C-O stretching vibration, located at 952 cm^{-1}, shifts to higher wavenumbers (blue shift), while the red shift occurs in CH_2 rocking vibration, located at 841 cm^{-1}.

The slightly visible asymmetric and symmetric stretching vibrations of the C-H bond of PVDF-HFP are subsequently located at 3024 and 2985 cm^{-1} [43], and the deformation vibration is observed at 1400 cm^{-1} [16]. These bands are absent in the tPE spectra. Furthermore, the characteristic vibrational bands of the crystalline phase (α-phase) of PVDF-HFP are assigned at 974, 795, 762, 613 and 532 cm^{-1}, whereas the bands at 871 cm^{-1} and 841 cm^{-1} identify the amorphous phase (β-phase) of PVDF-HFP. All characteristic bands for the crystalline phase are also absent in ternary polymer electrolyte systems, due to the interactions of all components in the electrolyte and amorphization of crystalline domains.

The bands of the C-H stretching vibration of the imidazolium ring are visible in the range of 3200–3100 cm^{-1} [43]. The bands at 1346 cm^{-1} and 1326 cm^{-1} show the asymmetric SO_2 stretching vibration and C-SO_2-N vibration of the LiTFSI and EMIMTFSI. Furthermore, the band at 1193 cm^{-1} represents the symmetric stretching vibration of the CF_3 group of LiTFSI and the C-H vibration of the imidazolium ring of EMIMTFSI, which shift to red in the spectra of the polymer electrolytes. The significant band at 1133 cm^{-1} shows the C-SO_2-N vibration in $TFSI^-$ anions of LiTFSI and EMITFSI, while the asymmetric S-N-S stretching vibration of the conducting salt and IL is assigned at 1051 cm^{-1} [43]. For more information, other vibration bands are listed in Table S1 in the Supporting Information.

When only PVDF-HFP is mixed with LiTFSI and EMIMTFSI, the resulting spectrum is dominated by features of EMIMTFSI and LiTFSI, with only a tiny contribution of PVDF-HFP. The mixture of PEO with LiTFSI and EMIMTFSI (1:1:1), representing a very viscous and glutinous liquid, was not suitable for measurements.

The IR spectrum of the polymer blend results in a combination of spectra of the individual polymers. The intensities of these bands are either absent or decrease strongly in the presence of the conducting salt and the ionic liquid. This could be explained by possible interactions of the lithium ions and/or the cations of the ionic liquid with the polymer chains taking place. Furthermore, most of the vibrational bands in the spectra of the ternary polymer electrolytes show broadening in comparison to the individual components. It can be assumed that interactions of the Li^+ and $EMIM^+$ cations with the polymer chains also take place. Therefore, the excitation of the vibrations is impeded. A possible explanation for the band shift is the coordination of the PEO chains with the lithium ions and the complexation of the PVDF-HFP chains with the positively charged imidazolium ions. Here, the fluorine atoms, which have the highest electronegativity and a high electron density, are coordinated by the positively charged N-atoms of the imidazolium ring [47]. The complexation could lead to an increase in ionic conductivity. In addition, the crystalline phases of both polymers are suppressed when the ionic liquid is added, which also supports the ionic mobility.

Thermogravimetry and differential thermal analysis were performed to investigate the thermal properties of the ternary polymer electrolytes under consideration. Figure 5 shows the thermal behaviour of the polymer electrolytes in a temperature range from room temperature to 600 °C.

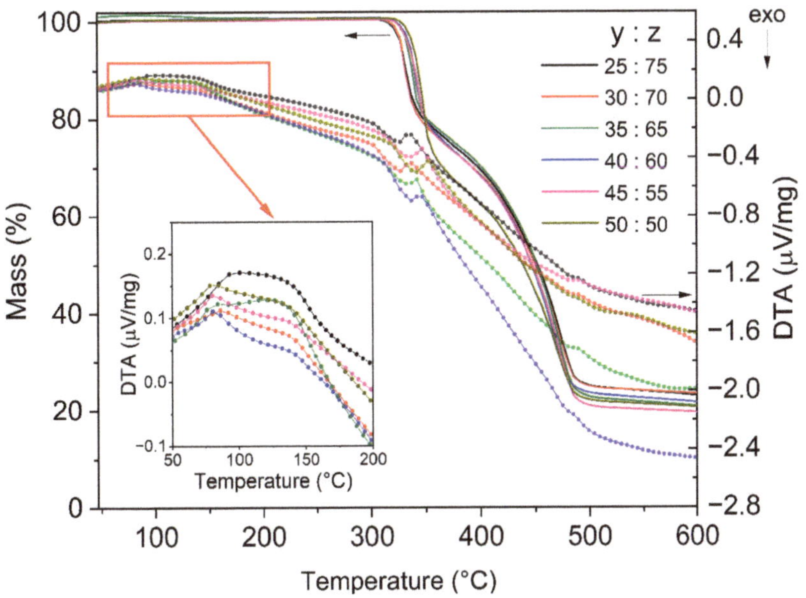

Figure 5. Results of TG/DTA analysis of ternary polymer $PEO_y/PVDF\text{-}HFP_z$/LiTFSI/EMIMTFSI electrolytes.

In the temperature range below 200 °C, two endothermic signals appear in the DTA curves, the first one at about 77 °C, and the second one at about 137 °C. These two features can be explained by melting of the remaining crystalline domains in both polymers. According to Kurzweil et al. [31], the melting point of PEO is 65 °C and of PVDF-HFP is 135 °C. The small deviation could be explained by interaction of the polymers with other components of the electrolyte systems.

The first mass loss of the ternary polymer electrolytes occurs at around 320 °C, which is visible in the TG curve, followed by the second one at 350 °C. According to Xue et al. [26], structural decomposition of PEO with a mass loss takes place at 194 °C, of PVDF-HFP at 388 °C, and of LiTFSI at 356 °C. EMIMTFSI is thermally stable up to 358 °C [48]. Therefore, the decomposition process results from contributions of all components. The mass loss between 320 °C and 350 °C is about 20% for almost all electrolyte compositions, with the exception of the polymer electrolyte tPE $PEO_{50}PVDF\text{-}HFP_{50}$, which shows a higher value of 28%. Presumably, this electrolyte is thermally less stable than other compositions due to the large amount of PEO. Decomposition of polymers is accompanied by a strong endothermic signal. This is caused by a combination of the decomposition processes of the CF_3 groups of the PVDF-HFP, ethyl and methyl groups from the imidazolium ring and chain scission of the PEO. The second stage, representing the main decomposition process of the entire structure of the polymer electrolytes, takes place in the temperature range from about 350 °C to 490 °C. Thereby, an endothermic DTA signal is visible here as well.

Although no mass loss can be detected in the TG curve below 200 °C, a change in the polymer structure must have already occurred at this temperature, as the crystalline phases of both polymers have reached their melting points.

3.2. Electrochemical Studies

For a better understanding how interactions between the polymers and other electrolyte components can impact on the performance of batteries, tPE were electrochemically investigated using electrochemical impedance spectroscopy, cyclic voltammetry, and galvanostatic cycling with potential limitation.

Ionic conductivity. The internal resistance in the tPE systems can be determined using the EIS technique. Figure 6a represents the Nyquist plots of all electrolyte compositions and their resulting circuit model. The component 0.35PEO-0.65PVDF-HFP/1LiTFSI/1EMIMTFSI has the lowest total resistance, in contrast to the electrolyte composition 0.25PEO-0.75PVDF-HFP/1LiTFSI/1EMIMTFSI. In addition, the total resistance changes greatly, according to the temperature change. Figure S1 shows that the Nyquist plot of the component 0.35PEO-0.65PVDF-HFP/1LiTFSI/1EMIMTFSI changed significantly at 20 °C, before and after heating at 80 °C.

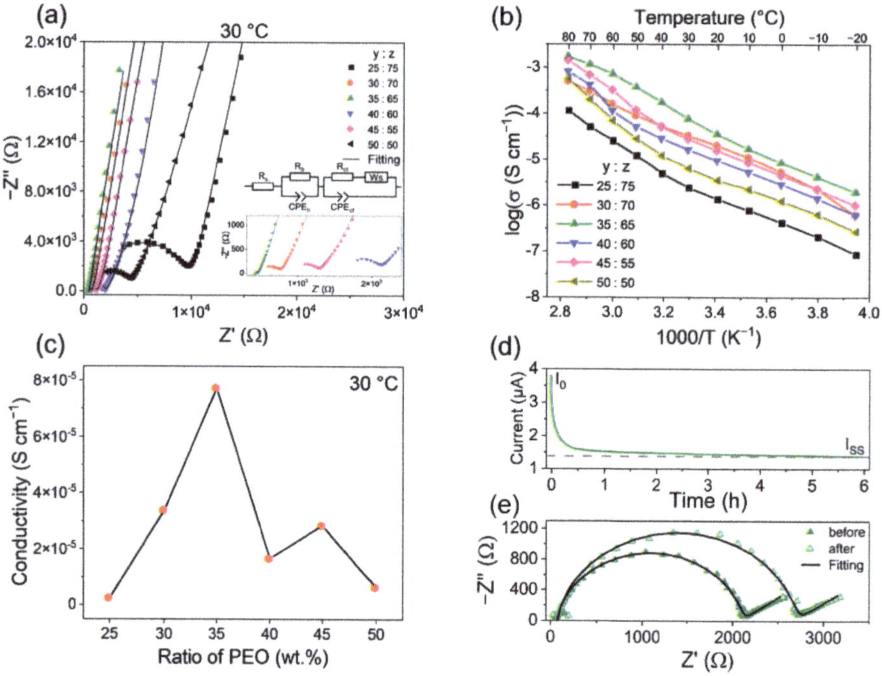

Figure 6. (**a**) Nyquist plots of polymer electrolytes tPE $PEO_yPVDF\text{-}HFP_z$ at 30 °C and the equivalent circuit model with R_s—series resistance, R_b—bulk resistance, R_{ct}—charge-transfer resistance, CPE—constant phase element, W_s—Warburg short element; (**b**) Ionic conductivity of polymer electrolytes as a logarithmical plot against temperature; (**c**) Relationship between ionic conductivity and mass ratio of PEO in the ternary electrolyte; (**d**) Chronoamperogram of tPE $PEO_{35}PVDF\text{-}HFP_{65}$ at 25 °C, and (**e**) its Nyquist plots before and after polarisation at 25 °C.

The ionic conductivity σ was calculated using the following function from the total resistance, which was determined by the EIS results [26,49].

$$\sigma = \frac{D}{R \cdot A}$$

Here, D is the thickness of the characterised polymer electrolytes, R is the total resistance, and A represents the surface area of the polymer electrolytes. Figure 6b represents the ionic conductivity of polymer electrolytes tPE $PEO_yPVDF\text{-}HFP_z$ as a logarithmical plot against temperature. Two distinct regions with different slopes below and above 30 °C can be separated, which still follow Arrhenius-type behaviour. The presence of two ranges can be explained by a pre-melting process of residual crystalline parts of the electrolyte, which usually occurs far below the melting points of individual polymers and more intensively with a high salt content [50].

Both temperature regions, between −20 °C and 30 °C and between 30 °C and 80 °C, were fitted using the Arrhenius equation, and the activation energy (E_A), which describes the lowest energy required for ion transport in the electric field [26], was calculated for each electrolyte composition.

$$\sigma(T) = \sigma_0 \cdot e^{-\frac{E_A}{kT}}$$

Here, σ_0 is the pre-exponential factor, k is the Boltzmann constant, and T is the temperature in Kelvin.

Table 2 shows ionic conductivity values at 30 °C as well as calculated activation energies for two temperature ranges, between −20–30 °C and 30–80 °C.

Table 2. Thickness of ternary polymer electrolytes 1(PEO_yPVDF-HFP_z)-1 LiTFSI-1 EMIMTFSI (y, z in wt%), ionic conductivity and their activation energies in two temperature ranges. For each polymer electrolyte, the thickness was measured at four different points, while the ionic conductivity was determined three times. The values in parentheses represent standard deviations. The activation energy values were calculated using a linear regression model $y = a + b*x$ based on the least squares method; the values in parentheses depict standard deviations.

tPE PEO_yPVDF-HFP_z y	z	Thickness (μm)	Ionic Conductivity at 30 °C (S cm^{-1})	E_{A1} (kJ mol^{-1}) −20–30 °C	R^2 Value (E_{a1})	E_{A2} (kJ mol^{-1}) 30–80 °C	R^2 Value (E_{a2})
25	75	118(6)	2.40(8) × 10^{-6}	42.2(9)	0.9979	69(2)	0.9973
30	70	129(3)	3.40(7) × 10^{-5}	50(3)	0.9815	50(2)	0.9922
35	65	187(2)	7.7(3) × 10^{-5}	47(1)	0.9960	56(3)	0.9895
40	60	178(6)	1.70(2) × 10^{-5}	41.8(6)	0.9990	72(7)	0.9597
45	55	182(5)	2.80(9) × 10^{-5}	42.4(2)	0.9999	73(3)	0.9919
50	50	134(2)	6.3(3) × 10^{-6}	40(1)	0.9961	81(5)	0.9801

The correlation between the ionic conductivity and the amount of PEO in the polymer electrolyte at 30 °C is shown in Figure 6c. Initially, the ionic conductivity increases with increasing mass proportion of PEO and then tends to decrease. Here, the tPE $PEO_{35}PVDF$-HFP_{65} composition reached the highest ionic conductivity of 7.7 × 10^{-5} S cm^{-1}. This composition may be expected to have the highest specific capacity and cyclic stability. The thickness of the electrolytes also shows a maximum dependence on the composition. The non-linearity in properties must be associated with structure of the electrolyte, including interaction of polymer chains, and its porosity, which should by directly proportional to its thickness, since the total weight of the electrolyte is nearly the same for all compositions. Activation energy for ionic movement in the low temperature range is lower for all electrolyte compositions than in the high temperature range. Apparently, the pre-melting process of crystalline domains in the polymer structure impedes ionic transport.

Lithium-ion transference number. Potentiostatic electrochemical impedance spectroscopy (PEIS) was combined with chronoamperometry to calculate the lithium transference number. This potentiostatic polarisation method was established by Bruce and Vincent for solid binary electrolytes, where the current changes with time in response to applying a small voltage amount ($\Delta V \leq 10$ mV) until the cell reaches the steady state and the impedance can be measured before and after potentiostatic polarisation [51]. The transference number t_+ can be calculated by the following equation [52]:

$$t_{+} = \frac{I_{SS}(\Delta V - I_0 R_0)}{I_0(\Delta V - I_{SS} R_{SS})}$$

where I_0 and I_{SS} represent the initial and steady state current, which can be interpreted from the chronoamperogram (Figure 6d), ΔV is the applied polarisation voltage of 10 mV, and R_0 and R_{SS} stand for the interfacial resistance before and after polarisation, obtained from the Nyquist plots (Figure 6e), respectively. All determined values and the calculated transference number for three electrolyte systems tPE $PEO_yPVDF\text{-}HFP_z$ are shown in Table 3. The Li^+ transference number slightly decreases with the increase in PEO amount in the ternary polymer film.

Table 3. Determined values and calculated Li^+ transference numbers of three tPE $PEO_yPVDF\text{-}HFP_z$ compositions at 25 °C.

tPE $PEO_yPVDF\text{-}HFP_z$ (y:z)	I_0 (μA)	I_{SS} (μA)	R_0 (Ω)	R_{SS} (Ω)	t_+
25:75	5.20	1.69	1437	1816	0.12
35:65	3.80	1.38	2017	2602	0.13
50:50	4.60	1.26	1560	1910	0.10

Electrochemical stability. Figure 7 shows the electrochemical stability window of tPE $PEO_{35}PVDF\text{-}HFP_{65}$ from the cyclic voltammetry between 0–5 V for asymmetrical Al/tPE/Li and symmetrical SS/tPE/SS cell configuration, −2–5 V for asymmetrical SS/tPE/Li cell with a scan rate of 0.1 mV s^{-1}.

For the Al/tPE/Li cell (Figure 7a), a noticeable redox process takes place in the first cycle. This could be caused by the fact that EMIMTFSI is not stable at potentials below 1 V [37]. Starting from the second cycle, the polymer electrolyte system has wide electrochemical stability between 1 V to 5 V. In addition, a Li–Al alloy can form at the potential below 0.5 V. Figure 7b represents the asymmetrical cell configuration with a stainless steel chip at the cathode and lithium chip at the anode. In the first scan, the Li deposition and stripping appear at −0.36 V and 0.22 V. However, some irreversibility and a comparatively smaller peak upon Li stripping are observed due to the redox instability of the $EMIM^+$ cations (similar to the case in Figure 7a), which affect the formation process of the solid–electrolyte interface [53]. After the initial cycle, the system exhibits reversibility, but a small peak is observed at 2.4 V, which may be an undesirable reaction between the stainless steel and lithium interface. Figure 7c shows wide electrochemical stability between 0 V and around 4 V in the symmetrical cell configuration with the stainless steel electrodes.

Half cells. The charge-discharge curves for the first cycle and cycle stability are shown in Figure 8a,b for Li/LFP cells, and Figure 8d,e for Li/LTO cells.

In the first cycle, both half cells with LFP and LTO show a relatively high capacity, which is close to the theoretical value (170 mAh g^{-1} for Li/LFP, and 175 mAh g^{-1} for Li/LTO cells). Note that the capacity values in Li/LFP cells do not follow the composition dependence of the ionic conductivity. Thus, the Li/LFP half cells with the lower mass fraction of PEO tend to have a higher specific capacity, except for the tPE $PEO_{35}PVDF\text{-}HFP_{65}$ composition. The cell polarisation, defined as a difference in potentials between the charge and discharge curve at half of the capacity value of the Li/LFP cells, is the lowest for 25% PEO:75% PVDF-HFP, at 165 mV, and increases to 489 mV at higher PEO mass fractions. Considering cycle stability, the discharge capacity of the compositions PEO:PVDF-HFP = 25:75, 35:65, and 50:50 showed around 92% after 50 cycles, compared to the discharge capacity of each composition in the first cycle. Here, the question may arise again of why the long-term performance is not in accordance with the dependence of the ionic conductivity on the polymer electrolyte composition. While the polymer compositions PEO:PVDF-HFP = 25:75, 30:70, 35:65, and 50:50 showed good stability, the discharge capacities of PEO:PVDF-HFP = 40:60, and 45:55 reduced significantly after 50 cycles. Thus, the discharge capacities were only about 7% and 3.4% of their discharge capacities in the first cycle. A possible explanation for this random relationship and the strongly decreasing capacity could be weak mechanical stability of the LFP electrode–electrolyte interface, leading to contact loss of LFP particles, either to the current collector or to the polymer electrolyte surface due to the volume change in the LFP structure during the lithiation and delithiation. Indeed, $LiFePO_4$ undergoes a two-phase transformation with the Li-free $FePO_4$ phase, having a 7% smaller unit cell volume [54]. The interface between the polymer electrolyte and LFP, shown in Figure 7c, features some cracks in the LFP electrode. These areas are definitely less active in the electrochemical process.

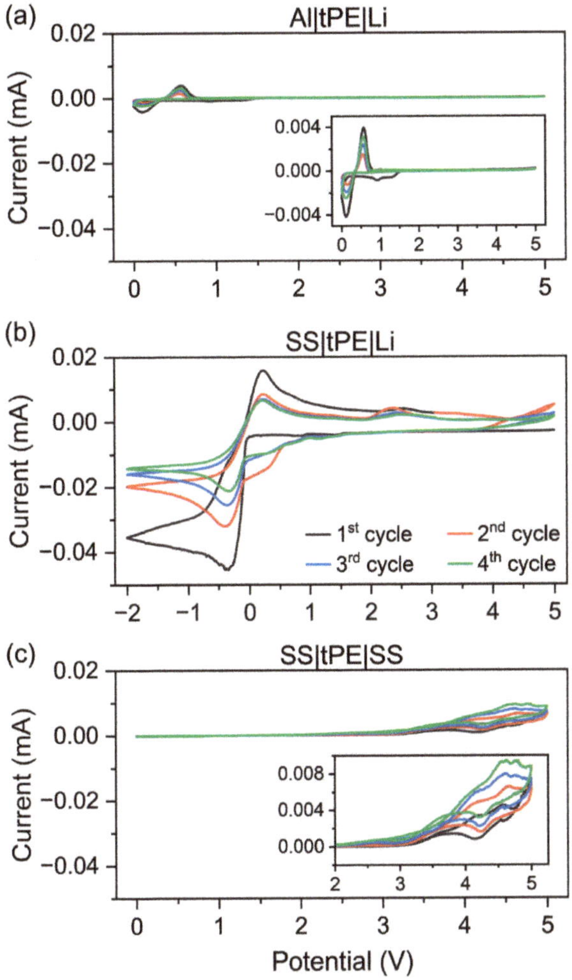

Figure 7. Cyclic voltammograms for tPE $PEO_{35}PVDF\text{-}HFP_{65}$ in asymmetric cell (**a**) Al/tPE/Li, (**b**) stainless steel/tPE/Li, and (**c**) symmetric cell setup with stainless steel at 0.1 mV s^{-1}.

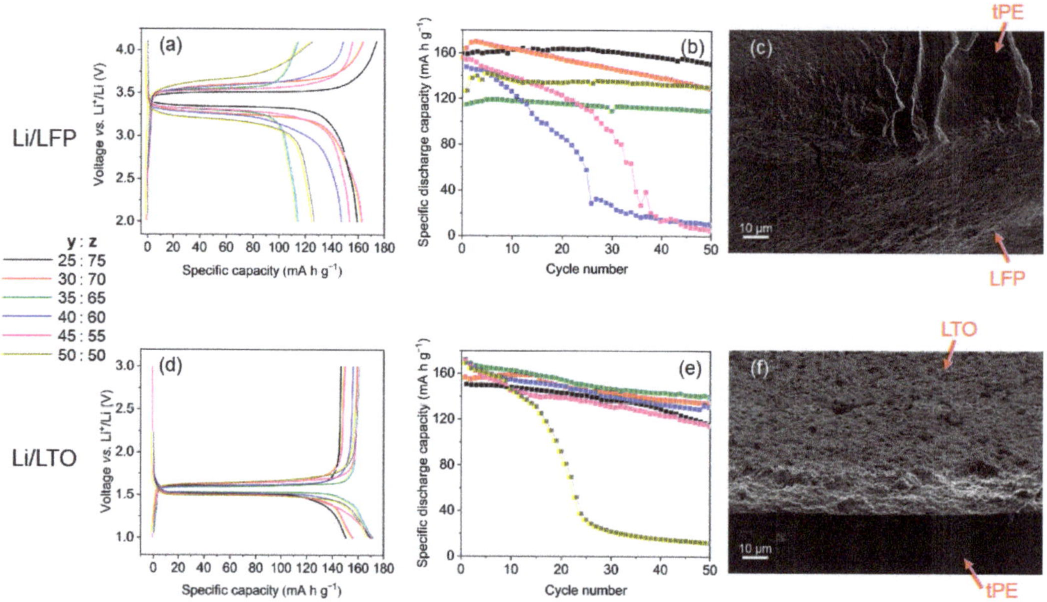

Figure 8. (**a**) Charge-discharge curve at 25 °C in the 1st cycle (0.1 C) and (**b**) cycle stability of Li/LFP cells; (**c**) cutting edge of tPE $PEO_{35}PVDF\text{-}HFP_{65}$/LFP under SEM; (**d**) charge-discharge curve at 25 °C in the 1st cycle (0.1 C) and (**e**) cycle stability of Li/LTO cells; (**f**) cutting edge of tPE $PEO_{35}PVDF\text{-}HFP_{65}$/LTO under SEM; y and z represent the mass ratio (%) of PEO and PVDF-HFP in polymer electrolytes.

The possible corrosion of the lithium salt LiTFSI on the aluminium current collector, reported in the studies [55,56], plays here an inferior role, since it should have the same impact for all compositions.

In contrast to Li/LFP half cells, Li/LTO half cells have a significantly smaller cell polarisation of 73 mV. This increases as the percentage of PEO in the electrolyte system increases, but the difference is less pronounced than for Li/LFP cells. The lowest polarisation of 73 mV occurs in the tPE $PEO_{35}PVDF\text{-}HFP_{65}$ composition, while the highest one of 220 mV belongs to the tPE $PEO_{50}PVDF\text{-}HFP_{50}$ composition, correlating with the ionic conductivity curve vs. composition in Figure 6b. The greater the polarisation, the greater the internal resistance in the electrolyte system. Thus, the internal resistance tends to increase with the increasing weight of PEO. In addition, the Li/LTO half cells represented a very high cycle stability, except for the composition with 50% PEO and 50% PVDF-HFP. The crystallinity of the entire polymer electrolyte system may increase as the crystalline phase of the PEO increases, which can lead to decreasing capacity and stability. In this case, tPE $PEO_{35}PVDF\text{-}HFP_{65}$ showed not only the highest specific discharge capacity, but also the best cycle stability after 50 cycles. Figure 8f depicts SEM images of the layer interface of LTO and the polymer electrolyte. LTO as a "zero-strain" material, demonstrating a very small volume change upon reversible lithiation [57], which is advantageous for interface stability. Rate performance of the tPE $PEO_{35}PVDF\text{-}HFP_{65}$ is shown in Figure 9a. After cycling with different C-rates, from 0.2 C to 2 C, the cell demonstrates high cycle stability at 0.2 C during further 50 cycles.

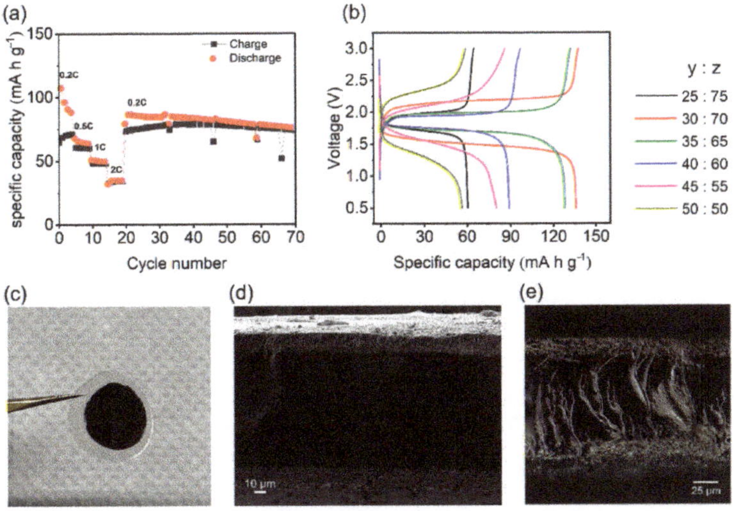

Figure 9. (**a**) Rate performance of the Li/tPE $PEO_{35}PVDF\text{-}HFP_{65}$/LTO cell; (**b**) full cells LTO/tPE/LFP in the first cycle; (**c**) tPE $PEO_{35}PVDF\text{-}HFP_{65}$ with electrode materials after 50 cycles; the sandwich LTO/tPE $PEO_{35}PVDF\text{-}HFP_{65}$/LFP; (**d**) under SEM and (**e**) under an optical microscope.

Full cells. Here, $Li_4Ti_5O_{12}$ served as the anode and $LiFePO_4$ was the cathode. The mass loadings of both active materials were nearly the same (LFP:LTO = ~1:1.2 *w*/*w*), and for all GCPL measurements, the results refer to the LFP cathode. One separate formation cycle for the cathode and anode was performed prior to electrochemical tests of full cells.

Figure 9b represents the first cycle of the full cell LTO/LFP with the polymer electrolytes under investigation at 25 °C. The highest specific discharge capacity was 137 mAh g^{-1}, measured for the tPE $PEO_{30}PVDF\text{-}HFP_{70}$ composition. However, this electrolyte composition has a large internal resistance, which can be deduced from the cell polarization of 830 mV. The second-best specific discharge capacity belongs to the polymer electrolyte with a 35% PEO and 65% PVDF-HFP mass ratio. This composition achieved a discharge capacity of 129 mAh g^{-1}, and the corresponding cell polarisation was 640 mV. The electrolyte compositions at y:z = 40:60 and 45:55 reached only about half of the theoretical value. The polymer electrolytes with tPE $PEO_{25}PVDF\text{-}HFP_{75}$ and tPE $PEO_{50}PVDF\text{-}HFP_{50}$ compositions achieved similar specific capacities of only around 60 mAh g^{-1}. However, there were differences between the electrolytes with 25% PEO and 50% PEO concerning polarisation. Furthermore, it is noticeable that the specific capacities decreased with increasing amounts of PEO. The large addition of PEO increases the crystalline part of the entire electrolyte system at room temperature, which limits the ionic transport. To better illustrate how the sandwich of the investigated ternary polymer electrolytes looks after electrochemical measurements, the battery cell with tPE $PEO_{35}PVDF\text{-}HFP_{65}$ after disassembling is shown in Figure 9c. Here, the mechanical integrity of the layers is preserved. This sandwich was also studied under SEM (Figure 9d) and optical microscopy (Figure 9e).

The SEM picture demonstrates a homogeneous structure of the polymer electrolyte in the LTO/tPE $PEO_{35}PVDF\text{-}HFP_{65}$/LFP sandwich, which is significant for high ionic mobility. Under the optical microscope, the polymer fibres are also visible in the electrolyte, which can increase polarisation and facilitate ageing of battery cells.

The cycling stability for 50 cycles is shown in Figure 10. Due to the very low mass ratio of PEO, the tPE $PEO_{25}PVDF\text{-}HFP_{75}$ achieved a lower performance, but the mechanical stability as well as cycle stability were very high, due to the support of the PVDF-HFP chains.

The electrolyte tPE $PEO_{30}PVDF\text{-}HFP_{70}$ has a higher initial capacity but becomes unstable after 50 cycles. This could be caused by the large internal resistance, including the contact resistance. Another possible explanation would be that the interface between the electrode and polymer electrolyte becomes unstable during long-term cycling. The battery cells with the PEO weight percentage of 35% and 40% show a more stable cycling. The capacity loss corresponds to 27% for tPE $PEO_{35}PVDF\text{-}HFP_{65}$ and 48% for tPE $PEO_{40}PVDF\text{-}HFP_{60}$. The full cells with a higher PEO mass fraction of 45% and 50% revealed a significantly lower stability (see Figure S2 in the Supporting Information). Two facts can play an important role here. First, the mechanical integrity of the interface, especially on the LFP/tPE side, is low. Second, the movement of the lithium ions is inhibited by very strong complexation with the PEO polymer chains, whereby not enough mobile ions can participate in the ion transport. For almost all polymer electrolytes, the Coulombic efficiencies reached about 100% and remained constant.

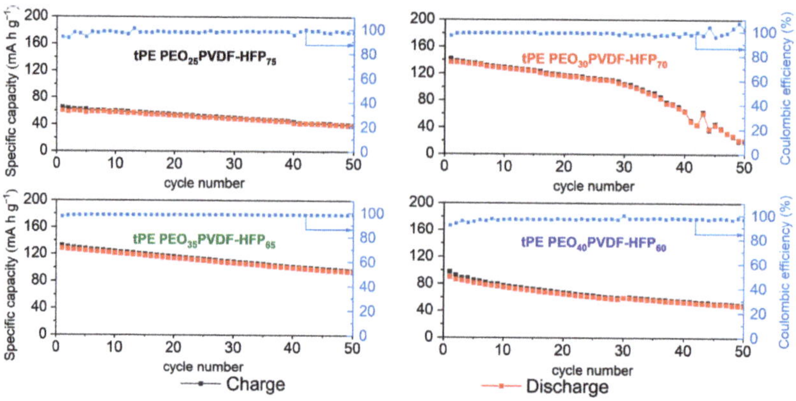

Figure 10. Cycle stability of full cells of LTO/tPE $PEO_yPVDF\text{-}HFP_z$/LFP at 0.1 C at 25 °C.

3.3. *Discussion*

In lithium-ion batteries with a conventional liquid electrolyte, containing organic solvents and $LiPF_6$ as a lithium salt, heat generation in the cell occurs, either due to water penetration or a short circuit. Water addition immediately causes dissociation of $LiPF_6$ with formation of PF_5 and HF. The PF_5 molecules attack solvent molecules, while HF reacts with metallic parts of the battery, producing highly flammable H_2. Altogether, this leads to the combustion of large amounts of active free radicals and thermal runaway of the battery cell [58]. Although PEO is readily flammable, formation of the complex polymer structure with the much less flammable PVDF-HFP and more fire-resistant ionic liquid significantly increases the stability of the whole system. Water addition to the cell system is not expected to be such a detrimental factor, as in case of the liquid $LiPF_6$-containing electrolytes. However, decreasing the PEO portion in the resulting electrolyte formulation is desirable, despite the promising characteristics of only PEO in a polymer electrolyte composed of 20 wt% PEO/40 wt% LiTDI/40% EMIMTFSI [59].

A nonmonotoneous impact of the PEO amount in the ternary **y**PEO/**z**PVDF-HFP):LiTFSI:EMITFSI electrolyte on the electrochemical behaviour in half LFP/Li and LTO-Li cells rcertainly results from the morphology change of the entire electrolyte film, as well as from a change in the coordination mechanism of Li-cations. The optimal composition corresponds to 35% PEO/65% PVDF-HFP:1 LiTFSI:1 EMITFSI for the operation temperatures of 25° and 60 °C (see Figure S3 in Supplementary Materials). This composition represents a compromise between a facilitated lithium-ion mobility along PEO polymer chains, and a high amount of detrimental crystalline domains in the overall polymer electrolytes. Moreover, an increased proportion of PEO interrupts the smoothness of the polymer film surface and can also result in weakened mechanical integrity of the interface between the electrode and polymer electrolyte. According to [60], in which room-temperature tensile behaviour of dried PVDF-HFP/PEO membranes with the PEO-composition varied between 0-16% by weight, and wet PVDF-HFP/PEO membranes after absorption of 1M $LiBF_4$ electrolyte solution of EC and PC (1:1 by weight), was studied, the addition of PEO resulted in significantly lower ultimate tensile strength and modulus in both dried and wet polymer films. The explanation for this was the development of more and larger voids in the sample. The lower stability of the PEO-rich membrane against strain/stress could have a negative impact for the LFP/polymer electrolyte interface during long-term battery cycling. Since the $LiFePO_4$ cathode material has a noticeable variation in the unit cell volume between the lithiated and lithium-free structure, a weaker mechanical contact at the beginning of cycling becomes worse with time. The cell polarisation, as a difference between the charging and discharging curve at the half of the capacity value, is also enhanced in half and full cells by the electrolytes containing more PEO. We also tested LFP/Li and LTO/Li cells with various electrolyte compositions at 60 °C. At this temperature, close to the melting point of PEO, variation of its amount in the polymer electrolyte should have a higher impact on the electrochemical behaviour of the battery cell. Indeed, the cell polarisation was much higher in case of the LFP/Li cell for all electrolyte compositions in comparison to the LTO/Li cell, which definitely relates to the bigger volume change in the LFP material upon (de)lithiation.

4. Conclusions

In metal-ion batteries, polymer electrolytes based on PVDF-HFP/PEO/EMIMTFSI exhibit less risk for flammability than pure PEO, since PVDF-HFP and EMIMTFSI are much less flammable. Although the estimated costs for the polymer electrolyte film in a battery cell are approximately two times higher than the conventional LP30 electrolyte with a Whatman separator, the increased stability makes polymer electrolytes advantageous for a stationary battery application and scaling up process. We prepared and characterised 1(PEO_yPVDF-HFP_z)-1 LiTFSI-1 EMIMTFSI ternary polymer electrolytes of different compositions ($25 < y < 50$, $75 > z > 50$), which were electrochemically investigated in Li/LFP and Li/LTO half cells as well as in LTO/LFP full cells at 25 °C. The role of the PEO amount for the electrochemical performance was evaluated, and the optimal electrolyte composition was suggested. Despite the somewhat low specific capacity in Li/LFP half cells, the electrolyte formulation containing 35% PEO/65% PVDF-HFP:1 LiTFSI:1 EMITFSI exhibits the highest overall ionic conductivity of 7.7×10^{-5} S cm^{-1}, one of the highest specific capacities and the best long-term stability in LTO/LFP full cells. This composition accumulates a facilitated lithium-ion mobility along PEO polymer chains and a low amount of disruptive crystalline domains in the overall structure of the polymer electrolyte film. To further enhance the electrochemical performance of a full battery cell with the optimal electrolyte composition, mechanical stability of the electrode–electrolyte interface should be increased.

Supplementary Materials: The following supporting information can be downloaded at: https://www.mdpi.com/article/10.3390/batteries11020045/s1, SI contains analysis of IR spectra, EIS data, cycling stability results and SEM image for some electrolyte compositions. Figure S1: Nyquist-plots of tPE PEO_{35}PVDF-HFP_{65} at 20 °C, recorded for materials before and after heating to 80 °C. Figure S2: Left: Cycle stability of full cells LTO/tPE PEO_yPVDF-HFP_z/LFP with a high mass fraction of PEO at 0.1C at 25 °C. Right: SEM image of the sandwich LTO/tPE PEO_{50}PVDF-HFP_{50}/LFP after cycling, reflecting a low mechanical integrity. Figure S3: Charge-discharge curves of half-cells LFP/tPE $PEO_yPVDF-HFP_z$/Li (left) and LTO/tPE $PEO_yPVDF-HFP_z$/Li (right) with various electrolyte compositions at 60 °C. Table S1: Vibrations assigning in IR spectra [16,29,43,45,46,61].

Author Contributions: H.B.T.N.: Conceptualization, Methodology, Investigation, Data curation, Visualization, Writing—original draft. L.D.: Investigation, Validation, Formal Analysis. B.P.: Investigation, Formal analysis, Validation. T.S.: Investigation, Formal analysis. H.B.A.N.: Investigation, Formal analysis. D.M.: Conceptualization, Writing—review and editing, Supervision, Project administration, Funding acquisition. All authors have read and agreed to the published version of the manuscript.

Funding: This work was funded by the German Research Foundation (DFG), project number 462115051.

Data Availability Statement: The original contributions presented in this study are included in the article and Supplementary Materials. Further inquiries can be directed to the corresponding authors.

Acknowledgments: The authors are deeply grateful to Nikolai Gaponik (Dresden University of Technology, Germany) for a fruitful discussion.

Conflicts of Interest: Authors declare no conflicts of interest.

References

1. Goodenough, J.B.; Kim, Y. Challenges for rechargeable Li batteries. *Chem. Mater.* **2010**, *22*, 587–603. [CrossRef]
2. Xue, Z.; He, D.; Xie, X. Poly(ethylene oxide)-based electrolytes for lithium-ion batteries. *J. Mater. Chem. A Mater.* **2015**, *3*, 19218–19253. [CrossRef]
3. Osada, I.; de Vries, H.; Scrosati, B.; Passerini, S. Ionic-Liquid-Based Polymer Electrolytes for Battery Applications. *Angew. Chem.* **2016**, *128*, 500–513. [CrossRef] [PubMed]
4. Long, L.; Wang, S.; Xiao, M.; Meng, Y. Polymer electrolytes for lithium polymer batteries. *J. Mater. Chem. A Mater.* **2016**, *4*, 10038–10069. [CrossRef]
5. Costa, C.M.; Nunes-Pereira, J.; Rodrigues, L.C.; Silva, M.M.; Ribelles, J.L.G.; Lanceros-Méndez, S. Novel poly(vinylidene fluoride-trifluoroethylene)/poly(ethylene oxide) blends for battery separators in lithium-ion applications. *Electrochim. Acta* **2013**, *88*, 473–476. [CrossRef]
6. Prabakaran, P.; Manimuthu, R.P. Enhancement of the electrochemical properties with the effect of alkali metal systems on PEO/PVdF-HFP complex polymer electrolytes. *Ionics* **2016**, *22*, 827–839. [CrossRef]
7. Croce, F.; Appetecchi, G.B.; Persi, L.; Scrosati, B. Nanocomposite polymer electrolytes for lithium batteries. *Nature* **1998**, *394*, 456–458. [CrossRef]
8. Christie, A.M.; Lilley, S.J.; Staunton, E.; Andreev, Y.G.; Bruce, P.G. Increasing the conductivity of crystalline polymer electrolytes. *Nature* **2005**, *433*, 50–53. [CrossRef] [PubMed]
9. Vallée, A.; Besner, S.; Prud'Homme, J. Comparative study of poly(ethylene oxide) electrolytes made with $LiN(CF_3SO_2)_2$, $LiCF_3SO_3$ and $LiClO_4$: Thermal properties and conductivity behavior. *Electrochim. Acta* **1992**, *37*, 1579–1583. [CrossRef]
10. Kim, G.T.; Appetecchi, G.B.; Carewska, M.; Joost, M.; Balducci, A.; Winter, M.; Passerini, S. UV cross-linked, lithium-conducting ternary polymer electrolytes containing ionic liquids. *J. Power Sources* **2010**, *195*, 6130–6137. [CrossRef]
11. Itoh, T.; Miyamura, Y.; Ichikawa, Y.; Uno, T.; Kubo, M.; Yamamoto, O. Composite polymer electrolytes of poly(ethylene oxide)/$BaTiO_3$/Li salt with hyperbranched polymer. *J. Power Sources* **2003**, *119–121*, 403–408. [CrossRef]
12. An, Y.; Cheng, X.; Zuo, P.; Liao, L.; Yin, G. Improved properties of polymer electrolyte by ionic liquid PP1.3TFSI for secondary lithium ion battery. *J. Solid State Electrochem.* **2012**, *16*, 383–389.
13. Jiang, Z.; Carroll, B.; Abraham, K.M. Studies of some poly(vinylidene fluoride) electrolytes. *Electrochim. Acta* **1997**, *42*, 2667–2677. [CrossRef]
14. Cao, J.H.; Zhu, B.K.; Xu, Y.Y. Structure and ionic conductivity of porous polymer electrolytes based on PVDF-HFP copolymer membranes. *J. Memb. Sci.* **2006**, *281*, 446–453. [CrossRef]
15. Liu, W.; Zhang, X.K.; Wu, F.; Xiang, Y. A study on PVDF-HFP gel polymer electrolyte for lithium-ion batteries. *IOP Conf. Ser. Mater. Sci. Eng.* **2017**, *213*, 012036. [CrossRef]
16. Herzog-Arbeitman, A.; Maletti, S.; Oswald, S.; Schmeida, T.; Giebeler, L.; Mikhailova, D. A Highly Conductive Gel Polymer Electrolyte for Li-Mg Hybrid Batteries. *ACS Appl. Energy Mater.* **2021**, *4*, 1906–1914. [CrossRef]
17. Miao, R.; Liu, B.; Zhu, Z.; Liu, Y.; Li, J.; Wang, X.; Li, Q. PVDF-HFP-based porous polymer electrolyte membranes for lithium-ion batteries. *J. Power Sources* **2008**, *184*, 420–426. [CrossRef]
18. Groce, F.; Gerace, F.; Dautzemberg, G.; Passerini, S.; Appetecchi, G.B.; Scrosati, B. Synthesis and characterization of highly conducting gel electrolytes. *Electrochim. Acta* **1994**, *39*, 2187–2194. [CrossRef]

19. Bohnke, O.; Frand, G.; Rezrazi, M.; Rousselot, C.; Truche, C. Fast ion transport in new lithium electrolytes gelled with PMMA. 1. Influence of polymer concentration. *Solid State Ion.* **1993**, *66*, 97–104. [CrossRef]
20. Passerini, S.; Rosolen, J.M.; Scrosati, B. Plasticized carbon electrodes of interest for lithium rocking chair batteries. *J. Power Sources* **1993**, *45*, 333–341. [CrossRef]
21. Pistoia, G.; Antonini, A.; Wang, G. Impedance study on the reactivity of gel polymer electrolytes towards a lithium electrode. *J. Power Sources* **1996**, *58*, 139–144. [CrossRef]
22. Dias, F.B.; Plomp, L.; Veldhuis, J.B.J. Trends in polymer electrolytes for secondary lithium batteries. *J. Power Sources* **2000**, *88*, 169–191. [CrossRef]
23. Elamin, K.; Shojaatalhosseini, M.; Danyliv, O.; Martinelli, A.; Swenson, J. Conduction mechanism in polymeric membranes based on PEO or PVdF-HFP and containing a piperidinium ionic liquid. *Electrochim. Acta* **2019**, *299*, 979–986. [CrossRef]
24. Das, S.; Ghosh, A. Charge Carrier Relaxation in Different Plasticized PEO/PVDF-HFP Blend Solid Polymer Electrolytes. *J. Phys. Chem. B* **2017**, *121*, 5422–5432. [CrossRef] [PubMed]
25. Kim, K.M.; Park, N.G.; Ryu, K.S.; Chang, S.H. Characteristics of PVdF-HFP/TiO_2 composite membrane electrolytes prepared by phase inversion and conventional casting methods. *Electrochim. Acta* **2006**, *51*, 5636–5644. [CrossRef]
26. Xue, X.; Zhang, X.; Liu, Y.; Chen, S.; Chen, Y.; Lin, J.; Zhang, Y. Boosting the Performance of Solid-State Lithium Battery Based on Hybridizing Micron-Sized LATP in a PEO/PVDF-HFP Heterogeneous Polymer Matrix. *Energy Technol.* **2020**, *8*, 2000444. [CrossRef]
27. Shalu; Chaurasia, S.K.; Singh, R.K. Crystallization behaviour of a polymeric membrane based on the polymer PVdF-HFP and the ionic liquid $BMIMBF_4$. *RSC Adv.* **2014**, *4*, 50914–50924.
28. Zhang, P.; Li, R.; Huang, J.; Liu, B.; Zhou, M.; Wen, B.; Xia, Y.; Okada, S. Flexible poly(vinylidene fluoride- co - hexafluoropropylene)-based gel polymer electrolyte for high-performance lithium-ion batteries. *RSC Adv.* **2021**, *11*, 11943–11951. [CrossRef] [PubMed]
29. Manjunatha, H.; Damle, R.; Kumaraswamy, G.N. Modification of polymer electrolyte blend PEO/PVDF–HFP by low-energy O+ ion irradiation to improve electrolyte behavior. *Polym. Bull.* **2022**, *79*, 3929–3950. [CrossRef]
30. Ue, M.J. Mobility and Ionic Association of Lithium and Quaternary Ammonium Salts in Propylene Carbonate and γ-Butyrolactone. *Electrochem. Soc.* **1994**, *141*, 3336–3342. [CrossRef]
31. Kurzweil, P.; Dietlmeier, O.K. *Elektrochemische Speicher—Superkondensatoren, Batterien, Elektrolyse-Wasserstoff, Rechtliche Rahmenbedingungen*; Springer Vieweg: Wiesbaden, Germany, 2018.
32. Aurbach, D.; Talyosef, Y.; Markovsky, B.; Markevich, E.; Zinigrad, E.; Asraf, L.; Gnanaraj, J.S.; Kim, H.J. Design of electrolyte solutions for Li and Li-ion batteries: A review. *Electrochim. Acta* **2004**, *50*, 247–254. [CrossRef]
33. Heider, U.; Oesten, R.; Jungnitz, M. Challenge in manufacturing electrolyte solutions for lithium and lithium ion batteries quality control and minimizing contamination level. *J. Power Sources* **1999**, *81–82*, 119–122. [CrossRef]
34. Kawamura, T.; Okada, S.; Yamaki, J.I. Decomposition reaction of LiPF6-based electrolytes for lithium ion cells. *J. Power Sources* **2006**, *156*, 547–554. [CrossRef]
35. Wiemers-Meyer, S.; Winter, M.; Nowak, S. Mechanistic insights into lithium ion battery electrolyte degradation-a quantitative NMR study. *Phys. Chem. Chem. Phys.* **2016**, *18*, 26595–26601. [CrossRef]
36. McEwen, B.; Ngo, H.L.; LeCompte, K.; Goldman, J.L. Electrochemical Properties of Imidazolium Salt Electrolytes for Electrochemical Capacitor Applications. *J. Electrochem. Soc.* **1999**, *146*, 1687–1695. [CrossRef]
37. Garcia, B.; Lavallée, S.; Perron, G.; Michot, C.; Armand, M. Room temperature molten salts as lithium battery electrolyte. *Electrochim. Acta* **2004**, *49*, 4583–4588. [CrossRef]
38. Bonhôte, P.; Dias, A.P.; Papageorgiou, N.; Kalyanasundaram, K.; Grätzel, M. Hydrophobic, Highly Conductive Ambient-Temperature Molten Salts. *Inorg. Chem.* **1996**, *35*, 1168–1178. [CrossRef]
39. Noda, A.; Hayamizu, K.; Watanabe, M. Pulsed-gradient spin-echo ^{1}H and ^{19}F NMR ionic diffusion coefficient, viscosity, and ionic conductivity of non-chloroaluminate room-temperature ionic liquids. *J. Phys. Chem. B* **2001**, *105*, 4603–4610. [CrossRef]
40. Appetecchi, G.B.; Henderson, W.; Villano, P.; Berrettoni, M.; Passerini, S. PEO-LiN$(SO_2CF_2CF_3)_2$ Polymer Electrolytes: I. XRD, DSC, and Ionic Conductivity Characterization. *J. Electrochem. Soc.* **2001**, *148*, A1171–A1178. [CrossRef]
41. Meyer, W.H. Polymer Electrolytes for Lithium-Ion Batteries. *Adv. Mater.* **1998**, *10*, 439–448. [CrossRef]
42. Stephan, A.M. Review on gel polymer electrolytes for lithium batteries. *Eur. Polym. J.* **2006**, *42*, 21–42. [CrossRef]
43. Shalu; Singh, V.K.; Singh, R.K. Development of ion conducting polymer gel electrolyte membranes based on polymer PVdF-HFP, BMIMTFSI ionic liquid and the Li-salt with improved electrical, thermal and structural properties. *J. Mater. Chem. C Mater.* **2015**, *3*, 7305–7318. [CrossRef]
44. Diddens, D.; Heuer, A. Lithium ion transport mechanism in ternary polymer electrolyte-ionic liquid mixtures: A molecular dynamics simulation study. *ACS Macro Lett.* **2013**, *2*, 322–326. [CrossRef] [PubMed]
45. Pritam; Arya, A.; Sharma, A.L. Dielectric relaxations and transport properties parameter analysis of novel blended solid polymer electrolyte for sodium-ion rechargeable batteries. *J. Mater. Sci.* **2019**, *54*, 7131–7155. [CrossRef]

46. Prabakaran, P.; Manimuthu, R.P.; Gurusamy, S.; Sebasthiyan, E. Plasticized polymer electrolyte membranes based on PEO/PVdF-HFP for use as an effective electrolyte in lithium-ion batteries. *Chin. J. Polym. Sci.* **2017**, *35*, 407–421. [CrossRef]
47. Tian, X.; Jiang, X.; Zhu, B.; Xu, Y. Effect of the casting solvent on the crystal characteristics and pervaporative separation performances of P(VDF-co-HFP) membranes. *J. Memb. Sci.* **2006**, *279*, 479–486. [CrossRef]
48. An, Y.; Zuo, P.; Cheng, X.; Liao, L.; Yin, G. Preparation and properties of ionic-liquid mixed solutions as a safety electrolyte for lithium ion batteries. *Int. J. Electrochem. Sci.* **2011**, *6*, 2398–2410. [CrossRef]
49. Yu, B.; Zhou, F.; Wang, C.; Liu, W. A novel gel polymer electrolyte based on poly ionic liquid 1-ethyl 3-(2-methacryloyloxy ethyl) imidazolium iodide. *Eur. Polym. J.* **2007**, *43*, 2699–2707.
50. Choi, B. Optical microscopy study on the crystallization in PEO-salt polymer electrolytes. *Solid State Ion.* **2004**, *168*, 123–129. [CrossRef]
51. Bruce, P.G.; Vincent, C.A. Steady state current flow in solid binary electrolyte cells. *J. Electroanal. Chem. Interfacial Electrochem.* **1987**, *225*, 1–17. [CrossRef]
52. Evans, J.; Vincent, C.A.; Bruce, P.G. Electrochemical measurement of transference numbers in polymer electrolytes. *Polymer* **1987**, *28*, 2324–2328. [CrossRef]
53. Zhu, C.; Cheng, H.; Yang, Y. Electrochemical Characterization of Two Types of PEO-Based Polymer Electrolytes with Room-Temperature Ionic Liquids. *J. Electrochem. Soc.* **2008**, *155*, A569. [CrossRef]
54. Yamada, A.; Koizumi, H.; Nishimura, S.; Sonoyama, N.; Kanno, R.; Yonemura, M.; Nakamura, T.; Kobayashi, Y. Room-temperature miscibility gap in Li_xFePO_4. *Nat. Mater.* **2006**, *5*, 357–360. [CrossRef] [PubMed]
55. Krause, L.J.; Lamanna, W.; Summerfield, J.; Engle, M.; Korba, G.; Loch, R.; Atanasoski, R. Corrosion of aluminum at high voltages in non-aqueous electrolytes containing perfluoroalkylsulfonyl imides; new lithium salts for lithium-ion cells. *J. Power Sources* **1997**, *68*, 320–325. [CrossRef]
56. Péter, L.; Arai, J. Anodic dissolution of aluminium in organic electrolytes containing per ¯ uoroalkylsulfonyl imides. *J. Appl. Electrochem.* **1999**, *29*, 1053–1061. [CrossRef]
57. Scharner, S.; Weppner, W.; Schmid-Beurmann, P. Evidence of Two-Phase Formation upon Lithium Insertion into the $Li_{1.33}Ti_{1.67}O_4$ Spinel. *J. Electrochem. Soc.* **1999**, *146*, 857–861. [CrossRef]
58. Chawla, N.; Bharti, N.; Singh, S. Recent Advances in Non-Flammable Electrolytes for Safer Lithium-Ion Batteries. *Batteries* **2019**, *5*, 19. [CrossRef]
59. Polu, A.R.; Kareem, A.A.; Rasheed, H.K. Thermal, electrical and electrochemical properties of ionic liquid-doped poly(ethylene oxide)–LiTDI polymer electrolytes for Li-ion batteries. *J. Solid State Electrochem.* **2023**, *27*, 409–416. [CrossRef]
60. Chung, N.K.; Kwon, Y.D.; Kim, D. Thermal, mechanical, swelling, and electrochemical properties of poly(vinylidene fluoride)-co-hexafluoropropylene/poly(ethylene glycol) hybrid-type polymer electrolytes. *J. Power Sources* **2003**, *124*, 148–154. [CrossRef]
61. Kiefer, J.; Fries, J.; Leipertz, A. Experimental Vibrational Study of Imidazolium-Based Ionic Liquids: Raman and Infrared Spectra of 1-Ethyl-3-methylimidazolium Bis(Trifluoromethylsulfonyl)imide and 1-Ethyl-3-methylimidazolium Ethylsulfate. *Appl. Spectrosc.* **2007**, *61*, 1306–1311. [CrossRef]

Disclaimer/Publisher's Note: The statements, opinions and data contained in all publications are solely those of the individual author(s) and contributor(s) and not of MDPI and/or the editor(s). MDPI and/or the editor(s) disclaim responsibility for any injury to people or property resulting from any ideas, methods, instructions or products referred to in the content.

batteries

MDPI

Article

Monodisperse Hierarchical N-Doped Carbon Microspheres with Uniform Pores as a Cathode Host for Advanced K–Se Batteries

Hyun-Jin Kim [1], Jeong-Ho Na [1] and Seung-Keun Park [1,2,3,*]

[1] Department of Intelligent Energy and Industry, Chung-Ang University, 84 Heukseok-ro, Dongjak-gu, Seoul 06974, Republic of Korea; robbit1212@cau.ac.kr (H.-J.K.); abcd1298@cau.ac.kr (J.-H.N.)
[2] Department of Advanced Materials Engineering, Chung-Ang University, 4726 Seodong-daero, Daedeok-myeon, Anseong 17546, Republic of Korea
[3] Western Seoul Center, Korea Basic Science Institute, 150 Bugahyeon-ro, Seodaemun-gu, Seoul 03759, Republic of Korea
* Correspondence: skpark09@cau.ac.kr

Abstract: K–Se batteries offer high energy density and cost-effectiveness, making them promising candidates for energy storage systems. However, their practical applications are hindered by Se aggregation, sluggish ion diffusion, and significant volumetric expansion. To address these challenges, monodisperse hierarchical N-doped carbon microspheres (NCHS) with uniformly sized pores were synthesized as cathode hosts. The flower-like microstructure, formed by the assembly of two-dimensional building blocks, mitigated Se aggregation and facilitated uniform distribution within the pores, enhancing Se utilization. Nitrogen doping, introduced during synthesis, strengthened chemical bonding between selenium and the carbon host, suppressed side reactions, and accelerated reaction kinetics. These synergistic effects enabled efficient ion transport, improved electrolyte accessibility, and enhanced redox reactions. Additionally, the uniform particle and pore sizes of NCHS effectively mitigated volumetric expansion and surface accumulation, ensuring long-term cycling stability and superior electrochemical performance. Se-loaded NCHS (Se@NCHS) exhibited a high discharge capacity of 199.4 mA h g^{-1} at 0.5 C after 500 cycles with 70.4% capacity retention and achieved 188 mA h g^{-1} at 3.0 C, outperforming conventional carbon hosts such as Super P. This study highlights the significance of structural and chemical modifications in optimizing cathode materials and offers valuable insights for developing high-performance energy storage systems.

Keywords: K–Se batteries; carbon microsphere; hierarchical structure; nitrogen doping; cathode host

1. Introduction

Over recent decades, lithium-ion batteries (LIBs) have been widely adopted as secondary energy storage devices, particularly in electric vehicles and portable electronics [1]. However, constraints related to energy density and the escalating costs of LIB technology have driven efforts to explore alternative battery systems [2]. Due to the widespread availability of K in the Earth's crust, potassium-ion batteries (KIBs) have emerged as a promising alternative. [3]. Despite this advantage, KIBs suffer from low gravimetric and volumetric energy densities, primarily due to the large ionic size and high atomic weight of K ions. Current studies on KIBs cathode materials has focused on Prussian blue analogs, layered metal oxides, polyanionic compounds, and organic materials, yet none have achieved adequate capacity for practical applications. Se, known for its high electrical conductivity (1×10^{-3} S m^{-1}), theoretical capacity of 675 mA h g^{-1}, and volumetric capacity of

 https://doi.org/10.3390/batteries11030101

3253 mA h cm^{-3}, has emerged as a potential cathode material for advanced KIBs [4]. The K–Se battery operates according to the reaction Se + 2M^{+} + 2e^{-} ↔ M_2Se (where M = Li, Na, K), offering a notable theoretical gravimetric capacity, making it a promising contender for next-generation energy storage solutions.

However, utilizing Se as a cathode material in K–Se batteries introduces critical challenges. The substantial large ionic radius of K^+ ions leads to sluggish ion transport [5]. Additionally, the shuttle effect, arising from the dissolution of polyselenide intermediates during redox reactions, decreases Coulombic efficiency and shortens cycle life [6]. The solid-state reaction between K and Se also leads to considerable volumetric expansion during charge/discharge cycles, causing mechanical degradation of the active material [7]. Several approaches have been proposed to address these issues. Yu et al. demonstrated that suppressing the formation of soluble polyselenide intermediates in carbonate-based electrolytes effectively alleviates the shuttle effect [7]. Despite such advances, Se cathodes still face challenges such as limited active material utilization, poor rate performance, and rapid capacity decay.

Unmodified commercial carbon materials such as Super P and Ketjen Black have been widely utilized as Se-loading hosts due to their availability and cost-effectiveness [8]. However, their non-uniform particle size distribution and small particle dimensions limit their performance [9]. Non-uniform particle sizes hinder the uniform distribution of Se within the internal pores of the host, leading to its accumulation in interparticle voids and reducing its utilization efficiency within the electrode structure [9]. This non-uniform dispersion limits the interaction between the active material and electrolyte, leading to a decline in total electrode capacity. Additionally, small particle sizes increase surface energy and promote particle aggregation [9], obstructing the pore structure and exacerbating non-uniform Se loading. Consequently, the aggregation and surface accumulation of Se degrade electrochemical performance and lead to rapid capacity fading. The combination of non-uniform and small particle sizes further restricts selenium from fully penetrating the internal pores of the carbon host, causing it to accumulate primarily on the surface or within interparticle voids [10]. Consequently, Se utilization is significantly diminished, severely limiting the energy density and efficiency of active materials in the battery. To overcome these limitations, extensive research has been conducted on host materials with a well-defined pore distribution. Yu et al. demonstrated that a uniform pore structure prevents the localized accumulation of Se and facilitates its homogeneous dispersion within the host material [11]. This approach enhances Se loading efficiency and utilization while improving interactions between the active material and electrolyte. Additionally, it contributes to increased electrode capacity and mitigates performance degradation caused by Se aggregation and surface accumulation [10].

However, aggregation issues stemming from non-uniform and small particle sizes cannot be fully resolved solely through pore design. Achieving a uniform particle size in carbon hosts is critical for improving performance. Uniformly synthesized flower-like microstructures constructed from two-dimensional (2D) building blocks offer several advantages for Se-based batteries [12]. The high surface-to-volume ratio of these microstructures significantly increases the number of electrochemical reaction sites, thereby enhancing active material utilization and reaction kinetics. Moreover, the three-dimensional (3D) architecture formed by the assembly of 2D building blocks improves electrolyte accessibility and optimizes ion diffusion pathways [12], resulting in enhanced Se utilization and long-term cycling stability. Additionally, the uniformly synthesized flower-like structure effectively mitigates aggregation issues caused by nonuniform and small particle sizes.

In this study, Se-loaded N-doped carbon spheres (Se@NCHS) with a uniform flower-like hierarchical architecture were developed as a cathode material for K–Se batteries. The uniform microstructure and precisely distributed pores formed along the periphery of the flower-like architecture effectively address the key challenges in Se-based batteries, significantly enhancing electrochemical performance. The uniformly distributed pores formed through the Mo etching process along the periphery of the flower-like structure minimize Se aggregation and surface accumulation while promoting efficient electrolyte infiltration and ion transport [11,13]. Furthermore, the interconnected framework constructed from 2D building blocks optimizes electrolyte access and ion diffusion pathways, ensuring long-term cycling stability. The flower-like microstructure with a uniform particle size mitigates aggregation issues caused by nonuniform and small particle sizes, ensuring stable and efficient performance. N doping, naturally introduced during synthesis, strengthens chemical bonding between Se and the carbon host, suppresses side reactions, and accelerates reaction kinetics, thereby improving Coulombic efficiency [14]. By combining structural and chemical modifications, the Se@NCHS electrode exhibits superior Se utilization, prolonged cycling stability, and enhanced electrochemical properties, making it a promising candidate for next-generation K–Se batteries.

2. Materials and Methods

2.1. Preparation of Mo@PDA, NCHS, Se@NCHS, and Se@Super P

The NCHS were prepared according to the method described in a previous study [12]. To synthesize the Mo@PDA spheres, 80 mL of distilled water and 160 mL of ethanol were mixed in a 250 mL beaker, followed by the addition of 0.3 g of ammonium molybdate ($(NH_4)_6Mo_7O_{24}$, SAMCHUN, Seoul, Republic of Korea) and 0.3 g and dopamine hydrochloride ($(HO)_2C_6H_3CH_2CH_2NH_2 \cdot HCl$, Sigma-Aldrich). Then, 0.7 mL of ammonium hydroxide solution ($NH_3 \cdot H_2O$, SAMCHUN, Seoul, Republic of Korea) was added, and the mixture was stirred for 6 h. The product was then washed with EtOH and dried in an oven for 24 h. Hierarchical NCHS samples were subsequently synthesized by heat-treating the Mo@PDA spheres in a quartz tube furnace, followed by etching. First, MoOx/NC spheres were obtained by heat-treating the Mo@PDA spheres at 750 °C for 4 h in an Ar atmosphere with a heating rate of 5 °C min^{-1}. The remaining Mo was etched using aqua regia prepared by mixing hydrochloric acid (HCl, SAMCHUN, Seoul, Republic of Korea) and nitric acid (HNO_3, SAMCHUN, Seoul, Republic of Korea) in a 3:1 ratio with stirring for 24 h. The resulting powder was filtered, thoroughly washed with distilled water, and dried in a vacuum oven at 70 °C to obtain the final NCHS. A 2:3 weight ratio of NCHS to commercial Se powder (Selenium, SAMCHUN, Seoul, Republic of Korea) was manually mixed, followed by heat treatment at 260 °C for 12 h in an argon-filled tube furnace, allowing Se to infiltrate the carbon matrix through melt-diffusion. The final product was named Se@NCHS. For comparison, Se@Super P spheres were synthesized by manually mixing commercial carbon (Super P) with Se in the same weight ratio as Se@NCHS and performing melt diffusion in a tube furnace under identical conditions.

2.2. Characterization

The surface morphology and structural properties of the prepared samples were examined via scanning electron microscopy (SEM, HITACHI S-5000, Hitachi Ltd., Tokyo, Japan) and transmission electron microscopy (TEM, JEM-2100F, Korea Basic Institute Daegu Center, Daegu, Republic of Korea) analysis. Thermogravimetric analysis (TGA, TG209F1, Korea Basic Institute Daegu Center, Daegu, Republic of Korea) was conducted with a TG209F1 analyzer over a temperature range of 30–600 °C at a heating rate of 10 °C min^{-1} under an N_2 atmosphere. The specific surface areas of the samples were determined through the Brunauer–Emmett–Teller (BET, BELSORP-mini II, Korea Basic Institute Daegu Center, Daegu, Republic of Korea) method, while pore size distribution was assessed using the Barrett–Joyner–Halenda (BJH) and Horváth–Kawazoe (HK) methods. Additionally, Raman spectroscopy (XPERRAM-S, Nanobase, Seoul, Republic of Korea) was employed to characterize the carbon structure of the samples, with measurements conducted over the 100–2000 cm^{-1} range.

2.3. Electrochemical Measurement

The synthesized samples were evaluated using a 2032-type coin-cell battery. For the preparation of working electrodes, the sample powder was blended with carbon black (Super P) and carboxymethyl cellulose (CMC) in deionized (DI) water, which acted as an ion-free medium, at a weight ratio of 8:1:1. The resulting mixture was evenly coated onto an Al foil, with each electrode containing approximately 1.2 mg cm^{-2} of the active material. Potassium metal served as both the counter and reference electrode, while a glass filter was employed as the separator. The electrolyte solution was composed of 3 M potassium bis(fluorosulfonyl)imide (KFSI) dissolved in a 1:1 volume ratio of ethylene carbonate (EC) and diethyl carbonate (DEC). Cyclic voltammetry (CV) tests were conducted at a scan rate of 0.1 mV s^{-1}. Electrochemical impedance spectroscopy (EIS) measurements were carried out using a ZIVE-SP1 potentiostat, covering a frequency range of 0.1 to 100 kHz. Galvanostatic charge-discharge tests were performed on a battery test station (Wonatech, Korea) within a voltage window of 0.5–3.0 V (K^+/K).

3. Results and Discussion

Figure 1 illustrates the synthesis process of Se@NCHS. Mo@PDA spheres were initially synthesized as uniform particles with an average size of approximately 1.3 μm, as observed in SEM images (Figure S1a,b). During synthesis, dopamine underwent polymerization, and 2D building blocks self-assembled into spheres with a unique flower-shaped morphology. The Mo@PDA spheres were then carbonized at 750 °C under an argon atmosphere, resulting in the formation of MoO_x@NC. MoO_x@NC particles, approximately 1.4 μm in size, exhibited no significant structural deformation of the carbon observed post-carbonization (Figure S1c,d).

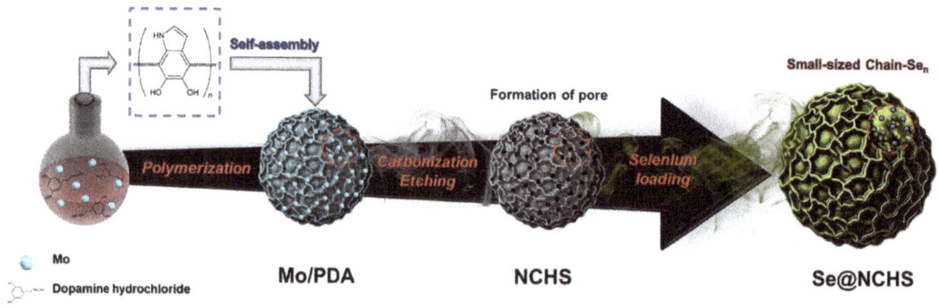

Figure 1. Schematic of Se@NCHS synthesis process.

After etching with aqua regia, the obtained NCHS retained its structure and size (Figure 2a,b). TEM images (Figure 2c,d) revealed a flower-shaped morphology with a large central core, indicating that the petal-like external structure had a higher pore density compared to the inner core. This unique flower-shaped morphology enhances the electrolyte contact area, thereby improving electrochemical performance [11,14]. High-resolution TEM (HR-TEM) images (Figure 2e,f) confirmed that no distinct lattice fringes were present within the carbon matrix of NCHS, while the selected area electron diffraction (SAED) pattern (Figure 2g) exhibited no lattice characteristics corresponding to metallic Mo. These findings confirm the complete removal of MoO_x, resulting in a porous amorphous carbon matrix. Elemental mapping via energy-dispersive X-ray spectroscopy (EDS) (Figure 2h) demonstrated that the carbon spheres primarily consisted of C and N, verifying the successful formation of N-doped carbon material. The NCHS material features a flower-shaped microstructure with uniformly distributed pores and nitrogen-induced defects, which facilitate additional ion diffusion channels and significantly enhance rate performance [15]. Furthermore, the flower-shaped design improves electrolyte accessibility and expands the reaction surface area, thereby maximizing electrochemical efficiency [11,14]. Additionally, the hierarchical architecture, developed through aqua regia etching, ensures a well-organized porous network across the flower-like surface, further optimizing charge transport [13]. These structural benefits collectively result in superior electrochemical performance compared to conventional carbon hosts [16].

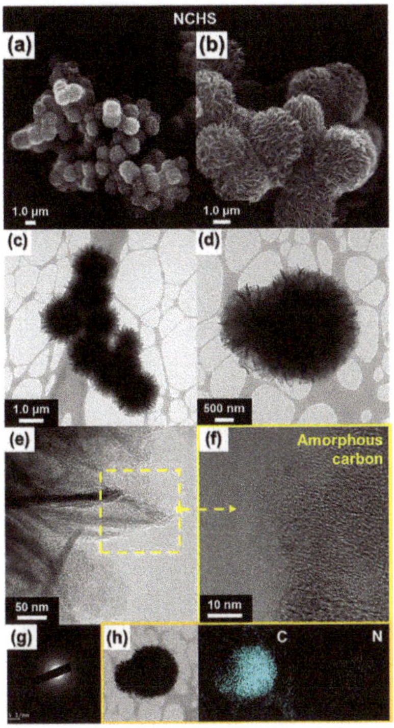

Figure 2. Morphology and structure characterizations of NCHS: (**a**,**b**) SEM images; (**c**,**d**) TEM images; (**e**–**g**) HR-TEM image and inlet SAED pattern, and (**h**) EDX mapping image.

Se@NCHS was synthesized via a melt-diffusion method, effectively embedding Se into the well-distributed pores and the flower-like architecture of NCHS. SEM images (Figure 3a,b) revealed that the Se@NCHS, approximately 1.2 μm in size, retained its structure without significant deformation after Se loading. Notably, no residual Se powder was observed on the outer surface of the carbon after Se loading, indicating the successful encapsulation of Se within the NCHS structure. HR-TEM images, along with SAED patterns, showed no distinct lattice fringes (Figure 3c–e) or lattice structures corresponding to Se, confirming the preservation of the amorphous carbon structure and atomic-scale confinement of Se within the pores of NCHS. The elemental mapping via energy-dispersive X-ray spectroscopy (EDX) of Se@NCHS (Figure 3f) revealed a homogeneous dispersion of C, N, and Se. The unique flower-shaped structure formed by the assembly of 2D building blocks provides a larger contact area than spherical carbon structures [11,14]. This increased surface area allowed for the uniform distribution of Se within the carbon matrix and created additional adsorption within the pores, contributing to the observed results. This structure enabled high surface Se loading, with no unincorporated or overgrown Se observed, and showed that more Se was loaded on the surface than within the internal pores. This distribution is attributed to the hierarchical structure of the NCHS, which was synthesized via aqua regia etching and is characterized by uniformly distributed pores across its flower-shaped surface. This porous architecture enabled Se to efficiently infiltrate the internal pores, thereby improving its utilization. Additionally, nitrogen within the Se@NCHS introduces polar sites with a strong affinity for charged Se species, effectively suppressing undesirable side reactions in the electrolyte [14]. These abundant active sites facilitated K^+ ion transport, enhanced electronic conductivity, promoted the conversion of Se species, accelerated reaction kinetics, and significantly improved redox capacity [14].

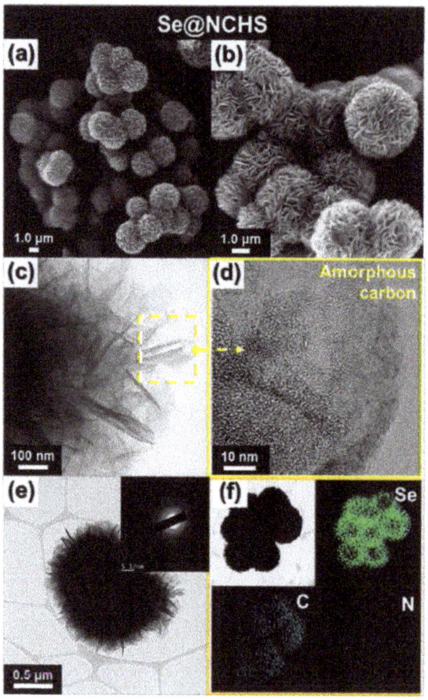

Figure 3. Morphology and structure characterizations of Se@NCHS: (**a**,**b**) SEM images; (**c**,**d**) TEM images; (**e**) HR-TEM image and inlet SAED pattern, and (**f**) EDX mapping image.

The N_2 absorption–desorption isotherms of NCHS, Super P, Se@NCHS, and Se@Super P were analyzed to assess the surface properties of the synthesized materials, as illustrated in Figure 4a. The adsorption-desorption characteristics of NCHS exhibited a type III isotherm, demonstrating substantial nitrogen absorption at low pressure levels and a distinct hysteresis loop at elevated pressures, validating its porous structure [16]. A steep increase in nitrogen adsorption at low relative pressures ($P/P_0 < 0.01$) indicated the presence of micropores. The BET surface area of NCHS was measured at 341.94 $m^2\ g^{-1}$, which was substantially larger than that of Super P (55.71 $m^2\ g^{-1}$). The micropore distribution, determined using the HK method, is presented in Figure 4b. The NCHS sample exhibited a distinct micropore peak at 0.9 nm, whereas Super P displayed a relatively low micropore concentration. The micropore volumes of NCHS and Super P, determined through t-plot analysis, were 0.10 $cm^3\ g^{-1}$ and 0.003 $cm^3\ g^{-1}$, respectively. The structural characteristics of these materials are outlined in Table S1.

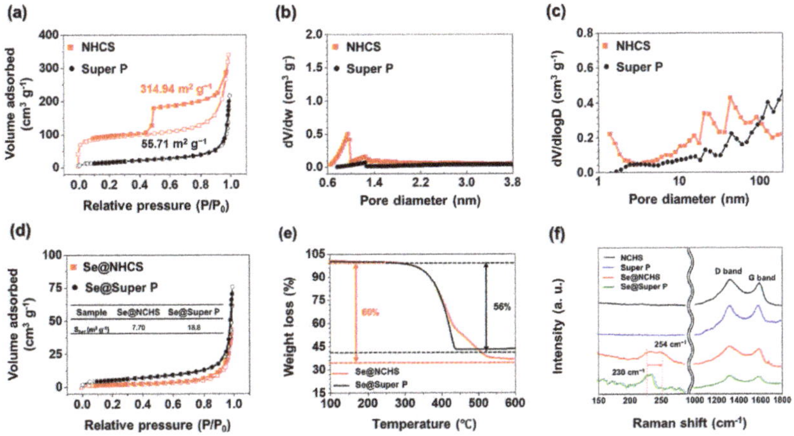

Figure 4. (**a**) N_2 adsorption-desorption isotherm curve, (**b**) HK method differential pore volume plot, (**c**) BJH pore-size distribution graph of NCHS and Super P, (**d**) N_2 adsorption-desorption isotherm curve, (**e**) TGA profiles of Se@NCHS and Se@Super P and (**f**) the Raman spectra of NCHS, Super P, Se@NCHS and Se@Super P.

The BJH pore size distribution curves for NCHS and Super P samples are shown in Figure 4c. Unlike Super P, NCHS has a hierarchical structure comprising mesopores and micropores. The mesopores formed by etching metallic Mo help mitigate volumetric changes in Se [13], enhance capacity retention, and provide stable cycling performance. Furthermore, the micropores generated by N-doping-induced defects are essential for facilitating the formation of chain-like Se_n, thereby improving electrochemical activity [15]. After Se loading, the N_2 adsorption–desorption isotherms (Figure 4d) show a significant reduction in surface areas for Se@NCHS and Se@Super P, measured at 7.69 and 18.78 $cm^3\ g^{-1}$, respectively. This reduction corresponds to a notable decrease in pore volume. The substantial decrease in the surface area of NCHS suggests that Se effectively penetrated the internal pores, blocking them and thereby reducing the available surface area. In contrast, the lower reduction rate observed in Super P indicates that Se did not fully infiltrate its internal pores, with residual Se remaining outside the pore structure [11,15].

TGA was performed to determine Se loading (Figure 4e). The Se content in Se@NCHS was measured at 60 wt.%, indicating that all the Se successfully infiltrated the internal pores during impregnation. The delayed weight loss at higher temperatures suggests that Se was deeply embedded within the inner pores, requiring additional time for sublimation or evaporation [11].

In contrast, Se@Super P exhibited a Se content of 56 wt.%, indicating that the loaded 60 wt% Se did not fully penetrate the internal pores of Super P. Unlike Se@NCHS, Se@Super P exhibited rapid weight loss at lower temperatures, which can be attributed to aggregation issues caused by non-uniform and smaller particle sizes [8]. These limitations restricted the infiltration of Se into the internal pores, resulting in its accumulation on the surface or within interparticle voids.

Raman spectroscopy was employed to analyze the structural properties of Se and carbon hosts in the samples (Figure 4f). The Raman spectra of Se@Super P and Se@NCHS offered key insights into how the carbon matrix affects the structural evolution of Se. For Se@Super P, a distinct peak appeared at 230 cm^{-1}, corresponding to the stretching vibration mode of cyclic Se_8 molecules [17,18]. Conversely, the Se@NCHS sample exhibited a peak shift from 230 cm^{-1} to 254 cm^{-1}, signifying a partial transformation of cyclic Se_8 into chain-like Se_n structures [12]. This shift implies that the microporous structure of NCHS plays a crucial role in facilitating this structural transition during the Se melting-diffusion process. Importantly, the presence of the 254 cm^{-1} peak confirms that the well-developed micropores in NCHS effectively suppress the stabilization of cyclic Se_8 while promoting its structural conversion into chain-like Se_n [18]. Since chain-like Se_n demonstrates superior electrochemical properties compared to cyclic Se_8, this transformation is a key factor in enhancing Se utilization for energy storage applications. Additionally, the structural features of the carbon hosts were examined through the D-band (1321 cm^{-1}) and G-band (1592 cm^{-1}) peaks, which correspond to disordered sp^3-hybridized carbon and graphitic sp^2-hybridized carbon, respectively [19]. The I_D/I_G ratio for NCHS was significantly higher than that of Super P, suggesting that KOH activation generated additional structural defects [20]. These defects are beneficial as strengthen the interaction between the carbon matrix and Se, thereby enhancing the confinement and stabilization of Se within the micropores [21]. Interestingly, after Se loading, the I_D/I_G values for both the Se@NCHS and Se@Super P samples remained unchanged compared to those of their pristine carbon hosts. This finding verifies that the Se melt-diffusion process preserves the structural stability of the carbon framework, further highlighting the durability of NCHS as a host material for Se [21].

CV tests were performed to examine the charge-discharge mechanism of Se@NCHS as a cathode material for K–Se batteries (Figure 5). The measurements were conducted at a scan rate of 0.1 mV s^{-1} within a voltage window of 0.5–3.0 V (vs. K^+/K) (Figure 5a). This voltage range does not include the 0.01–0.5 V region, where carbon interacts with K^+ ions, confirming that carbon primarily functions to enhance conductivity and maintain structural integrity [22]. During the first CV scan of Se@NCHS, three cathodic peaks were observed at 0.7, 1.54, and 2.12 V. The 0.7 V peak significantly diminished in subsequent cycles, most likely due to irreversible capacity loss caused by electrolyte decomposition and the formation of a solid electrolyte interphase (SEI) layer [23]. In contrast, anodic peaks appeared at 1.89, 2.01, and 2.31 V, corresponding to the reversible transition from K_2Se to Se [23]. The first charge-discharge profile of Se@NCHS (Figure 5b) exhibits discharge and charge capacities of 658.0 and 463.0 mA h g^{-1}, respectively, resulting in a coulombic efficiency of 70.4%. The initial decline in coulombic efficiency is mainly due to the irreversible formation of the SEI layer and side reactions involving K^+ ions with functional groups on the carbon surface. [24]. According to prior research [18], suggest that Se reduction in carbonate-based electrolytes follows a stepwise mechanism: Se $\rightarrow$ K_2Se_2 $\rightarrow$ K_2Se. Consequently, the cathodic peaks at 1.54 and 2.12 V in the first CV scan of Se@NCHS correspond to the formation of K_2Se (Se $\rightarrow$ K_2Se_2 $\rightarrow$ K_2Se) [18]. An expanded view of the first discharge plateau for Se@NCHS (Figure 5c) shows that the initial plateau at 2.12 V aligns with the first CV cycle, indicating that the electrochemical reactions are consistent, particularly in the stepwise reduction process of Se [18,21]. This finding confirms that Se is effectively incorporated into the NCHS matrix, demonstrating that the conversion reactions of Se proceeded efficiently within the electrode structure, enabling a high utilization rate of the active material.

The cycling durability of Se@NCHS and Se@Super P electrodes at a current density of 0.5 C over 500 cycles is presented in Figure 6a. While the Se@Super P electrode exhibited negligible capacity, the Se@NCHS sample retained a discharge capacity of 199.4 mA h g^{-1}. This notable variation is mainly due to the enhanced utilization of chain-like Se (Se_n), enabled by the hierarchical structure of NCHS [25]. The minimal capacity retention in Se@Super P can be explained by the voltage window (0.5–3.0 V) selected for K–Se batteries, which is consistent with the range used in CV tests and lies outside the potassium ion interaction range (0.1–0.5 V). Additionally, the coulombic efficiency stabilized at approximately

99% by the 50th cycle and remained consistent in the following cycles (indicated by a black arrow). To examine the carbon host's contribution to cathode capacity, galvanostatic measurements were conducted on NCHS without Se. As shown in the first charge/discharge profile of NCHS (Figure S2a), a substantial decline in capacity occurred after the initial cycle, resulting in a coulombic efficiency of 64.12%, primarily caused by irreversible side reactions. The cycling performance data (Figure S2b) indicate that the capacity of NCHS dropped to nearly zero under conditions identical to Se@NCHS, verifying that the carbon matrix has minimal influence on the overall capacity of K–Se batteries.

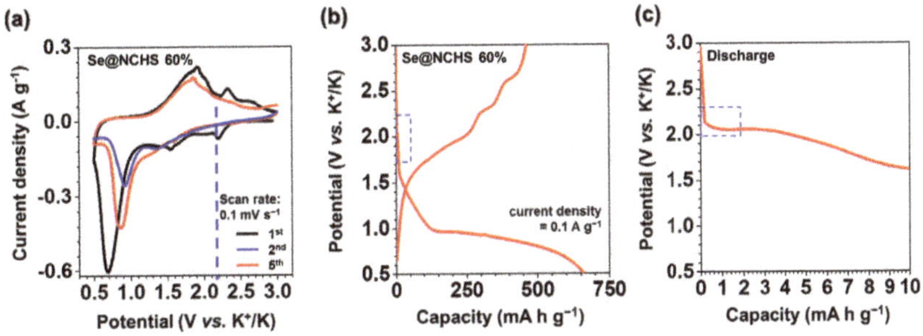

Figure 5. Charge–discharge mechanism of Se@NCHS electrodes; (**a**) CV curve, (**b**) charge-discharge 1st profiles at 0.1 C and (**c**) 1st discharge profile at 0.1 C.

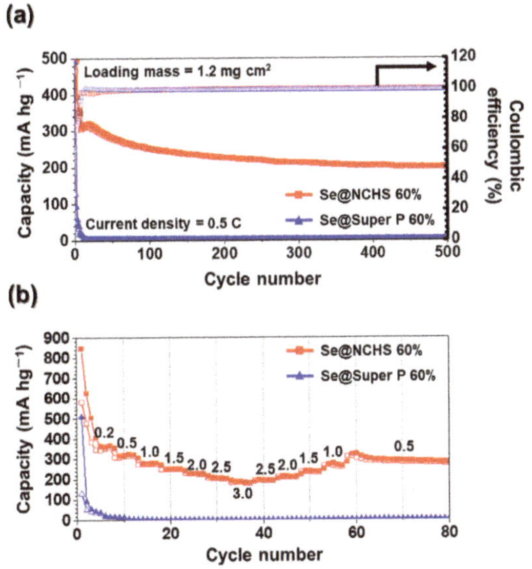

Figure 6. Electrochemical properties of Se@NCHS and Se@Super P electrodes; (**a**) Long-term cycle stability and efficiency of charge transfer at 0.5 C current rate and (**b**) Rate-dependent performance evaluated at varying current densities.

The rate capability of the Se@NCHS and Se@Super P electrodes at various current densities is depicted in Figure 6b. The capacity decline observed during the initial 0.1 C is primarily associated with the stabilization of the electrode-electrolyte interface and structural rearrangements within the electrode. To ensure the accuracy and reproducibility of the data, the discharge capacity of the Se@NCHS electrode was assessed at current densities of 0.2, 0.5, 1.0, 1.5, 2.0, 2.5, and 3.0 C, yielding respective values of 478, 351, 313, 277,

250, 226, 204, and 188 mA h g^{-1}. When the current density reverted to 0.5 C, the capacity retention remained high, reaching 282 mA h g^{-1}. In contrast, the Se@Super P electrode exhibited a sharp decline in capacity, dropping from 53 to 1 mA h g^{-1} as the current density increased from 0.1 to 3.0 C. These results confirm that the enhanced electrochemical performance of Se@NCHS electrode stems from the synergistic effects of its structural architecture and morphological attributes. The hierarchical framework of the Se@NCHS electrode effectively accommodates Se volume fluctuations during cycling and facilitates the formation of chain-like Se (Se_n), thereby significantly improving Se utilization [18,21]. Furthermore, its distinct flower-shaped architecture provides a greater electrolyte contact area compared to spherical carbon, effectively shortening ion diffusion pathways [11].

To examine the influence of the hierarchical architecture and morphology on electrode resistance variations during cycling, ex situ EIS measurements were performed on Se@NCHS and Se@Super P electrodes. In the Nyquist plot, the semicircle radius represents the impedance of the electrode, with a larger semicircle indicating higher resistance [26]. In the fresh state, the Se@Super P electrode (Figure 7a) exhibited a smaller semicircle compared to Se@NCHS, suggesting that the high electrical conductivity of Super P nanoparticles facilitated more efficient electron and ion transport during the initial EIS analysis [26]. Over the course of cycling, the semicircle radius of the Se@NCHS electrode progressively decreased, while that of the Se@Super P electrode exhibited an increased in R_{ct} after cycling (Figure 7a–c). Prior studies indicate that Rct tends to decrease as electronic conductivity increases, primarily due to the intrinsically high conductivity of carbon-based materials [27]. The Se@NCHS electrode demonstrated a continuous reduction in R_{ct} over successive cycles, which can be attributed to enhanced electrolyte penetration and shortened ion diffusion paths, facilitated by its distinct flower-like structure of NCHS. These findings highlight that the hierarchical structure of the Se@NCHS electrode effectively mitigates volume expansion during charge/discharge cycles, thereby preserving structural integrity and stabilizing electrode interfaces. Figure 7d presents the correlation between phase angle ($\omega^{-1/2}$) and real impedance (Z') for both electrodes after the 100 cycles in the low-frequency region. The Se@NCHS electrode displayed a lower slope than Se@Super P electrode, indicating a higher K^+ diffusion rate. This enhancement is attributed to the uniform dispersion of active materials within the electrode, facilitated by its hierarchical structure, which effectively minimizes interfacial resistance at the electrode-electrolyte interface.

Consequently, the decrease in impedance detected in EIS measurements during cycling can be associated with the mitigation of volume expansion and improved ion transport. These findings indicate that the hierarchical structure of NCHS, defined by its uniform pore distribution and precisely controlled particle size, is essential for reducing resistance and enhancing long-term electrochemical performance. To further explore the effect of pore architecture on ion movement, galvanostatic intermittent titration technique (GITT) tests were performed (Figure 8). Moreover, the potassium-ion diffusion coefficient (D_K) for Se@NCHS and Se@Super P was derived using Fick's second law to assess ion diffusion behavior. In GITT, a longer reaction time indicates deeper ion penetration into the electrode, allowing for a more complete reaction with the entire active material and facilitating uniform utilization of active sites. The longer reaction time observed for Se@NCHS compared to Se@Super P suggests higher structural stability, enabling the electrode to maintain a greater capacity during charge-discharge cycles. This structural robustness was credited to the hierarchical structure of the Se@NCHS electrode, which effectively accommodates volume variations, thereby enhancing overall electrochemical performance. Furthermore, Se@NCHS exhibited significantly higher ion diffusion rates during both charge and discharge processes than Se@Super P. This improvement was attributed to the hierarchical structure, which provided additional ion pathways through the uniform pore distribution created by etching and nitrogen-doping-induced defects, thereby enabling faster ion diffusion. Additionally, the unique flower-shaped morphology formed by the assembly of 2D building blocks increased the contact area with the electrolyte relative to spherical carbon spheres, effectively shortening ion diffusion pathways [14,15]. The variation in the diffusion coefficient during the discharge of Se@NCHS can be attributed to the electrochemical reaction of Se, in which the transformation from cyclo-Se_8 to chain-like Se_n affects the diffusion coefficient. This phase transition alters the diffusion pathway, resulting in significant fluctuations in diffusion coefficients within a specific voltage range. In contrast, no significant variation in the diffusion coefficient was observed for Se@Super P, due to the non-uniform and small particle sizes of Super P, which limit Se penetration into the internal pores and result in Se accumulation on the surface, minimizing obstruction of the diffusion pathway. These findings provide strong evidence that the hierarchical structure, uniform pore distribution, and particle size of NCHS contribute to enhanced electrochemical performance.

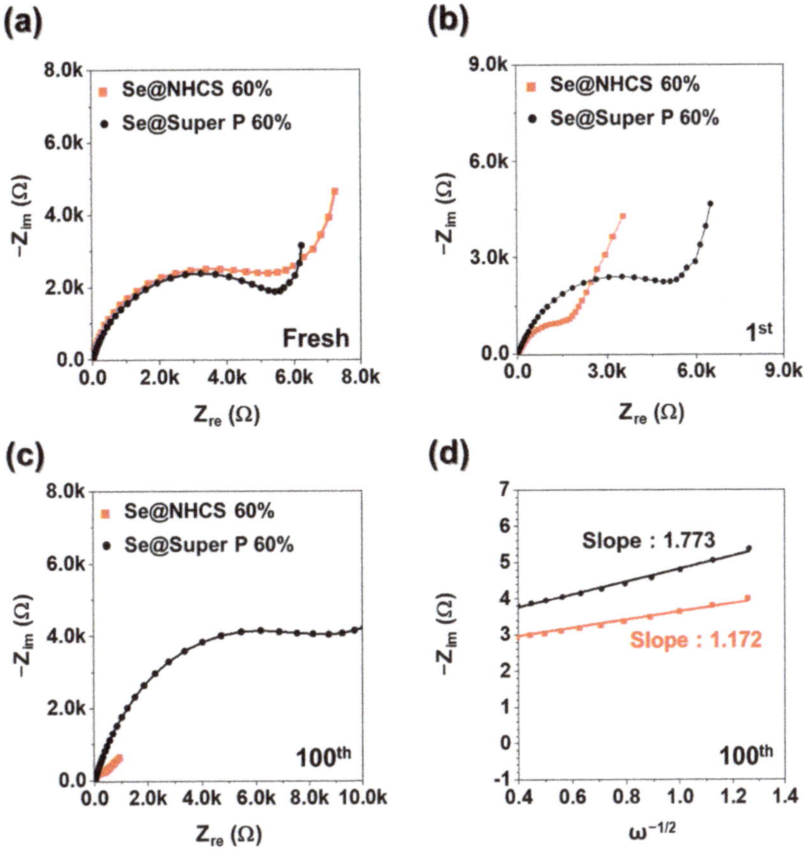

Figure 7. EIS Nyquist plots of Se@NCHS and Se@Super P electrodes; (**a**) at fresh cells, (**b**) after the first cycle, (**c**) after 100 cycles and (**d**) Z_{re}–$w^{-1/2}$ relationship after the 100 cycles.

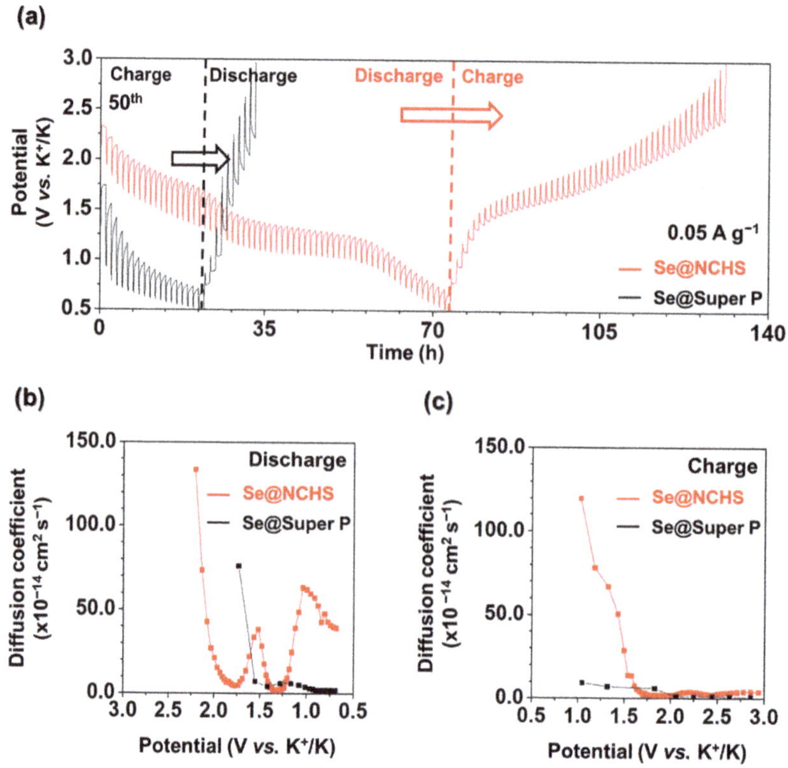

Figure 8. GITT analysis of Se@NCHS and Se@Super P electrodes: (**a**) GITT profiles, the calculated ion diffusion coefficient values (**b**) during discharging process and (**c**) during charging process.

4. Conclusions

This study demonstrated that an NCHS, synthesized through a simple process, significantly enhanced the electrochemical performance of K–Se batteries. The uniform pore distribution formed on the outer surface of the flower-shaped structure during Mo etching minimized Se aggregation and surface accumulation, thereby addressing key challenges in Se-based batteries. The uniformly synthesized flower-like microstructures constructed from 2D building blocks improved electrolyte accessibility, optimized ion diffusion pathways, and ensured long-term cycling stability. Additionally, the uniform particle size of NCHS mitigated aggregation issues, further improving performance. N doping, naturally introduced during the synthesis process, strengthened the chemical bonding between Se and the carbon host, improved interactions, suppressed side reactions, and played a key role in enhancing electrochemical performance. This unique flower-like hierarchical structure contributed significantly to the potential of the material as a cathode host for K–Se batteries, providing valuable insights into the role of pore architecture and doping in electrochemical performance.

Supplementary Materials: The following supporting information can be downloaded at: https://www.mdpi.com/article/10.3390/batteries11030101/s1, Figure S1: SEM images of Mo@PDA and MoO_X@NC; Figure S2: The initial galvanostatic charge-discharge curves and cycling stability of NCHS electrodes; Table S1: Textural Properties of NCHS and Super P; Table S2: Electrochemical properties of various nanostructured Se cathode host for K–Se batteries reported in previous literatures.

Author Contributions: H.-J.K.: Conceptualization, Investigation, Data curation, Formal analysis, Writing—original draft. J.-H.N.; Conceptualization, Validation, Writing—review & editing. S.-K.P.: Supervision, Conceptualization, Funding acquisition, Writing—review & editing. All authors have read and agreed to the published version of the manuscript.

Funding: This research was supported by the Chung-Ang University Research Scholarship Grant in 2023 and the Materials/Parts Technology Development Program (No. RS-2024-00456324) funded by the Ministry of Trade, Industry & Energy (MOTIE, Korea).

Data Availability Statement: Dataset available on request from the authors.

Conflicts of Interest: The authors declare no conflicts of interest.

References

1. Nwanna, E.C.; Bitire, S.; Imoisili, P.E.; Jen, T.-C. An overview of the application of atomic layer deposition process for lithium-ion based batteries. *Int. J. Energy Res.* **2022**, *46*, 10499–10521. [CrossRef]
2. Zhang, Q.; Wang, C.; Xie, Z.; Zhou, Z. Defective/Doped Graphene-Based Materials as Cathodes for Metal–Air Batteries. *Energy Environ. Mater.* **2022**, *5*, 1103–1116. [CrossRef]
3. Liao, J.; Chen, C.; Hu, Q.; Du, Y.; He, Y.; Xu, Y.; Zhang, Z.; Zhou, X. A low-strain phosphate cathode for high-rate and ultralong cycle-life potassium-ion batteries. *Angew. Chem.* **2021**, *133*, 25779–25786. [CrossRef]
4. Xu, R.; Yao, Y.; Wang, H.; Yuan, Y.; Wang, J.; Yang, H.; Jiang, Y.; Shi, P.; Wu, X.; Peng, Z. Unraveling the nature of excellent potassium storage in small-molecule Se@peapod-like N-doped carbon nanofibers. *Adv. Mater.* **2020**, *32*, 2003879. [CrossRef]
5. Yang, J.; Ju, Z.; Jiang, Y.; Xing, Z.; Xi, B.; Feng, J.; Xiong, S. Enhanced capacity and rate capability of nitrogen/oxygen dual-doped hard carbon in capacitive potassium-ion storage. *Adv. Mater.* **2018**, *30*, 1700104. [CrossRef]
6. Tai, Z.; Zhang, Q.; Liu, Y.; Liu, H.; Dou, S. Activated carbon from the graphite with increased rate capability for the potassium ion battery. *Carbon* **2017**, *123*, 54–61. [CrossRef]
7. Yao, Y.; Chen, M.; Xu, R.; Zeng, S.; Yang, H.; Ye, S.; Liu, F.; Wu, X.; Yu, Y. CNT interwoven nitrogen and oxygen dual-doped porous carbon nanosheets as free-standing electrodes for high-performance Na-Se and K-Se flexible batteries. *Adv. Mater.* **2018**, *30*, 1805234. [CrossRef]
8. Chung, S.-H.; Yu, C.-H. Design and development of high-loading carbon-sulfur nanocomposite cathodes with drop-casting method. In *Electrochemical Society Meeting Abstracts*; The Electrochemical Society, Inc.: Philadelphia, PA, USA; p. 347.
9. Shrestha, S.; Wang, B.; Dutta, P. Nanoparticle processing: Understanding and controlling aggregation. *Adv. Colloid Interface Sci.* **2020**, *279*, 102162. [CrossRef]
10. Wang, H.; Jiang, Y.; Manthiram, A. Long cycle life, low self-discharge sodium–selenium batteries with high selenium loading and suppressed polyselenide shuttling. *Adv. Energy Mater.* **2018**, *8*, 1701953. [CrossRef]
11. Li, H.; Dong, W.; Li, C.; Barakat, T.; Sun, M.; Wang, Y.; Wu, L.; Wang, L.; Xia, L.; Hu, Z.-Y. Three-dimensional ordered hierarchically porous carbon materials for high performance Li-Se battery. *J. Energy Chem.* **2022**, *68*, 624–636. [CrossRef]
12. Deng, N.; Feng, Y.; Wang, G.; Wang, X.; Wang, L.; Li, Q.; Zhang, L.; Kang, W.; Cheng, B.; Liu, Y. Rational structure designs of 2D materials and their applications toward advanced lithium-sulfur battery and lithium-selenium battery. *Chem. Eng. J.* **2020**, *401*, 125976. [CrossRef]
13. Deng, K.; Erofeev, I.; Chowdhuri, A.R.; Philipsen, H.; Aabdin, Z.; Hartanto, A.W.; Tjiu, W.W.; Zhang, M.; Fernando, D.; Saidov, K. Nanoscale wet etching of molybdenum interconnects with organic solutions. *Small* **2024**, *20*, 2406713. [CrossRef] [PubMed]
14. Gao, F.; Yue, X.-A.; Xu, X.-Y.; Xu, P.; Zhang, F.; Fan, H.-S.; Wang, Z.-L.; Wu, Y.-T.; Liu, X.; Zhang, Y. A N/Co co-doped three-dimensional porous carbon as cathode host for advanced lithium–selenium batteries. *Rare Met.* **2023**, *42*, 2670–2678. [CrossRef]
15. Mo, R.; Rooney, D.; Sun, K.; Yang, H.Y. 3D nitrogen-doped graphene foam with encapsulated germanium/nitrogen-doped graphene yolk-shell nanoarchitecture for high-performance flexible Li-ion battery. *Nat. Commun.* **2017**, *8*, 13949. [CrossRef]
16. Zhao, C.; Hu, Z.; Luo, J. Porous carbon nanoplate/Se composite derived from potassium citrate as high-performance Li-Se battery cathode: A study on structure-function relation. *Colloids Surf. A Physicochem. Eng. Asp.* **2019**, *560*, 69–77. [CrossRef]
17. Zhao, Q.; Meng, Y.; Su, L.; Cen, W.; Wang, Q.; Xiao, D. Nitrogen/oxygen codoped hierarchical porous Carbons/Selenium cathode with excellent lithium and sodium storage behavior. *J. Colloid Interface Sci.* **2022**, *608*, 265–274. [CrossRef]
18. Wu, X.; Chen, X.; Wu, H.; Xie, B.; Wang, D.; Wang, R.; Zhang, X.; Piao, Y.; Diao, G.; Chen, M. Encapsulation of Se in dual-wall hollow carbon spheres: Physical confinement and chemisorption for superior Na–Se and K–Se batteries. *Carbon* **2022**, *187*, 354–364. [CrossRef]
19. Dong, L.-l.; Chen, W.-G.; Deng, N.; Zheng, C.-H. A novel fabrication of graphene by chemical reaction with a green reductant. *Chem. Eng. J.* **2016**, *306*, 754–762. [CrossRef]

20. Fu, H.; Zhang, J.; Zhao, L.; Huang, Y.; Chen, B. Investigations of NO reduction by coal-based activated carbon with KOH activation: Performance and mechanism. *Chemosphere* **2024**, *346*, 140506. [CrossRef]
21. Zhang, G.; Wu, G.; Li, J.; Wang, Y.; Xu, S.; Niu, X.; Wu, R.; Jiang, J.; Blackwood, D.J.; Chen, J.S. KOH-activated hollow carbon spheres with surface functionalization for high-capacity and long-cycle-life lithium-selenium batteries. *J. Colloid Interface Sci.* **2024**, *674*, 852–861. [CrossRef]
22. Alvin, S.; Cahyadi, H.S.; Hwang, J.; Chang, W.; Kwak, S.K.; Kim, J. Revealing the intercalation mechanisms of lithium, sodium, and potassium in hard carbon. *Adv. Energy Mater.* **2020**, *10*, 2000283. [CrossRef]
23. Zhou, L.; Cui, Y.; Kong, D.; Feng, W.; Gao, X.; Yan, Y.; Ren, H.; Hu, H.; Xue, Q.; Yan, Z. Amorphous Se species anchored into enclosed carbon skeleton bridged by chemical bonding toward advanced K-Se batteries. *J. Energy Chem.* **2021**, *61*, 319–326. [CrossRef]
24. Zhou, X.; Wang, L.; Yao, Y.; Jiang, Y.; Xu, R.; Wang, H.; Wu, X.; Yu, Y. Integrating conductivity, captivity, and immobility ability into N/O dual-doped porous carbon nanocage anchored with CNT as an effective Se host for advanced K-Se battery. *Adv. Funct. Mater.* **2020**, *30*, 2003871. [CrossRef]
25. Cheng, L.; Ma, C.; Lu, W.; Wang, X.; Yue, H.; Zhang, D.; Xing, Z. A graphitized hierarchical porous carbon as an advanced cathode host for alkali metal-selenium batteries. *Chem. Eng. J.* **2022**, *433*, 133527. [CrossRef]
26. Allison, A.; Andreas, H.A. Minimizing the nyquist-plot semi-circle of pseudocapacitive manganese oxides through modification of the oxide-substrate interface resistance. *J. Power Sources* **2019**, *426*, 93–96. [CrossRef]
27. Youn, D.H.; Heller, A.; Mullins, C.B. Simple synthesis of nanostructured Sn/nitrogen-doped carbon composite using nitrilotriacetic acid as lithium ion battery anode. *Chem. Mater.* **2016**, *28*, 1343–1347. [CrossRef]

Disclaimer/Publisher's Note: The statements, opinions and data contained in all publications are solely those of the individual author(s) and contributor(s) and not of MDPI and/or the editor(s). MDPI and/or the editor(s) disclaim responsibility for any injury to people or property resulting from any ideas, methods, instructions or products referred to in the content.

batteries

MDPI

Article

Surface Fluorination for the Stabilization in Air of Garnet-Type Oxide Solid Electrolyte for Lithium Ion Battery

Michael Herraiz [1], Saida Moumen [1], Kevin Lemoine [1], Laurent Jouffret [1], Katia Guérin [1], Elodie Petit [1], Nathalie Gaillard [1], Laure Bertry [2], Reka Toth [3], Thierry Le Mercier [2], Valérie Buissette [2] and Marc Dubois [1,*]

1 Institut de Chimie de Clermont-Ferrand (UMR 6296), Université Clermont Auvergne, CNRS, BP 10448, F-63000 Clermont-Ferrand, France; michael.herraiz@uca.fr (M.H.); saida.moumen@doctorant.uca.fr (S.M.); kevin.lemoine@uca.fr (K.L.); laurent.jouffret@uca.fr (L.J.); katia.araujo_da_silva@uca.fr (K.G.); elodie.petit@uca.fr (E.P.); nathalie.gaillard@uca.fr (N.G.)

2 Centre de Recherche & Innovation, Syensqo, 52 Rue de la Haie Coq, 93300 Aubervilliers, France; thierry.le-mercier@syensqo.com (T.L.M.); valerie.buissette@syensqo.com (V.B.)

3 Solvay, 9 rue des Cuirassiers, 69003 Lyon, France; reka.toth@solvay.com

* Correspondence: marc.dubois@uca.fr

Abstract: After reviewing the state of the art of the fluorination of inorganic solid electrolytes, an application of gas/solid fluorination is given and how it can be processed. Garnet-type oxide has been chosen. These oxides with an ideal structure of chemical formula $A_3B_2(XO_4)_3$ are mainly known for their magnetic and dielectric properties. Certain garnets may have a high enough Li^+ ionic conductivity to be used as solid electrolyte of lithium ion battery. The surface of LLZO may be changed in contact with the moisture and CO_2 present in the atmosphere that results in a change of the conductivity at the interface of the solid. LiOH and/or lithium carbonate are formed at the surface of the garnet particles. In order to allow for handling and storage under normal conditions of this solid electrolyte, surface fluorination was performed using elemental fluorine. When controlled using mild conditions (temperature lower or equal to 200 °C, either in static or dynamic mode), the addition of fluorine atoms to LLZO with $Li_{6,4}Al_{0,2}La_3Zr_2O_{12}$ composition is limited to the surface, forming a covering layer of lithium fluoride LiF. The effect of the fluorination was evidenced by IR, Raman, and NMR spectroscopies. If present in the pristine LLZO powder, then the carbonate groups disappear. More interestingly, contrary to the pristine LLZO, the contents of these groups are drastically reduced even after storage in air up to 45 days when the powder is covered with the LiF layer. Surface fluorination could be applied to other solid electrolytes that are air sensitive.

Keywords: garnet-type oxides; lithium ion battery; solid electrolyte; fluorination; surface treatment

1. The State of the Art

The potential of solid electrolytes in the energy field for Li ion batteries (LiBs) has been the subject of considerable interest. This novel class of electrolytes has demonstrated a multitude of benefits for Li ion batteries, effectively circumventing a substantial challenge inherent to liquid electrolytes. One such issue is safety, given the flammable nature of the electrolyte used in Li ion batteries. Meanwhile, solid electrolytes have demonstrated considerable efficacy in limiting this issue by impeding the growth of Li dendrites on the Li anode [1,2]. Furthermore, solid electrolytes are distinguished by their high energy density and high ionic conductivity, which can approach that of liquid electrolytes [3,4]. A variety of solid electrolyte types have been developed and tested for use in LiBs. Notably, organic solid electrolytes utilizing polymers such as polyethylene oxide (PEO), polyacrylonitrile (PAN), polyethylene carbonate (PEC), polypropylene carbonate (PPC), and polymethyl methacrylate (PMMA) have been developed [5–7]. Furthermore, inorganic solid electrolytes are gaining popularity for use in LiBs due to their high ionic conductivities. This category encompasses numerous subcategories, including oxides such as LISICON, garnets LLMO (M = Zr, Ta, Nb), perovskites ABO_3 (A = Li, Ca, Sr, La/B = Al or Ti), LIPON, LATP, and sulfides like thio-LISICON and argyrodites [8–11]. Each category offers valuable performance characteristics for LiBs. A third category of solid electrolytes has recently been developed, designated as hybrid solid electrolytes. The mixture comprises organic and inorganic materials, which are combined in order to take advantage of the characteristics of both [12–14]. Garnet-type oxides have an ideal structure with the chemical formula $A_3B_2(XO_4)_3$ and generally crystallize in a centered cubic lattice belonging to the $Ia\bar{3}d$ space group. The cationic sites A, B, and X have an oxygen coordination of VIII, VI, and IV, respectively. Synthetic garnets are mainly known for their magnetic and dielectric properties. However, it has been observed that some garnets can exhibit sufficiently high Li^+ ionic conductivity for use as solid electrolytes in lithium batteries [15–18]. In 2007, for example, teams succeeded in preparing a new garnet with the formula $Li_7La_3Zr_2O_{12}$ (LLZO) and achieved a total conductivity of the order of 3.10^{-4} S/cm [19]. Other studies have also shown that ionic conductivity is highest when the garnet has a cubic rather than a tetragonal structure. Other teams have shown that ionic conductivity is enhanced when the LLZO-type garnet includes another chemical element such as aluminum or niobium. Because of their high conductivity, LLZO-type garnets can be used as solid electrolytes in lithium batteries.

Despite the numerous advantages exhibited by solid electrolytes in Li ion batteries, there are certain properties that require further enhancement if solid electrolytes are to become the dominant technology in the field of energy storage. The use of solid electrolytes is constrained by several limitations, particularly in regard to their handling and the range of applications for which they are suitable. The surface of LLZO-type garnets can be modified by contact with moisture and CO_2 present in the atmosphere, leading to a drastic change in conductivity at the solid interface [16,20]. Indeed, LiOH or lithium carbonate is observed to form on the surface of garnet particles when they are in contact with an ambient atmosphere [21]. It would therefore be useful to have garnets with a suitable ionic conductivity for use as a solid electrolyte in a lithium battery that can be stored and handled under normal conditions.

The most prevalent challenge associated with solid electrolytes pertains to their chemical, thermal, and electrochemical stability, which is largely attributed to interfacial side reactions [22–28]. The chemical stability of certain solid electrolytes may be compromised when they are exposed to ambient air. This reaction results in the degradation of the electrolyte, which manifests itself in various ways. For instance, sulfide solid electrolytes generate toxic gases such as hydrogen sulfide (H_2S) [22,29], while oxide solid electrolytes undergo a H^+/Li^+ exchange [13,30]. Additionally, garnet-type LLZO electrolytes formed layers of impurities at the surface, including Li_2CO_3 and LiOH. It is therefore imperative that this type of electrolyte be handled in an inert environment [22,31]. Furthermore, electrochemical stability is a crucial property that must be evaluated. Some solid electrolytes exhibit a narrow electrochemical window or unstable interfaces between the electrolyte and the electrode, stemming from the formation of parasitic reactions at the interface during charge/discharge cycles [22,32]. Mechanically, properties of the material may be compromised by the formation of lithium dendrites, the development of cracks, and the propagation of fractures [22,33]. Ultimately, thermal stability is a crucial factor in battery safety. During the cycling process, certain parasitic reactions may occur, resulting in the decomposition of the electrolyte and an increase in temperature. An increase in temperature can result in the decomposition of the electrolyte, which may subsequently lead to short circuits [22,34,35]. Therefore, solid electrolytes must possess sufficient thermal resistance of functioning under thermal conditions.

A novel strategy has recently been devised with the objective of enhancing the properties of solid electrolytes, with a particular focus on stability. The doping of materials has been identified as a means of enhancing the performance of batteries. Among the various doping types, fluorination, also known as halogenation, is worthy of particular mention. Fluorination can be defined as the introduction of fluorine ions into the structure of a solid electrolyte. Many research works have demonstrated the beneficial effects of fluorination on solid electrolytes. Indeed, the incorporation of fluorine ions results in the formation of a passivation layer of LiF at the interface between the solid electrolyte and the lithium anode [35,36]. This interface is designated the Solid Electrolyte Interface (SEI), and it plays a pivotal role in optimizing battery performance. The formation of this layer enables the blocking of lithium dendrite growth, thereby ensuring the longevity of batteries and preventing safety issues [37]. Moreover, numerous studies have demonstrated the distinction between solid electrolytes prior to and following fluorination. The formation of the LiF layer allows for the creation of a broad and stable electrochemical window following fluorination. The electrolyte structure's electrochemical stability under high voltages can be obtained by introducing fluorine, which is known for its strong oxidation resistance [38,39]. Solid electrolytes were more stable during battery cycling than normal electrolytes as was it confirmed in many studies, such as the case of $Li_2ZrCl_{6-x}F_x$ [40], $Li_3YBr_{5.7}F_{0.3}$ [41], and more other electrolytes. Furthermore, the formation of the SEI layer provides protection for the solid electrolyte, which has beneficial mechanical properties due to the increased roughness and robustness of the layer [42]. Additionally, the fluorination of solid electrolytes enhances the ionic conductivity. As evidenced by the findings of Lu et al., the ionic conductivity of $Li_{6.25}Ga_{0.25}La_3Zr_2O_{12}$ (LGLZO) increased from 5.43×10^{-4} S/cm to 1.28×10^{-3} S/cm after fluorination [43].

The effect of surface fluorination was also investigated. Li et al. demonstrated that the introduction of LiF into LLZO increases the stability of the electrolyte against moisture and CO_2. The LiF layer protects the LLZO garnet surface from the formation of Li_2CO_3, hence the critical role of LiF in protecting the electrolyte surface [44], as well as enhancing interfacial reactions with lithium metal. This improves resistance to dendrite penetration and supports higher current densities without short circuiting [45]. It was also demonstrated that the fluorination of the electrode surface results in the formation of a passive layer of LiF at the interface. This layer played a pivotal role in stabilizing and protecting the electrolyte from decomposition [46,47]. Moreover, electrochemical characterization was conducted on LiF-coated Li metal and Li metal prior to fluorination. During the cycling process, the anode with the LiF coating demonstrated stability. However, the anode lacking LiF coating is not stable, and the formation of parasitic reactions on the surface results in short circuits [47]. Furthermore, the stabilizing effect of the LiF layer was also evident in other findings by fluorinating the surface of an anode material, $Li_4Ti_5O_{12}$. The formation of the LiF layer protected the anode surface from carbonates, reduced interfacial reactions between the electrode/electrolyte, and reduced the generation of gases like CO_2 during cycling [48]. These advantages were previously proven by fluorinating the surface of graphite anode for Li ion batteries. Surface analyses revealed a significant reduction in oxygen and hydrogen content, which positively impacts electrochemical performances by reducing unwanted side reactions [49]. The same benefits of the LiF layer were obtained in the research works of Xiong et al. The formation of the LiF layer on the surface of the

MXene/TiO_2 membrane was efficient in suppressing lithium dendrites and reinforcing the SEI [50].

Surface fluorination has emerged as a key technique to enhance the performance and durability of lithium ion battery (LiB) anodes and cathodes. For anodes, surface fluorination creates a protective lithium fluoride (LiF) layer that stabilizes the Solid Electrolyte Interphase (SEI), reduces side reactions, and prevents dendrite formation. Similarly, cathodes benefit from surface fluorination by improving interfacial stability, mitigating transition metal dissolution, and enhancing high-voltage cycling performance. Studies have demonstrated significant improvements in capacity retention and cycle life through controlled fluorination methodologies.

Surface fluorination typically involves gas/solid or plasma-assisted techniques using elemental fluorine (F_2) or fluorine-containing gases such as XeF_2. For example, the fluorination of $LiNi_xCo_xMn_{1-2x}O_2$ (NCM) cathodes at rates of 2–5 atomic percent has been shown to produce an optimal LiF layer, balancing protection and ionic conductivity. Similarly, fluorinated graphite anodes with coverage rates around 10–15% have demonstrated enhanced electrochemical stability [51]. These methodologies involve moderate temperatures (below 200 °C) to prevent bulk fluorination, which could degrade the underlying material structure. The fluorination of NCM cathodes has led to an increase in capacity retention by 20–30% over 500 cycles. Additionally, high-voltage cathodes such as $LiCoO_2$ (LCO) have shown improved performance, with a reduction in capacity fade by up to 50% due to the stabilizing LiF layer at the interface [52].

For graphite anodes, fluorination has reduced initial capacity losses by 15–25% while extending cycle life. The LiF layer formed during fluorination reduces the decomposition of organic electrolytes and enhances lithium ion transport through the SEI. The fluorination of SiO_x materials also leads to the formation of an LiF-rich SEI layer, which enhances the cycling stability of lithium ion batteries [53]. In the case of LTO, the surface fluorination of $Li_4Ti_5O_{12}$ (LTO) has been investigated by the decomposition of solid XeF_2 [54]. The surface fluorination of LTO passivates its surface, reducing side reactions with the electrolyte and improving the overall electrochemical stability. For instance, the fluorination process was found to influence the Ti^{4+} catalytic properties, thereby affecting gas formation within the cells. Surface fluorination did not alter the bulk structure of LTO but modified its surface reactivity, leading to enhanced electrochemical performance.

The aim of the present work is to provide a proof of concept of the surface fluorination applied to a solid electrolyte, i.e., garnets, without degrading their physicochemical properties. As a consequence, their ionic conductivity as a solid electrolyte in lithium batteries (not measured in the present work) is expected to be unchanged.

In practice, surface fluorination is not easy to achieve experimentally, as the active phase of LLZO (cubic structure) is extremely sensitive to processing conditions, as it is not thermodynamically stable at room temperature and is highly sensitive to humidity. Direct liquid fluorination would appear to be out of the question, hence the need to study the feasibility of gaseous fluorination. A heterogenous gas/solid fluorination with elemental fluorine (F_2) is considered. The gas/solid fluorination process using molecular fluorine is designed to be easy to implement without adding any material other than LiF in very small quantities. It therefore differs from alternative stabilization approaches such as ALD coating, ZrO_2 shell formation, or Li_3PO_4 interfacial layers requiring additives or complex methods.

2. A New Approach for the Stabilization of Inorganic Solid Electrolyte

2.1. Material

The garnet used is an Al-doped LLZO obtained by the method described in ref. [21,55]. The starting LLZO was obtained according to the 2-step procedure described in patent US20220278355. In brief, a La/Zr/Al precursor is first obtained by co-precipitation of nitrate salts with ammonia in water, followed by calcination in the presence of a templating agent. The precursor is then impregnated by a $LiNO_3$ aqueous solution and calcined successively at 900 °C and 1000 °C. The cations have the following relative composition determined by ICP: $Li_{6.97}Al_{0.22}La_3Zr_{1.98}$.

2.2. Fluorination

For the fluorination, 99.9% pure elemental fluorine (F_2) was used (with an HF impurity level < 0.5 vol% and an O_2/N_2 impurity of around 0.5 vol%). First of all, the reactivity of LLZO towards elemental fluorine was evaluated under an initial pressure of pure F_2 gas equal to 200 mbar. The pressure in the closed reactor (static mode) was controlled (Figure 1). After 200 mbar of fluorine gas was injected into the reactor without LLZO pressure initially under primary vacuum (10^{-3} mbar), the temperature increased in the system, resulting in a pressure increase. After this variation was demonstrated, the same experiment was carried out with LLZO (500 mg). The difference in terms of pressure evolution with and without LLZO was therefore related to the consumption of F_2 molecules or the decomposition of the sample with gas evolution. Here, for the same quantity of fluorine injected into the reactor, the response with the sample is slightly lower, giving the first indication of reactivity. The gas flows—both fluorine and inert gas for purging—are carefully controlled by mass flow controllers. Each fluorination experiment is repeated to ensure reproducibility, and pressure vs. time and pressure vs. temperature profiles are systematically compared across runs. Additionally, we have successfully scaled up the process from a few hundred milligrams to several grams of material without observing changes in air stability, confirming the robustness of the method.

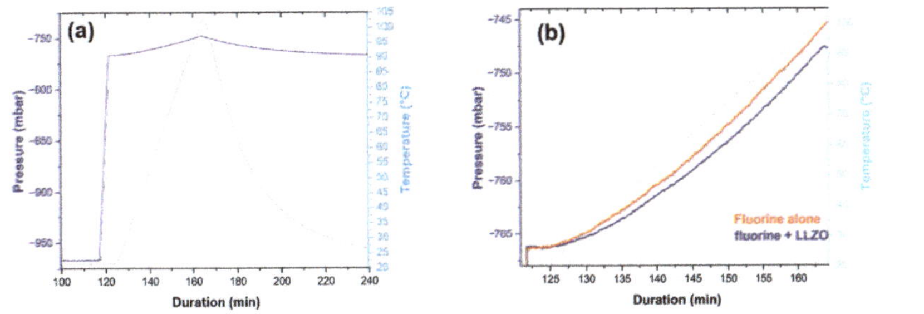

Figure 1. (**a**) Temperature and pressure curves for fluorine in the reactor as a function of time, and (**b**) zoom in on the difference in temperature behavior of fluorine with and without a sample.

In order to cover a wide range of fluorination conditions, several samples were treated:

- Static fluorination at room temperature (RT) with a F_2 pressure of 300 mbar for 1 h (**Stat-1 h**);
- Dynamic fluorination at RT for 1 h (**Dyn-1 h**);
- Dynamic fluorination at RT for 2 h (**Dyn-2 h**).

For each treatment, 500 mg of LLZO was deposited in a passivated nickel boat as a bed of powder with a thickness of less than 2 mm. The plate was inserted into the 1 L reactor at 25 °C. After the fluorination, the reactor was flushed with N_2 (50 mL/min) for an hour and a half. When the temperature is different than room temperature, the reactor was left to cool freely under N_2 flux.

2.3. Physicochemical Characterization

Powder X-ray diffraction (XRD) patterns were obtained using a Malvern Panalytical Xpert Pro diffractometer equipped with a Cu source monochromated in the Kα wavelength (1.541874 Å)with a $K\alpha_2/K\alpha_1$ intensity ratio of 0.50. Scans were collected in the range of 10° to 90°, with a step size of 0.033° and counting time of 200 s. Diffraction patterns were analyzed using X'Pert HighScore Plus Software 4.9 [56].

Fourier-transform infrared (FTIR) spectra were recorded on a Nicolet 6700 FT-IR (Thermo scientific, Waltham, MA, USA) spectrometer in transmission mode at 4 cm^{-1} resolution, and 256 scans were taken for each spectrum.

The samples were analyzed by Raman spectroscopy on a T64000 spectrometer from the company Jobin-Yvon (Edison, NJ, USA) equipped with a confocal microscope. The spectra were recorded after storing for 2 months in a sealed flask under ambient conditions. The incident laser used was an ionized argon solid laser operating at 532 nm. The incident power of the laser was 100 mW. The Raman analyses were carried out in the range of 250–1500 cm^{-1} with acquisition times of 60 s for each window and a spectral width of 500 cm^{-1}, repeated twice.

Then, ^{19}F and ^{7}Li solid-state nuclear magnetic resonance (NMR) spectra were acquired on a Bruker AVANCE spectrometer where 2.5 zirconia rotors rotating at 30 kHz were used. For MAS (magic angle spinning) spectra, a simple sequence was performed with a single $\pi/2$ pulse length of 4.0 and 3.5 μs for ^{19}F and ^{7}Li, respectively. The ^{7}Li NMR was performed at a frequency of 116.59 MHz and LiCl was used as the reference. The ^{19}F NMR was carried out with a frequency of 282.2 MHz, and the spectra were externally referenced to $CFCl_3$. For the determination of the proportion of fluorine by NMR, poly(vinylidene fluoride) (PVDF SOLEF 1015/1001) was used as an internal standard. The quantification was carried out using a Bruker 400 MHz Avance Neo spectrometer equipped with a 1.9 mm probe.

For Scanning Electron Microscopy–Energy Dispersive Spectroscopy (SEM–EDS) analysis, the powder was embedded in a Epofix resin, which polymerized at room temperature over 24 h. After polymerization, the solid containing the LLZO powder undergoes a section on a microtome setup (Reichert and Jung Ultracut E model) under dry conditions, making the section of some solid particles accessible. Then, the surface of the preparation was treated by a platinum sputtering under secondary vacuum in a Cressington 208HR sputter coater. The deposited thickness was a few nanometers. The preparation was introduced in SEM FEG LEO 1525. SEM–EDS analysis was performed at 8 kV with diaphragm 60 μm and working distance 8.5 mm. The EDS spectrum was analyzed by an Oxford SDD 80 mm^2 detector X Max N cooled by Peltier effect. Data treatment was conducted under AZTEC software V4.4 after bean optimization on a silicon standard. Line profiles are acquired at magnification 2000 with 500 data points for a length that is typically 20 mm. Under these conditions at 8 kV, the analyzed volume was <1 μm^3. The absolute intensity of the elements F (K lines) and La (M lines) was measured as a function of the position on the line profile.

For X-ray Photoelectron Spectroscopy (XPS), the powdered samples were pressed on an indium pellet. The XPS instrument was a THERMO K-alpha+ with a monochromatized AlKα X-ray source. The data treatment software was Avantage 3.1. The atomic concentrations are obtained from high-resolution spectra for each element. XPS analysis was conducted on samples as is and with Ar^+ ions. In order to give an order of magnitude of the sputtered thickness, we referred to the sputtering rate of SiO_2 in the following results. The analytical and instrumental specifications were as follows: take-off angle of 90°, depth of analysis lower than 10 nm (average 3 nm), spot diameter of 400 μm, all elements were detectable except H and He, sensitivity limits from 0.1% to 0.5at., quantitative analysis with accuracy of 10 to 20%, and precision 2 to 5% (relative) depending on the concentration.

3. Results

3.1. Fluorination Is Limited to the Outmost Surface

To gain insight into the structure before and after fluorination, X-ray diffraction analyses of the powders were carried out (Figure 2). The overall structure of the sample appears unaffected by the treatment, with the same diffraction peaks, both in terms of width at half-height and position. Only a low-intensity peak can be observed for Dyn-2 h at 28° that could correspond to LaF_3 in a very small quantity.

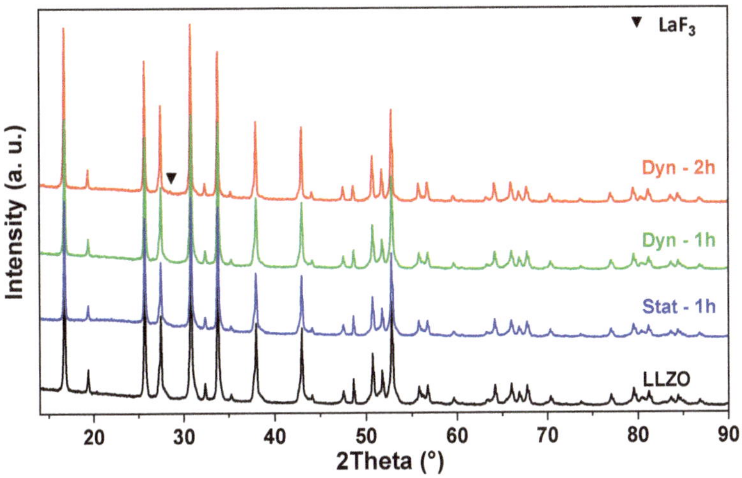

Figure 2. X-ray diffractograms of various LLZO powders fluorinated under different conditions; from top to bottom, Dyn-2 h, Dyn-1 h, Stat-1 h, and LLZO.

In view of this result, it is then necessary to resort to a characterization technique more suited to the detection of elements present in a matrix, in the case of the solid-state NMR study for the ^{19}F nucleus. As the initial compound does not contain fluorine, this element is only introduced through the process used. Solid-state NMR on ^{19}F nuclei is very useful, not only for highlighting the formation of LiF but also the destruction of the inorganic matrix. In this case, fluorides are formed, which cause resonance bands to appear in the NMR spectrum.

Figure 3a,b shows the solid-state NMR spectra of the samples for, respectively, the ^{19}F nucleus, synonymous with the presence of fluorine within the sample, in small quantities (number of pulses at least 128), and ^{7}Li. According to the band resonance, the symmetrical signal at −204 ppm/$CFCl_3$ indicates the presence of Li-F bonds. Indeed, according to the XRD previously exploited, there is no formation of crystallized fluorinated phases visible in the results, which implies that the matrix has been fluorinated, in a partial way, without changing its structure. It is also interesting to note that for the longest dynamic fluorination time (2 h), additional bands appear (marked by red arrows). Given the very low fluorine content in **Stat-1 h**, the rotor signature can be seen (black arrow). These additional signals around −178 and −138 ppm reflect different environments for ^{19}F nuclei. Because of its proximity to the main LiF band, the one at −100 ppm may be an environment similar to that of LiF. This additional signature is only obtained for **Dyn-2 h**. A possible reactivity between lanthanum La and fluorine F would be visible by a resonance in the range of

[−30/−130 ppm], which is not necessarily visible in our case for all samples, with the exception of **Dyn-2 h**.

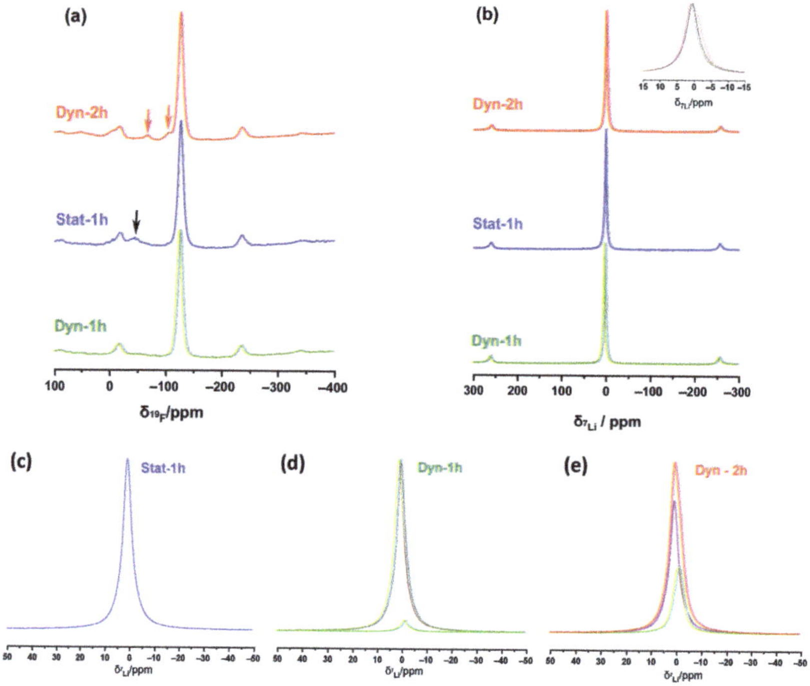

Figure 3. ^{19}F (**a**) and ^{7}Li (**b**) MAS NMR spectra (30 kHz) of LLZO fluorinated under different conditions: pulsed fluorination (20 mbar every 2 min/200 mbar total); static (300 mbar—1 h); dynamic fluorination (1 h); dynamic fluorination (2 h). (**c**–**e**) display the fits with Lorentzian lines of the ^{7}Li spectra.

The ^{7}Li spectra (Figure 3b) show a signal centered at around 0.7 ± 0.1 ppm. The fits of the spectra (Figure 3c–e) reveal two components for **Dyn-1 h** and **Dyn-2 h**, at 0.7± 0.1 ppm and 1.4 ± 0.1 ppm. In accordance with ^{19}F NMR data, this band is assigned to LiF. The second band at 0.74 ± 0.02 ppm is related to Li^+ of the LLZO lattice. The longer the dynamic fluorination duration, the higher the amount of LiF. As seen by ^{19}F NMR data, Lithium fluorine is also present in **Stat-1 h** but with very low content; the line is negligible in comparison with Li^+ in LLZO.

Previous results showed that, whatever the fluorination conditions, the LLZO structure was largely preserved, but that the fluorine reacted with the precursor, since the presence of Li-F was detected. The compound treated in dynamic mode for 2 h stood out for the presence of at least one additional minority phase.

Another issue is the instability in air of LLZO samples. Indeed, the product tends to undergo carbonation when stored under standard conditions. Once received, it was stored in a glove box, under an inert atmosphere, to preserve it and limit pollution before fluorination. It should be noted that fluorination is sometimes used to purify products, notably by degrading carbon layers on the surface through the formation of volatile fluorocarbons (CF_4).

Raman spectroscopy was used to monitor the presence of carbonate groups on the sample as a function of fluorination conditions. Figure 4 displays the evolution of the Raman response for samples **Dyn-1 h**, **Dyn-2 h**, and **Stat-1 h** in the [250–1500 cm^{-1}] range. Initially, the spectra of raw and treated LLZO shows several common bands, evidencing the maintaining of the structure. The vibration band at 354 and 508 cm^{-1} are characteristic of the deformation modes of the octahedral units based on ZrO_6^- [22,57–63]. The band at 648 cm^{-1} is assigned to the stretching modes of these same units [64]. The band at 1080 cm^{-1} changes in intensity according to the fluorination conditions [65]. It is assigned to the vibration of C–O bonds of CO_3^{2-} carbonate groups ($A_{1'}$ symmetrical stretch) [57–59]. This band decreases in intensity for **Stat-1 h** and **Dyn-1 h** and even disappears for **Dyn-2 h**. These data confirm the phenomenon of carbonation once the material has been returned to air. After fluorination, the results demonstrate the benefits of fluorination, since the structure stabilizes as fluorination progresses, with almost total limitation of the phenomenon after 2 h of treatment in dynamic mode. It should be noted that the fluorinated sample **Dyn-2 h** was stored in air for almost 3 months. No carbonation was visible using Raman spectroscopy. Carbonate ions were removed from LLZO by prolonged fluorination (2 h).

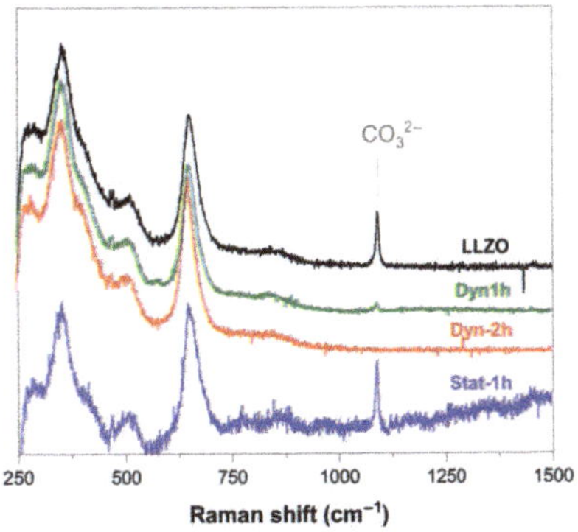

Figure 4. Raman spectra of fluorinated samples in dynamic (**top**) and static (**bottom**) modes.

To confirm LLZO decarbonation and stabilization, IR spectroscopy in ATR mode was carried out. The spectra are shown in Figure 5. The main vibrational modes are related to the carbonate ions E′ asymmetric stretching and $A_{2''}$ out-of-plane bending at 1409–1460 and 879 cm^{-1}, respectively [57,61,62,66]. These bands disappear almost completely for LLZO treated for 2 h in dynamic mode. The band at 846 and 680 cm^{-1} are assigned to lithium hydroxyl and water rocking vibrations, respectively. $LiOH.H_2O$ forms on the surface, which is a necessary intermediate step for Li_2CO_3 formation [60,67]. The weak bands for LiOH and H_2O disappear only after long-term dynamic treatment. At this stage, the LiF layer is enough of a covering to block hydrolysis and subsequently the formation of Li_2CO_3.

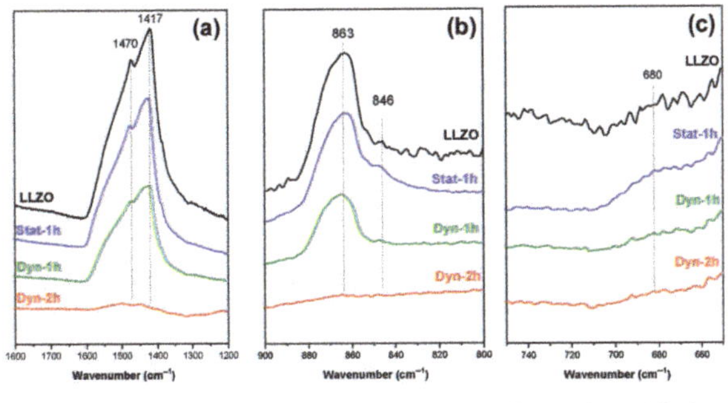

Figure 5. IR spectra (ATR mode) of fluorinated samples with dynamic and static modes in the 1200–1800 (**a**), 800–900 (**b**), and 650–750 cm^{-1} range (**c**).

IR and Raman spectroscopies indisputably indicate carbonate removal and protection against LLZO carbonation by fluorination. Fluorination must be sufficiently long in dynamic mode (2 h with excess of F_2 gas) for the phenomena to be completed.

The results of SEM–EDS analysis as reported in Figure 6 evidence the distribution of fluorine and lanthanum elements along the section of fluorinated LLZO solid particles. The fluorine atoms are concentrated on the surface of the particles, while their core is lanthanum-rich. The ratio F/Zr from XPS were expressed in function of material depth. F/Zr is equal to 1.0, 0.7, and 0.3 at the surface, at 15 nm, and 50 nm from the surface, respectively.

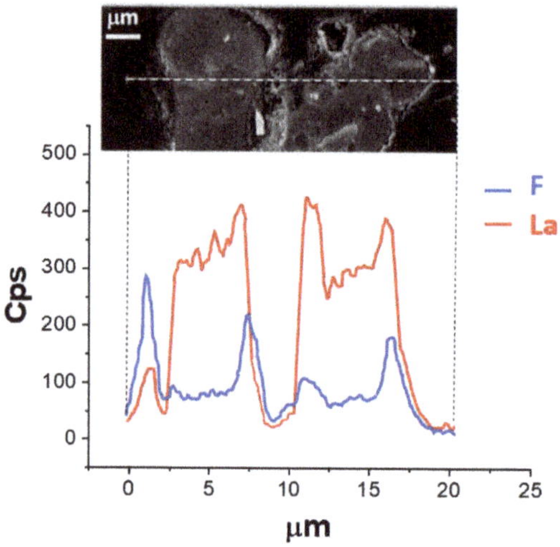

Figure 6. SEM–EDS analysis of **Dyn-2 h**.

Fluorination of LLZO results in a layer of LiF on the surface of the grain. This layer seems to be enough of a covering when the fluorination is performed for a long duration (2 h) in dynamic mode, i.e., with a continuous flux of F_2 molecules. The reaction with moisture and CO_2 from air is avoided thanks to this covering layer, which is evidenced in Figure 7.

Using PVDF SOLEF 1015/1001 as an internal standard for the quantification, the relative contents are 0.7, 3.3, and 0.9%F (by weight) for Dyn-1 h, Dyn-2 h, and Stat-1 h, respectively.

3.2. *Fluorination Allows for Long-Term Stabilization*

The air exposure-induced contamination impacts the interfacial resistance significantly when exposure time exceeds 24 h [22]. The following step was to investigate for how long the protection in air by fluorination is effective. The infrared spectra were recorded just after the fluorination t0 and according to the time of air exposure up to 45 days (Figure 7). The intensity increases of all the vibration bands for carbonates started as soon as the exposure. The band at 1570 cm^{-1} appears clearly only for the raw LLZO, especially after 11 days of exposure (Figure 7c); it can be assigned to symmetrical water H-O-H deformation [67]. On the contrary, all the vibration bands remained low-intensity for **Dyn-2 h,** whatever the air exposure duration. Only their wavenumbers are slightly shifted, i.e., 1437 and 1490 cm^{-1} instead of 1417 and 1471 cm^{-1} for the raw LLZO. The very weak band at 1572 cm^{-1} is not shifted. The long-term (45 days) stability in air is significantly enhanced thanks to the fluorination for 2 h in dynamic mode. Nevertheless, the LiF coating may be not completely hermetic to moisture since the vibration bands in the ranges associated with carbonates and hydroxide groups are very weak but observable.

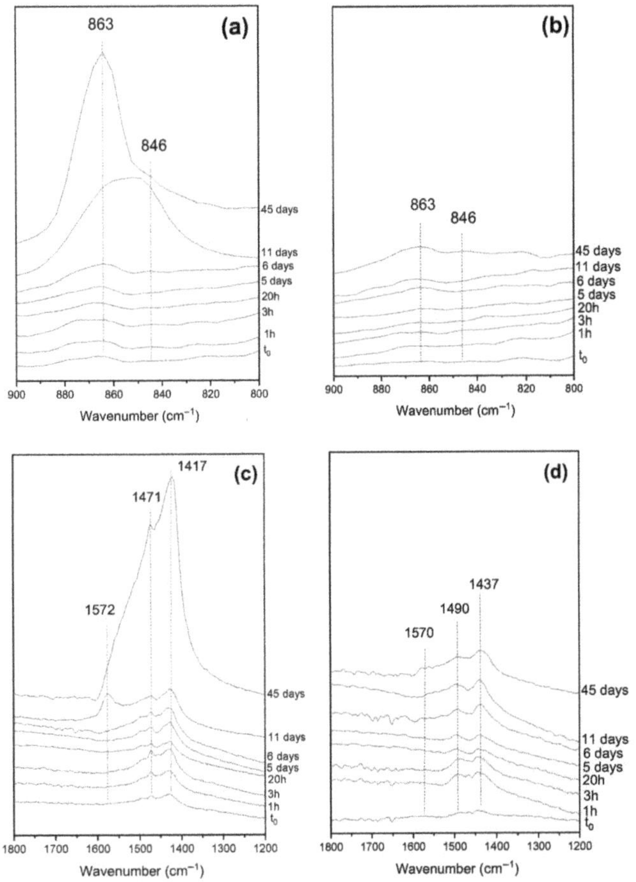

Figure 7. IR spectra (ATR mode) of raw LLZO (**a**,**c**) and Dyn-2 h (**b**,**d**) according to the time of exposure in air.

4. Conclusions and Perspectives

As the reactivity of LLZO with fluorine had never been demonstrated in the literature, several fluorination methods and conditions to quickly refocus the range were explored. Dynamic fluorination proved to be a serious option for purifying the precursor and stabilizing it under standard atmospheric conditions. Indeed, with the formation of carbonates as soon as the sample is exposed to air, precursor pollution can be reduced or even almost annihilated by dynamic fluorination, as shown by Raman and IR (ATR mode) spectroscopies. The structure appears intact after fluorination according to X-ray diffraction, but NMR spectroscopy for the ^{19}F and ^{7}Li nuclei shows the formation of Li-F or even a very few M-F bonds (M = La or Zr). Further analysis of the lithium will be useful to determine which atoms the fluorine has reacted with and whether the mobility of Li^+ ions is affected by the process. Because of the interest from an industrial standpoint to be able to handle the powder outside of a dry room for all stages prior to sintering, the surface fluorination of LLZO was carried out on the powder material. Nevertheless, the stability in air was only relevant for unpressed powders, as LLZO requires high-temperature annealing of 1000 °C to form a ceramic pellet with good conductivity. The maintaining of the LiF surface coating and its diffusion need to be evaluated. Moreover, the effect of surface fluorination on ionic conductivity as a solid electrolyte in lithium batteries must be investigated using electrochemical impedance spectroscopy. To continue further in this process, the structure or porosity of the LiF layer requires study, especially during long-term operation. Finally, although ionic conductivity measurements are beyond the scope of the present work, it is essential to determine whether the improved air stability achieved through surface fluorination also enhances the electrochemical performance of LLZO.

Author Contributions: Conceptualization, M.D., K.L., L.J., K.G., M.H., L.B., R.T., T.L.M. and V.B.; methodology, M.D., K.L., L.J., K.G., M.H., L.B., R.T., T.L.M. and V.B.; formal analysis, M.H., M.D., K.L., L.J., E.P., N.G., S.M. and K.G.; data curation, M.H., M.D., K.L., L.J., E.P., N.G., S.M. and K.G.; writing—original draft preparation, M.D., K.L., L.J., K.G., L.B., R.T., T.L.M. and V.B.; writing—review and editing, M.D., K.L., L.J., S.M., K.G., L.B., R.T., T.L.M. and V.B.; visualization, supervision, M.D., K.L., L.J., K.G., L.B., R.T. and T.L.M.; project administration, M.D. and K.G.; funding acquisition, M.D. and K.G. All authors have read and agreed to the published version of the manuscript.

Funding: This research received funding from SYENSQO.

Data Availability Statement: Data is contained within the article.

Conflicts of Interest: Authors Laure Bertry, Thierry Le Mercier and Valérie Buissette were employed by the company Syensqo; Author Reka Toth was employed by the company Solvay. The remaining authors declare that the research was conducted in the absence of any commercial or financial relationships that could be construed as a potential conflict of interest.

References

1. Cheng, X.-B.; Zhao, C.-Z.; Yao, Y.-X.; Liu, H.; Zhang, Q. Recent Advances in Energy Chemistry between Solid-State Electrolyte and Safe Lithium-Metal Anodes. *Chem* **2019**, *5*, 74–96. [CrossRef]
2. Meabe, L.; Aldalur, I.; Lindberg, S.; Arrese-Igor, M.; Armand, M.; Martinez-Ibañez, M.; Zhang, H. Solid-State Electrolytes for Safe Rechargeable Lithium Metal Batteries: A Strategic View. *Mater. Futures* **2023**, *2*, 33501. [CrossRef]
3. Wang, C.; Wang, C.; Li, M.; Zhang, S.; Zhang, C.; Chou, S.; Mao, J.; Guo, Z. Design of Thin Solid-State Electrolyte Films for Safe and Energy-Dense Batteries. *Mater. Today* **2024**, *72*, 235–254. [CrossRef]
4. Aurbach, D.; Talyosef, Y.; Markovsky, B.; Markevich, E.; Zinigrad, E.; Asraf, L.; Gnanaraj, J.S.; Kim, H.-J. Design of Electrolyte Solutions for Li and Li-Ion Batteries: A Review. *Electrochim. Acta* **2004**, *50*, 247–254. [CrossRef]
5. Xi, G.; Xiao, M.; Wang, S.; Han, D.; Li, Y.; Meng, Y. Polymer-Based Solid Electrolytes: Material Selection, Design, and Application. *Adv. Funct. Mater.* **2021**, *31*, 2007598. [CrossRef]
6. Chen, Z.; Kim, G.-T.; Wang, Z.; Bresser, D.; Qin, B.; Geiger, D.; Kaiser, U.; Wang, X.; Shen, Z.X.; Passerini, S. 4-V Flexible All-Solid-State Lithium Polymer Batteries. *Nano Energy* **2019**, *64*, 103986. [CrossRef]
7. Mauger, A.; Julien, C.M. Solid Polymer Electrolytes for Lithium Batteries: A Tribute to Michel Armand. *Inorganics* **2022**, *10*, 110. [CrossRef]
8. Manthiram, A.; Yu, X.; Wang, S. Lithium Battery Chemistries Enabled by Solid-State Electrolytes. *Nat. Rev. Mater.* **2017**, *2*, 16103. [CrossRef]
9. Knauth, P. Inorganic Solid Li Ion Conductors: An Overview. *Solid State Ion.* **2009**, *180*, 911–916. [CrossRef]
10. Zheng, F.; Kotobuki, M.; Song, S.; Lai, M.O.; Lu, L. Review on Solid Electrolytes for All-Solid-State Lithium-Ion Batteries. *J. Power Sources* **2018**, *389*, 198–213. [CrossRef]
11. Zhao, N.; Khokhar, W.; Bi, Z.; Shi, C.; Guo, X.; Fan, L.-Z.; Nan, C.-W. Solid Garnet Batteries. *Joule* **2019**, *3*, 1190–1199. [CrossRef]
12. Keller, M.; Varzi, A.; Passerini, S. Hybrid Electrolytes for Lithium Metal Batteries. *J. Power Sources* **2018**, *392*, 206–225. [CrossRef]
13. Cheng, Z.; Liu, T.; Zhao, B.; Shen, F.; Jin, H.; Han, X. Recent Advances in Organic-Inorganic Composite Solid Electrolytes for All-Solid-State Lithium Batteries. *Energy Storage Mater.* **2021**, *34*, 388–416. [CrossRef]
14. Bhardwaj, R.K.; Zitoun, D. Recent Progress in Solid Electrolytes for All-Solid-State Metal(Li/Na)–Sulfur Batteries. *Batteries* **2023**, *9*, 110. [CrossRef]
15. Chen, R.; Li, Q.; Yu, X.; Chen, L.; Li, H. Approaching Practically Accessible Solid-State Batteries: Stability Issues Related to Solid Electrolytes and Interfaces. *Chem. Rev.* **2020**, *120*, 6820–6877. [CrossRef] [PubMed]
16. Murugan, R.; Thangadurai, V.; Weppner, W. Fast Lithium Ion Conduction in Garnet-Type $Li_7La_3Zr_2O_{12}$. *Angew. Chemie Int. Ed.* **2007**, *46*, 7778–7781. [CrossRef] [PubMed]
17. Ramakumar, S.; Janani, N.; Murugan, R. Influence of Lithium Concentration on the Structure and Li^+ Transport Properties of Cubic Phase Lithium Garnets. *Dalt. Trans.* **2015**, *44*, 539–552. [CrossRef] [PubMed]

18. Awaka, J.; Kijima, N.; Hayakawa, H.; Akimoto, J. Synthesis and Structure Analysis of Tetragonal $Li_7La_3Zr_2O_{12}$ with the Garnet-Related Type Structure. *J. Solid State Chem.* **2009**, *182*, 2046–2052. [CrossRef]
19. Thangadurai, V.; Kaack, H.; Weppner, W.J.F. Novel Fast Lithium Ion Conduction in Garnet-Type $Li_5La_3M_2O_{12}$ (M = Nb, Ta). *J. Am. Ceram. Soc.* **2003**, *86*, 437–440. [CrossRef]
20. Cheng, L.; Crumlin, E.J.; Chen, W.; Qiao, R.; Hou, H.; Franz Lux, S.; Zorba, V.; Russo, R.; Kostecki, R.; Liu, Z.; et al. The Origin of High Electrolyte–Electrode Interfacial Resistances in Lithium Cells Containing Garnet Type Solid Electrolytes. *Phys. Chem. Chem. Phys.* **2014**, *16*, 18294–18300. [CrossRef] [PubMed]
21. Cheng, L.; Park, J.S.; Hou, H.; Zorba, V.; Chen, G.; Richardson, T.; Cabana, J.; Russo, R.; Doeff, M. Effect of Microstructure and Surface Impurity Segregation on the Electrical and Electrochemical Properties of Dense Al-Substituted $Li_7La_3Zr_2O_{12}$. *J. Mater. Chem. A* **2014**, *2*, 172–181. [CrossRef]
22. Sharafi, A.; Yu, S.; Naguib, M.; Lee, M.; Ma, C.; Meyer, H.M.; Nanda, J.; Chi, M.; Siegel, D.J.; Sakamoto, J. Impact of Air Exposure and Surface Chemistry on Li–$Li_7La_3Zr_2O_{12}$ Interfacial Resistance. *J. Mater. Chem. A* **2017**, *5*, 13475–13487. [CrossRef]
23. Lim, H.-D.; Park, J.-H.; Shin, H.-J.; Jeong, J.; Kim, J.T.; Nam, K.-W.; Jung, H.-G.; Chung, K.Y. A Review of Challenges and Issues Concerning Interfaces for All-Solid-State Batteries. *Energy Storage Mater.* **2020**, *25*, 224–250. [CrossRef]
24. Janek, J.; Zeier, W.G. A Solid Future for Battery Development. *Nat. Energy* **2016**, *1*, 16141. [CrossRef]
25. Chen, Y.; Qian, J.; Li, L.; Wu, F.; Chen, R. Advances in Inorganic Solid-State Electrolyte/Li Interface. *Chem. A Eur. J.* **2024**, *30*, e202303454. [CrossRef] [PubMed]
26. Afyon, S.; Kravchyk, K.V.; Wang, S.; Broek, J.V.D.; Hänsel, C.; Kovalenko, M.V.; Rupp, J.L.M. Building Better All-Solid-State Batteries with Li-Garnet Solid Electrolytes and Metalloid Anodes. *J. Mater. Chem. A* **2019**, *7*, 21299–21308. [CrossRef]
27. Lou, S.; Zhang, F.; Fu, C.; Chen, M.; Ma, Y.; Yin, G.; Wang, J. Interface Issues and Challenges in All-Solid-State Batteries: Lithium, Sodium, and Beyond. *Adv. Mater.* **2021**, *33*, 2000721. [CrossRef] [PubMed]
28. Xu, R.C.; Xia, X.H.; Zhang, S.Z.; Xie, D.; Wang, X.L.; Tu, J.P. Interfacial Challenges and Progress for Inorganic All-Solid-State Lithium Batteries. *Electrochim. Acta* **2018**, *284*, 177–187. [CrossRef]
29. Muramatsu, H.; Hayashi, A.; Ohtomo, T.; Hama, S.; Tatsumisago, M. Structural Change of Li_2S–P_2S_5 Sulfide Solid Electrolytes in the Atmosphere. *Solid State Ion.* **2011**, *182*, 116–119. [CrossRef]
30. Simon, D.; Kelder, E.; Wagemaker, M.; Mulder, F.; Schoonman, J. Characterization of Proton Exchanged $Li_4Ti_5O_{12}$ Spinel Material. *Solid State Ion.* **2006**, *177*, 2759–2768. [CrossRef]
31. Samson, A.J.; Hofstetter, K.; Bag, S.; Thangadurai, V. A Bird's-Eye View of Li-Stuffed Garnet-Type $Li_7La_3Zr_2O_{12}$ Ceramic Electrolytes for Advanced All-Solid-State Li Batteries. *Energy Environ. Sci.* **2019**, *12*, 2957–2975. [CrossRef]
32. Han, F.; Zhu, Y.; He, X.; Mo, Y.; Wang, C. Electrochemical Stability of Li_{10} GeP_2 S_{12} and Li_7 La_3 Zr_2 O_{12} Solid Electrolytes. *Adv. Energy Mater.* **2016**, *6*, 1501590. [CrossRef]
33. Famprikis, T.; Canepa, P.; Dawson, J.A.; Islam, M.S.; Masquelier, C. Fundamentals of Inorganic Solid-State Electrolytes for Batteries. *Nat. Mater.* **2019**, *18*, 1278–1291. [CrossRef] [PubMed]
34. Feng, X.; Ouyang, M.; Liu, X.; Lu, L.; Xia, Y.; He, X. Thermal Runaway Mechanism of Lithium Ion Battery for Electric Vehicles: A Review. *Energy Storage Mater.* **2018**, *10*, 246–267. [CrossRef]
35. von Aspern, N.; Röschenthaler, G.-V.; Winter, M.; Cekic-Laskovic, I. Fluorine and Lithium: Ideal Partners for High-Performance Rechargeable Battery Electrolytes. *Angew. Chemie Int. Ed.* **2019**, *58*, 15978–16000. [CrossRef] [PubMed]
36. Tan, J.; Matz, J.; Dong, P.; Shen, J.; Ye, M. A Growing Appreciation for the Role of LiF in the Solid Electrolyte Interphase. *Adv. Energy Mater.* **2021**, *11*, 2100046. [CrossRef]
37. Wang, T.; Duan, J.; Zhang, B.; Luo, W.; Ji, X.; Xu, H.; Huang, Y.; Huang, L.; Song, Z.; Wen, J.; et al. A Self-Regulated Gradient Interphase for Dendrite-Free Solid-State Li Batteries. *Energy Environ. Sci.* **2022**, *15*, 1325–1333. [CrossRef]
38. Tan, Y.-H.; Lu, G.-X.; Zheng, J.-H.; Zhou, F.; Chen, M.; Ma, T.; Lu, L.-L.; Song, Y.-H.; Guan, Y.; Wang, J.; et al. Lithium Fluoride in Electrolyte for Stable and Safe Lithium-Metal Batteries. *Adv. Mater.* **2021**, *33*, 2102134. [CrossRef] [PubMed]
39. Yang, F.; Liu, Y.; Liu, T.; Wang, Y.; Nai, J.; Lin, Z.; Xu, H.; Duan, D.; Yue, K.; Tao, X. Fluorinated Strategies Among All-Solid-State Lithium Metal Batteries from Microperspective. *Small Struct.* **2023**, *4*, 2200122. [CrossRef]
40. Ganesan, P.; Soans, M.; Cambaz, M.A.; Zimmermanns, R.; Gond, R.; Fuchs, S.; Hu, Y.; Baumgart, S.; Sotoudeh, M.; Stepien, D.; et al. Fluorine-Substituted Halide Solid Electrolytes with Enhanced Stability toward the Lithium Metal. *ACS Appl. Mater. Interfaces* **2023**, *15*, 38391–38402. [CrossRef] [PubMed]
41. Yu, T.; Liang, J.; Luo, L.; Wang, L.; Zhao, F.; Xu, G.; Bai, X.; Yang, R.; Zhao, S.; Wang, J.; et al. Superionic Fluorinated Halide Solid Electrolytes for Highly Stable Li-Metal in All-Solid-State Li Batteries. *Adv. Energy Mater.* **2021**, *11*, 2101915. [CrossRef]
42. Xu, R.; Zhang, X.-Q.; Cheng, X.-B.; Peng, H.-J.; Zhao, C.-Z.; Yan, C.; Huang, J.-Q. Artificial Soft–Rigid Protective Layer for Dendrite-Free Lithium Metal Anode. *Adv. Funct. Mater.* **2018**, *28*, 1705838. [CrossRef]
43. Lu, Y.; Meng, X.; Alonso, J.A.; Fernández-Díaz, M.T.; Sun, C. Effects of Fluorine Doping on Structural and Electrochemical Properties of $Li_{6.25}Ga_{0.25}La_3Zr_2O_{12}$ as Electrolytes for Solid-State Lithium Batteries. *ACS Appl. Mater. Interfaces* **2019**, *11*, 2042–2049. [CrossRef] [PubMed]

44. Li, Y.; Xu, B.; Xu, H.; Duan, H.; Lü, X.; Xin, S.; Zhou, W.; Xue, L.; Fu, G.; Manthiram, A.; et al. Hybrid Polymer/Garnet Electrolyte with a Small Interfacial Resistance for Lithium-Ion Batteries. *Angew. Chemie Int. Ed.* **2017**, *56*, 753–756. [CrossRef] [PubMed]
45. Dong, B.; Haworth, A.R.; Yeandel, S.R.; Stockham, M.P.; James, M.S.; Xiu, J.; Wang, D.; Goddard, P.; Johnston, K.E.; Slater, P.R. Halogenation of $Li_7La_3Zr_2O_{12}$ Solid Electrolytes: A Combined Solid-State NMR, Computational and Electrochemical Study. *J. Mater. Chem. A* **2022**, *10*, 11172–11185. [CrossRef]
46. Lin, D.; Liu, Y.; Chen, W.; Zhou, G.; Liu, K.; Dunn, B.; Cui, Y. Conformal Lithium Fluoride Protection Layer on Three-Dimensional Lithium by Nonhazardous Gaseous Reagent Freon. *Nano Lett.* **2017**, *17*, 3731–3737. [CrossRef] [PubMed]
47. Zhao, J.; Liao, L.; Shi, F.; Lei, T.; Chen, G.; Pei, A.; Sun, J.; Yan, K.; Zhou, G.; Xie, J.; et al. Surface Fluorination of Reactive Battery Anode Materials for Enhanced Stability. *J. Am. Chem. Soc.* **2017**, *139*, 11550–11558. [CrossRef] [PubMed]
48. Charles-Blin, Y.; Flahaut, D.; Guérin, K.; Dubois, M.; Monconduit, L.; Louvain, N.; Martinez, H. Surface Atomic Layer Fluorination of $Li_4Ti_5O_{12}$: Investigation of the Surface Electrode Reactivity and the Outgassing Behavior in LiBs. *Appl. Surf. Sci.* **2020**, *527*, 146834. [CrossRef]
49. Groult, H.; Nakajima, T.; Perrigaud, L.; Ohzawa, Y.; Yashiro, H.; Komaba, S.; Kumagai, N. Surface-Fluorinated Graphite Anode Materials for Li-Ion Batteries. *J. Fluor. Chem.* **2005**, *126*, 1111–1116. [CrossRef]
50. Xiong, C.; Wang, Z.; Peng, X.; Guo, Y.; Xu, S.; Zhao, T. Bifunctional Effect of Laser-Induced Nucleation-Preferable Microchannels and in Situ Formed LiF SEI in MXenes for Stable Lithium-Metal Batteries. *J. Mater. Chem. A* **2020**, *8*, 14114–14125. [CrossRef]
51. Breddemann, U.; Sicklinger, J.; Schipper, F.; Davis, V.; Fischer, A.; Huber, K.; Erickson, E.M.; Daub, M.; Hoffmann, A.; Erk, C.; et al. Fluorination of Ni-Rich Lithium-Ion Battery Cathode Materials by Fluorine Gas: Chemistry, Characterization, and Electrochemical Performance in Full-Cells. *Batter. Supercaps* **2021**, *4*, 632–645. [CrossRef]
52. Kim, S.; Kim, E.J.; Charles-Blin, Y.; Guérin, K.; Dubois, M.; Flahaut, D.; Martinez, H.; Deschamps, M.; Miller, D.N.; Irvine, J.T.S.; et al. Atomic Layer Fluorination of 5 V Class Positive Electrode Material $LiCoPO_4$ for Enhanced Electrochemical Performance. *Batter. Supercaps* **2020**, *3*, 1051–1058. [CrossRef]
53. Son, J.E.; Yim, J.-H.; Lee, J.-W. Fluorination of SiOx as an Effective Strategy to Enhance the Cycling Stability of Lithium-Ion Batteries. *Electrochem. Commun.* **2023**, *152*, 107517. [CrossRef]
54. Charles-Blin, Y.; Flahaut, D.; Ledeuil, J.-B.; Guérin, K.; Dubois, M.; Deschamps, M.; Perbost, A.-M.; Monconduit, L.; Martinez, H.; Louvain, N. Atomic Layer Fluorination of the $Li_4Ti_5O_{12}$ Surface: A Multiprobing Survey. *ACS Appl. Energy Mater.* **2019**, *2*, 6681–6692. [CrossRef]
55. Le Mercier, T.; Braida, M.; Buissette, V.; Amisse, R.; Toth, R.; Bertry, L. Method of Preparation of a Garnet-Type Inorganic Material. Patent WO2021023599A1, 11 February 2021.
56. Degen, T.; Sadki, M.; Bron, E.; König, U.; Nénert, G. The HighScore Suite. *Powder Diffr.* **2014**, *29*, S13–S18. [CrossRef]
57. Neises, J.; Scheld, W.S.; Seok, A.-R.; Lobe, S.; Finsterbusch, M.; Uhlenbruck, S.; Schmechel, R.; Benson, N. Study of Thermal Material Properties for Ta- and Al-Substituted $Li_7La_3Zr_2O_{12}$ (LLZO) Solid-State Electrolyte in Dependency of Temperature and Grain Size. *J. Mater. Chem. A* **2022**, *10*, 12177–12186. [CrossRef]
58. Jeong, W.; Park, S.S.; Yun, J.; Shin, H.R.; Moon, J.; Lee, J.-W. Tailoring Grain Boundary Structures and Chemistry of Li7La3Zr2O12 Solid Electrolytes for Enhanced Air Stability. *Energy Storage Mater.* **2023**, *54*, 543–552. [CrossRef]
59. Tsai, C.-L.; Ma, Q.; Dellen, C.; Lobe, S.; Vondahlen, F.; Windmüller, A.; Grüner, D.; Zheng, H.; Uhlenbruck, S.; Finsterbusch, M.; et al. A Garnet Structure-Based All-Solid-State Li Battery without Interface Modification: Resolving Incompatibility Issues on Positive Electrodes. *Sustain. Energy Fuels* **2019**, *3*, 280–291. [CrossRef]
60. Xia, W.; Xu, B.; Duan, H.; Tang, X.; Guo, Y.; Kang, H.; Li, H.; Liu, H. Reaction Mechanisms of Lithium Garnet Pellets in Ambient Air: The Effect of Humidity and CO_2. *J. Am. Ceram. Soc.* **2017**, *100*, 2832–2839. [CrossRef]
61. Zhang, Z.; Zhang, L.; Liu, Y.; Wang, H.; Yu, C.; Zeng, H.; Wang, L.; Xu, B. Interface-Engineered $Li_7La_3Zr_2O_{12}$-Based Garnet Solid Electrolytes with Suppressed Li-Dendrite Formation and Enhanced Electrochemical Performance. *ChemSusChem* **2018**, *11*, 3774–3782. [CrossRef] [PubMed]
62. Mohammadi, S.; Golmohammad, M.; SharifiRad, A.; Alizadeh, S.M. Surface Modification and Chemical Stability of Garnet LLZO Solid Electrolyte by ZnO Coating through a Facile and Practical Method. *Mater. Chem. Phys.* **2023**, *308*, 128235. [CrossRef]
63. Scheld, W.S.; Collette, Y.; Schwab, C.; Ihrig, M.; Uhlenbruck, S.; Finsterbusch, M.; Fattakhova-Rohlfing, D. Ga-Ion Migration during Co-Sintering of Heterogeneous Ta- and Ga-Substituted LLZO Solid-State Electrolytes. *J. Eur. Ceram. Soc.* **2025**, *45*, 116936. [CrossRef]
64. Zhu, Y.; Chon, M.; Thompson, C.V.; Rupp, J.L.M. Time-Temperature-Transformation (TTT) Diagram of Battery-Grade Li-Garnet Electrolytes for Low-Temperature Sustainable Synthesis. *Angew. Chemie Int. Ed.* **2023**, *62*, e202304581. [CrossRef] [PubMed]
65. Liu, M.; Zhang, M.; Liu, S.; Chen, H.; Li, B.; Li, G.; Zhang, S.; Wen, Y.; Qiu, J.; Chen, J.; et al. Direct Wetting of $Li_7La_3Zr_2O_{12}$ Electrolyte with Molten Li Anode and Its Application in Solid-State Lithium Batteries. *J. Power Sources* **2024**, *593*, 233977. [CrossRef]

66. Sun, J.; Zhao, N.; Li, Y.; Guo, X.; Feng, X.; Liu, X.; Liu, Z.; Cui, G.; Zheng, H.; Gu, L.; et al. A Rechargeable Li-Air Fuel Cell Battery Based on Garnet Solid Electrolytes. *Sci. Rep.* **2017**, *7*, 41217. [CrossRef] [PubMed]
67. Gennick, I.; Harmon, K.M. Hydrogen Bonding. VI. Structural and Infrared Spectral Analysis of Lithium Hydroxide Monohydrate and Cesium and Rubidium Hydroxide Hydrates. *Inorg. Chem.* **1975**, *14*, 2214–2219. [CrossRef]

Disclaimer/Publisher's Note: The statements, opinions and data contained in all publications are solely those of the individual author(s) and contributor(s) and not of MDPI and/or the editor(s). MDPI and/or the editor(s) disclaim responsibility for any injury to people or property resulting from any ideas, methods, instructions or products referred to in the content.

MDPI AG
Grosspeteranlage 5
4052 Basel
Switzerland
Tel.: +41 61 683 77 34

Batteries Editorial Office
E-mail: batteries@mdpi.com
www.mdpi.com/journal/batteries

Disclaimer/Publisher's Note: The title and front matter of this reprint are at the discretion of the Guest Editor. The publisher is not responsible for their content or any associated concerns. The statements, opinions and data contained in all individual articles are solely those of the individual Editor and contributors and not of MDPI. MDPI disclaims responsibility for any injury to people or property resulting from any ideas, methods, instructions or products referred to in the content.

www.ingramcontent.com/pod-product-compliance
Lightning Source LLC
Chambersburg PA
CBHW061615210926
51850CB00009B/2333
* 9 7 8 3 7 2 5 8 7 0 2 4 0 *